DON'T KNOW MUCH ABOUT®
MUCH ABOUT®

The Civil War

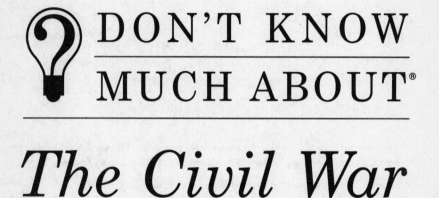

DON'T KNOW MUCH ABOUT®

The Civil War

EVERYTHING YOU NEED TO KNOW ABOUT AMERICA'S
GREATEST CONFLICT BUT NEVER LEARNED

Kenneth C. Davis

Perennial
An Imprint of HarperCollinsPublishers

First Avon Books edition published 1997.

Reissued in Perennial 2001, 2004.

The Library of Congress has catalogued the hardcover edition as follows:

Davis, Kenneth C.
   Don't know much about® the Civil War: everything you need to know about
America's greatest conflict but never learned / Kenneth C. Davis.
       p. cm.
   Includes bibliographical references and index.
   1. United States—History—Civil War, 1861–1865. 2. Slavery—United States
History. I. Title.
   E462.D38   1996   92-50352
   973.7—dc20   CIP

ISBN 0-380-71908-8 (pbk.)

06 07 08 09 10 FG/RRD 20 19 18

Acknowledgments

An individual writer's name goes on the cover, but every book is a collaboration. There are many people I must thank for assisting me in the production of this book. First, I wish to acknowledge the many great historians who have preceded me. Their names are in the Bibliography. I hope that readers of this book will turn to the vast, rich literature of the Civil War and the rest of American history.

With much gratitude, I want to single out some of the many people at William Morrow and Avon Books who have assisted me: Will Schwalbe, Beccy Goodhart, Sharyn Rosenblum, William Wright, Mike Greenstein, Bruce Brill, and Rachel Klayman. I'd like to also thank my friend Mark Levine, as well as Professor Matthew Murdzak, for their readings of the manuscript, and David Black for his help.

Among my many supportive friends, I wish to thank two points of light, Rosemary Altea and Joni Evans. Their encouragement, insight, and wisdom are precious gifts to me.

I trace my love of history to my parents, who understood that history and learning aren't found only in books, but in places like Gettysburg. So to them, again, my thanks. I am always grateful to my own children, Colin and Jenny, for their patience, love and just being. And to Joann, the biggest thank you for making it all possible.

The greatest surprise and pleasure of my work has been meeting with and hearing from teachers all over the country. No praise is more gratifying to me than the words of teachers and students who have found something useful, stimulating, or challenging in my books. We ask our teachers to do a critical and difficult job without giving them proper support—or credit. Most of the teachers I have encountered bring love, deep commitment, and sacrifice to their work. To the teachers of America, this book is dedicated.

Contents

————— ✳ —————

It is well that war is so terrible; else we should grow too fond of it.
—GENERAL ROBERT E. LEE

War is at best barbarism.... Its glory is all moonshine.
It is only those who have neither fired a shot, nor heard the
shrieks and groans of the wounded who cry aloud for blood,
more vengeance, more desolation.
—GENERAL WILLIAM T. SHERMAN

If I were to speak of war, it would not be to show you the glories of
conquering armies but the mischief and misery they strew in
their tracks; and how, while they marched on with tread of iron and
plumes proudly tossing in the breeze, some one must follow closely
in their steps, crouching to the earth, toiling in the rain and darkness,
shelterless themselves, with no thought of pride or glory, fame or
praise, or reward; hearts breaking with pity, faces bathed in tears
and hands in blood. This is the side which history never shows.
—CLARA BARTON

✻

Introduction

"Oh, I'm so interested in the Civil War. I have been ever since I read *Gone With the Wind*!"

I have lost count of the number of people who told me that while I was writing this book. Unfortunately, it seems that a great many Americans owe their understanding of the central event in our history to a sixty-year-old piece of fiction.

"Frankly, my dear," this notion terrifies me. It suggests that legions of Miss Mitchell's readers know as much about the Civil War as Scarlett O'Hara's slave Prissy knew about "birthing babies." It means that millions of people's notions of the Civil War era and slavery were shaped by a book in which all the white folks speak the King's English and slaves sound like this:

"Runned away? No'm, us ain' runned away. Dey done sont an' tuck us, kase us wuz de fo' bigges' an' stronges' han's at Tara.... Dey specially sont fer me, kase Ah could sing so good."

Perhaps their views of plantation life derive from this Pulitzer Prize–winning description of an idyllic Georgia barbecue: "Over behind the barns there was always another barbecue pit, where the house servants and the coachmen and maids of the guests had their own feast of hoecakes and yams and chitterlings, that dish of hog entrails so dear to negro hearts, and, in season, watermelons enough to satiate."

But maybe it wasn't Margaret Mitchell's prose and views on the war and slaves that did the trick. Perhaps they were among those

Americans who learned about the Civil War this way: "Slavery died out in the North because it was expensive and inefficient. *The African was too recently removed from his tropical home to endure the harsh winters,* and even on a farm his labor did not pay for itself. *There was no time to train his primitive fingers to perform all the varied tasks required of a Yankee handyman.*" (Emphasis added.)

Does that sound like 1859 plantation propaganda? It actually comes from a *1959* McGraw-Hill textbook called *The War Between the States*.

For a long time, I thought that Margaret Mitchell's famous novel was to blame for the obsession that so many Americans have with the Civil War period. I assumed that *Gone With the Wind*—still this country's most beloved book and movie—made us a nation consumed by a devastating war.

And consumed we are.

* Early in 1995, a man was killed in a parking lot after an argument that started over a Confederate flag hanging in the back of the dead man's pickup. This incident occurred not long after Virginia's hotly contested 1994 Senate race, in which Republican candidate Oliver North supported those who wish to retain the Confederate flag as an emblem in several southern state flags. It was another skirmish in an emotional war between those who see the "stars and bars" as an offensive racist symbol and those who honor it as a symbol of their heritage.

* Discussing affirmative action policies in 1995, Republican Senator Robert Dole questioned whether "future generations should have to pay" for slavery. Said Dole, "Some would say yes. I think it's a tough question."

* In June 1995 the Southern Baptists, who had split from the Baptist Church in 1845 over the issue of slaveowners becoming missionaries, apologized to African Americans for the sin of slavery.

* In 1994 the heavily armed Disney divisions were outflanked in their attempt to build an American history theme park and real estate development near the site of a major Civil War battlefield. In the Third Battle of Bull Run, the Fightin' Mouseketeers, hold-

ing their banners "High, High, High," were routed by a regiment of guerrilla historians and roving bands of historical preservationists.

* In Richmond, a bitter dispute ran along racial lines in 1995 over the placement of a statue honoring Arthur Ashe, the late tennis star and a native of the Virginia capital. Some traditionalists—who happened to be white—didn't think a statue honoring a tennis player—who happened to be black—belonged with monuments to the heroes of the Confederacy, Robert E. Lee and Stonewall Jackson. Richmond's blacks objected because the young tennis star wasn't welcome in that section of the city in the segregated days of his youth, when he was unable to play on public courts. After an ugly controversy, the statue was unveiled on Monument Drive.

* Two historians recently requested court permission to remove John Wilkes Booth's body from the Booth family plot in a Baltimore cemetery. They suggested that the body is not Booth's and that Lincoln's assassin might have actually escaped. (Quick. Call Oliver Stone!)

* A mock slave auction, reenacted in 1994 at Williamsburg, Virginia, set off a protest by African Americans, who thought the drama was degrading and too painful to repeat.

* In the Republican Congress elected in 1994, there was a concerted move to return certain powers to the states. About the same time, the Supreme Court issued several decisions questioning the extent of federal powers. The "states' rights" tune is on the Hit Parade once more.

And you thought the Civil War was over.

Back in 1970, I spent the summer in Wilmington, North Carolina, tutoring children under the auspices of a historic black church that had been a station on the Underground Railroad. A young New Yorker who had never been in the South, I got my first surprise from a highway billboard: THE KU KLUX KLAN WELCOMES YOU.

My next dose of culture shock came when we took a group of

black children on their first trip to the ocean. Several young men who preferred the days when the beaches were closed to blacks screamed ugly racial epithets at the children. My third eye-opener came during a trip to a prominent "white" church to hear its magnificent pipe organ. As the black children settled into the pews, an older member of the church stormed down the aisle. His face purple with anger, this good Christian man furiously demanded, "What are these"— the *N* word was on the tip of his tongue, but he caught himself and spit out—"*people* doing here?"

I, too, had thought the Civil War was over.

The Civil War lives on. In American politics. In our social consciousness. In popular memory. America's favorite movie hero of 1994, Forrest Gump, was named after a slave-trading Confederate general who was responsible for the massacre of black soldiers and helped found the Ku Klux Klan. A four-hour movie about the Battle of Gettysburg made a best-seller out of *The Killer Angels*, the novel on which the film was based. Gettysburg remains the most visited historic site in America outside Washington, D.C. In 1995 plans were announced for a multimillion-dollar museum on the site of Georgia's notorious Andersonville Prison. Dedicated to the prisoners of all wars, the museum will focus on the thousands of Union prisoners who were held and died there. And tens of thousands of "reenactors" keep this history alive by donning Civil War uniforms and recreating the battles of the war.

A sense of romanticism was present from the war's outset. Mark Twain once said that there was a Civil War because Southerners read too much Sir Walter Scott, the nineteenth-century author of *Ivanhoe* (1820). Twain meant that all of Scott's notions of chivalry, courtly love, and honor clung fast in the planter's world and had a great deal to do with the ill-fated Confederate "Cause."

Obviously, it was far more complex than that. But in thinking about the Civil War at the end of the twentieth century, it's important to remember the vast difference between our modern cultural attitudes and those of Americans in 1860. It *was* a far more romantic and patriotic age. The country was younger and emotionally closer to its founding. The ideals of the American Revolution were stronger, more palpable. Many men who fought in the Civil War were direct descendants of veterans of George Washington's army, and their grandfathers' swords hung over the fireplace. On both sides of the conflict, they saw themselves as guardians of the Revolutionary spirit, custodians of liberty. The very idea of the United States was still new, and an American identity had not yet been forged. Certainly, our modern cynicism did not yet exist. War was still glorious and heroic and offered boys one of the few opportunities to escape the routine of the farm. Before the Civil War, warfare was not as horrifying as we have come to view it today. The Civil War, the first "modern" war, had a lot to do with changing people's minds about the "glory" of war.

Even today, that sense of glory is part of the Civil War's tug. For me, it began in 1963, on a family camping trip to Gettysburg during the hundredth anniversary of the battle there. I still have my souvenir from that trip—a wooden pistol with the battle dates printed on the barrel: July 1, 2, 3, 1863. But more important is the recollection of Gettysburg that was permanently seared in my memory. To walk those green fields, climb the hills, squeeze into the narrow crevices in the rocks of Devil's Den, and feel the summer heat in the Peach Orchard is to touch history.

Many Americans share a fascination for the Civil War's supposed glory: a deep, passionate interest in a war that split the country, pitted families and former comrades against one another, and left lingering wounds for decades. No other period in our history has generated so many books—histories, biographies of the legendary and the infamous, guides to battle sites, genealogies, almanacs, battlefield atlases, diaries of generals, infantrymen, slaves, and plantation owners. One recent book even dealt with sex in the Civil War! The list goes on without signs of slowing. Add the magazines devoted to the war, videotaped tours of battlefields and reenactments of the battles themselves along with board games, puzzles, and computer software that allow people to "relive" the "War Between the States" and you have evidence of a popular obsession that defies the notion that

Americans aren't interested in history. But this intense fascination has also obscured the horrors of the war—the lives lost, the economic devastation, and the long-lasting effects of the Civil War on American politics, society, and life.

There was nothing civil about the Civil War.

It was a long, bloody, murderous affair. The fighting left more than 620,000 American soldiers dead and hundreds of thousands more crippled and maimed. As many as 50,000 civilians in the Confederacy may have died. It wreaked utter havoc on a good portion of the South—a scourge of economic devastation from which it took decades to recover. It left even deeper wounds on the national psyche.

For more than 130 years, writers, historians, sociologists, novelists, and poets have attempted to explain the Civil War in military, economic, social, and political terms. But the roots of this terrible war are still misunderstood by too many Americans. And there are all those who still know nothing at all. For in spite of all that has been said and written, far too many Americans still don't have a basic grasp of the Civil War, the central act in the drama of this nation. Much attention has recently been paid to the shocking deficits in American schooling and particularly in our knowledge of our nation's history. We are, as *New York Times* columnist Bob Herbert recently put it, "a nation of nitwits" when it comes to understanding our past. That is why, even with so many books already lining the shelves, here is another Civil War book.

The chief reason so many of us don't know and don't care about history—and the Civil War in particular—is plain boredom. Let's face it. Most of our schoolbooks were dull, dreary collections of dates and facts. A young woman, interested in my earlier book *Don't Know Much About® History*, asked me, "It's not just a bunch of battles, is it?" She is not alone. Many readers have dozed off under the influence of the weighty textbook histories, with their armies sweeping across flanks and wheeling to outmaneuver an enemy. Too many Civil War books look at the war with the clinical detachment of a description of a chess match or the official scorer's account of a baseball game.

Instead, I have tried to give the Civil War a human face while defining the essence of what the war was about and how it affected America. In *Don't Know Much About® History* and *Don't Know Much About® Geography*, I tried to humanize these subjects, so often viewed as boring, by showing that the participants in history were real people with human qualities. While even I confess it is tough to make "nullification" sound like fun, textbooks that emphasize facts over understanding have drained the lifeblood from history. I have tried to present the historical icons—Lincoln, Lee, Grant, Sherman, Frederick Douglass—as people. Like all of history, the Civil War was about people. And for every story of heroism, there is one of cowardice. For every example of sacrifice and devotion to duty, there are stories of corruption, money-grubbing, ambition, and double-dealing. But these tales remind us that the Civil War happened to real people. It's not just a bunch of dates and battles and speeches.

This is where even the most ardent Civil War buff will find what I call the hidden history of the period. Students of the Civil War may know every strategic detail about Gettysburg and what went wrong for the Confederates. They may not know that Robert E. Lee was suffering from severe diarrhea. Did this ailment, which was actually the leading killer of the Civil War affect Lee's judgment at this crucial moment? Is it widely known that disease was responsible for twice as many deaths as battle wounds? Do people know that Ulysses Grant once expelled all the Jews from the territory he commanded? Or that a wartime riot in New York City by Irish workingmen left 150 people dead and was far more savage than the recent riots in Los Angeles? Have they heard of the atrocities committed on both sides of the war, especially in Kansas, where civilians were murdered without remorse?

This touches on another reason for this book: the Civil War has been the spawning ground for many myths—old ones that often die hard. But new ones come along all the time. This is particularly true during the current debate over the study of American history. The grievous oversights and misguided teachings of the past—especially as they relate to blacks, women, and Native Americans—need to be addressed. For instance, during my school days in the 1960s, I never read the works of Frederick Douglass and heard precious little about him; never heard of Fort Pillow, where surrendering black soldiers were murdered by Confederate troops; and never learned about the exploits of the black troops depicted in the film *Glory*. We never heard about what half the country—the female half—was doing while the men were off fighting. In fact, the women were running the farms, taking jobs, or following their men into battle and nursing them. We did hear about the millions of immigrants who came to America during the war and were conscripted for service in a war they didn't understand. Sadly, we didn't hear the other side of the story—about the estimated one in twenty immigrant women who were forced into prostitution.

Of course, problems arise when revisionists play as fast and loose with the facts as the old historians did. There is little good in history standards that include the legendary escaped slave Harriet Tubman and pioneer feminists at the expense of George Washington or Robert E. Lee. That simply trades one deficient mythology for another. So my interest isn't in a "politically correct" view of the Civil War but one that is historically accurate.

Like most conflicts in history, the Civil War was essentially a struggle for power, which comes in many forms. These happen all the time in every level of life. In 1994 and 1995 Americans watched a power struggle that produced a seemingly unthinkable baseball strike and the cancellation of the World Series, which had previously survived two world wars and an earthquake. (While on the subject of the Civil War, baseball, and myths, let's get rid of the Abner Doubleday story once and for all. A Civil War general—and ancestor of the current owner of the New York Mets—Abner Doubleday did *not* invent baseball.) In a power struggle like this, what seems like a good idea on both sides goes downhill fast, like a child on a sled. Knowing he's going too fast, the child panics but then realizes there are no brakes.

The Civil War was no baseball strike. But there are parallels in the way these conflicts played out. Positions were staked out, leaving no room for retreat. Any sign of compromise was taken as a sign of weakness. Intractable, the two sides ride the brakeless sled. Before the Civil War, the two sides had marked their territory in a quest for power and were unwilling to budge in spite of the consequences. Once the sled was going downhill, it was impossible to stop.

The runaway sled in American history was slavery, and it is necessary to examine it to understand the power struggle that made the Civil War. For hundreds of years in America, slavery represented power, power over another human being. But far beyond that slavery meant economic, cultural, and political power.

When I was in school in the sixties, I recall that it became fashionable to say that the Civil War was not about slavery. Instead, "The War of Northern Aggression" was brought on by other issues, such as "states' rights," "nullification," and "preservation of the the Union." There are still people who cling to the idea that the Civil War was not fought over slavery. They criticize such modern historians as James M. McPherson or the documentary filmmakers Ric and Ken Burns as "neoabolitionists." These critics often point to the fact that soldiers on both sides were fighting neither to free the slaves nor defend slavery. Clearly, many Union men, like General William T. Sherman, had no argument with slavery and were downright racist in their attitudes. But this argument confuses two issues. The many reasons why men fought in the Civil War do not explain why there was a Civil War. A lot of young men picked up weapons back then for the same reasons that young men have gone to war for centuries— adventure, glory, duty, manhood, or "three squares" and a paycheck. Yes, many of those young men serving in the Civil War were not fighting for or against slavery, just as many of the young Americans who enlisted after Pearl Harbor didn't go to save the Jews from the Holocaust, rescue democracy in Europe, or save Asia from Japanese control. As most men on both sides saw it, their homeland had been attacked. "God and country" called and they answered.

Even so, the Civil War was clearly a war fought because of African slavery. All the other justifications come down to political differences, reflecting the social and cultural gulf between the free and slave states that might have been bridged if not for the deeply divisive issue of slavery. One way to approach this issue is to turn the question around: Without the division slavery caused, would the Confederate states still have left the Union?

Finally, the question of slavery raises what I think is the most important reason for this book. The Civil War never really ended. More than 130 years after the guns were silenced, we're still fighting over race in America, a dangerous fault line that threatens to split the country. It is clear that this deep-seated racial chasm is the most pernicious legacy of America's slave past and the Civil War. Until we all understand that chasm and how it has been part of the American experience since colonial times, this country will never heal the wounds, get over them, and move on.

*

Author's Note: The word *nigger* is an offensive vulgarity. And in the current age of "political correctness," it has become unfashionable to even mention such a word or teach a classic like Twain's *Huckleberry Finn* because of its use in that book. However, *nigger* is frequently used here in its historical context. I apologize to those who find its use painful or take offense.

DON'T KNOW MUCH ABOUT®

MUCH ABOUT®

The Civil War

chapter one

"*The Wolf by the Ears*"

As God Himself hath set them over you here in the nature of his stewards or overseers, He expects you will do everything for them, as you do for Himself: That you must be obedient and subject to them in all things, and do whatever they order you to do.

—THOMAS BACON
"SERMON TO NEGRO SLAVES" (1743)

We hold these truths to be self-evident, that all men are created equal, that they are endowed by their Creator with certain unalienable rights, that among these are life, liberty and the pursuit of happiness.

—THOMAS JEFFERSON
DECLARATION OF INDEPENDENCE (1776)

How is it that we hear the loudest yelps for liberty from the drivers of Negroes?

—SAMUEL JOHNSON (1775)

✳

E arly in April 1865 Charles Coffin, a correspondent for the *Boston Journal,* witnessed an extraordinary occurrence. Having reported on the Civil War from the earliest days, he had observed many significant events, from the First Battle of Bull Run four years earlier through Grant's final campaign in Virginia in 1865. Now he was present at the capture of Richmond and the arrival of President Abraham Lincoln in the fallen Confederate capital. Coffin recounted:

> No carriage was to be had, so the President, leading his son, walked to General Weitzels' headquarters—Jeff Davis's mansion. . . . The walk was long and the President halted a moment to rest.
>
> "May de good Lord bless you, President Linkum!" said an old Negro, removing his hat and bowing, with tears of joy rolling down his cheeks.
>
> The President removed his own hat and bowed in silence. It was a bow which upset the forms, laws, customs, and ceremonies of centuries of slavery.

So ended the Civil War. Of course, it would be a few more days before Robert E. Lee's final surrender at Appomattox Courthouse on April 9, 1865. And months more of dislocation, followed by years of bitter Reconstruction and decades of hatefulness between victor and vanquished. But in this brief moment, in the crumbling, burning remnants of the Confederate capital, the heart and symbol of the ruined Confederate Cause, the war came to a close. Hundreds of thousands were dead. A large part of the country was in ruins, smoldering. A deep sense of regional mistrust and racial hatred would sunder America for decades. But here, as the tall, somber president bowed to a former slave, the war was crystalized in an eternal moment of reconciliation: the doomed Lincoln, symbol of the Union, worn down by the years and the losses, slow to name slavery as the enemy but indomitable in his will to ultimately destroy it, and an aged slave, bent by years of relentless labor, glorying in the first flush of freedom.

"We have the wolf by the ears," an aging Thomas Jefferson had written to a friend forty-five years earlier. "And we can neither hold him, nor safely let him go. Justice is in one scale, and self-preservation in the other." Jefferson's "wolf" was, of course, slavery. And this big, bad wolf had been banging at America's door almost since the arrival of the English in America. It huffed and puffed and nearly blew the house down.

The United States was born out of a revolutionary idea that Jefferson (1743–1826) expressed eloquently in his Declaration of Independence: All men are created equal and are entitled to life, liberty, and the pursuit of happiness. These notions, along with the very radical statement that governments could rule only by the consent of the governed, formed the basis of the Great American Contradiction: How could a nation so constituted, dedicated to the proposition that all men are created equal and supposedly founded on the cornerstone of "liberty for all," maintain a system that enslaved other human beings? It was this contradiction—this Great Divide—that eventually split the country in two.

Only about one quarter of the people in the slave states kept slaves. And of the five and a half million whites living in the slave states in 1860, only forty-six thousand held more than twenty slaves. But to understand fully the Civil War, this Great Divide—this American Contradiction—must be understood. Its roots were deep, planted about the same time that the first English colonists were learning how to plant tobacco in Virginia.

Who Brought Slavery to America?

George Washington did it. Patrick "Give Me Liberty or Give Me Death" Henry did it. Thomas "All Men Are Created Equal" Jefferson did it. George Mason, who wrote the Virginia Declaration of Rights, on which the American Bill of Rights was based, was against it but did it anyway. Even good old Benjamin Franklin did it. In fact, many of America's Founding Fathers did it. They bought, kept, bred, and sold human beings.

When the Constitutional Convention met in Philadelphia in 1787 to debate the great issues facing a struggling, infant nation, seventeen of these Founding Fathers collectively held about fourteen hundred slaves. George Mason, John Rutledge, and George Washington were three of the largest slaveholders in America. Most of Washington's slaves actually belonged to his wife, Martha Custis Washington; they had belonged to her first husband. This is the part of Washington's story that gets left out when children learn the tale of the cherry tree. The moral subtext: Telling lies is wrong; keeping people in chains isn't so bad.

Slavery was as American as apple pie. It was a well-established American institution in the thirteen original colonies long before Washington, Patrick Henry, Franklin, and Jefferson were born, but it threatened to tip the great American ship of state from the republic's very beginnings. Both Washington and Jefferson expressed deep reservations about the practice of slavery and its future in America. Nevertheless, neither of them fretted sufficiently about human bondage on their plantations to do much about it. Granted, Washington freed his slaves in his will. Jefferson, who seems to have been brilliant about everything but his finances, couldn't afford that luxury for most of his slaves. He had to rely on the kindness of creditors to let five of his favored slaves have their freedom.

Of course, America had no monopoly on slavery. The institution was as old as civilization itself. Throughout human history, slavery has taken on many guises, and few civilizations have been built without some form of servitude. In his prize-winning book, *Freedom*, Orlando Patterson wrote, "Slaveholding and trading existed among the earliest and most primitive of peoples. The archaeological evidence reveals that slaves were among the first items of trade within, and between, the primitive Germans and Celts, and the institution was an established part of life, though never of major significance, in primitive China, Japan and the prehistoric Near East."

It is impossible to think of such major ancient civilizations as Egypt, Mesopotamia, Greece, Rome, or the Aztecs without acknowledging the role of slavery. Occasionally, slavery took on a variety of slightly more humane forms. In some societies, for instance, slaves could serve in armies, buy their freedom, or even rise to positions of rank and power. Remember the biblical story of Joseph and the pharaoh? Sold as a slave into Egypt, Joseph demonstrated his wisdom to save the country from famine and rose to become an adviser to the Egyptian court. But for the most part, slavery always meant a short life of brute labor or an existence of household servitude passed on to succeeding generations. The monuments of the great civilizations—from the Egyptian Pyramids to the Roman roads and Jefferson's prized Virginia estate, Monticello—were all constructed with slave labor.

But what made Anglo-American slavery so uniquely perverse was the institutionalization, or grotesque efficiency, of the African slave trade. Involving vast numbers of people and contradicting America's founding ideals so obviously, African slavery became America's "peculiar institution."

The history of Africans in America doesn't begin, however, with slave auctions in the British colonies. The first black men came to the Americas along with the earliest wave of Spaniard conquistadors who followed Columbus in the 1500s. These early European explorers included black natives of Spain and Portugal such as Pedro Alonso Niño, a navigator on Columbus's first voyage. Black men served with Cortés in Mexico and with Balboa at the discovery of the Pacific. And several Iberian blacks were prominent in the exploration of what is now the southwestern United States, including most notably Estebanico, who spent eight years wandering in what is now Mexico and New Mexico before being killed in a battle with Zuñi Indians. This history, long overlooked, is somewhat sketchy.

Better documented but still overlooked is the fact that the first "cargo" of twenty Africans to land in the future United States of America arrived at Jamestown, Virginia—known to generations of American schoolchildren as the first permanent English settlement in America—aboard a Dutch man-of-war in August 1619. These African captives had been taken off a Spanish slave ship en route to Spain's colonies by a polyglot group of pirates who preyed on Spanish slave and treasure ships. This moment escaped generations of history books. Of course, those same books hailed the glorious consequences of the opening of the House of Burgesses in Jamestown. Sure. It was the first legislative assembly in America. But a nasty little secret was that in the very same month those gentlemen planters met to begin inventing American democracy, they were laying the foundation for American slavery.

By the time the English arrived in America, African slavery was well established in Europe. As a sideline to the more famous voyages of discovery he sponsored, Prince Henry, lionized as "the Navigator," led Portugal in opening the African slave trade in 1444 with the blessing of Pope Nicholas V. Quick to recognize a profitable venture and desperate for labor in its mines and plantations in the New World, Spain joined in and dominated the African slave trade for almost a hundred years. Gradually, other European trading nations saw the profits to be made from this "black ivory." For about fifty years, the Dutch dominated the trade until they were supplanted by the English, who got an exclusive right to supply slaves—the Asiento—to Spain's colonies in 1713.

PRELUDE TO THE CIVIL WAR: 1619–1787

1619

August Twenty Africans, carried on a Dutch ship, are brought to Jamestown, Virginia, to be sold as indentured servants, not slaves, a fine distinction that probably escaped their notice.

1652

May Rhode Island outlaws slavery. The first colony to do so, it was founded by religious leader Roger Williams (1603–1683) on the grounds of religious freedom, the separation of church and state, and the compensation of the Indians for their lands. But slavery, because it was so profitable, was later permitted, and Newport became a major slave trading port.

The Netherlands permits African slaves to be brought into New Netherlands, its American colony, which later became New York. The Dutch colony enacts laws governing the treatment of slaves, who may only be whipped with the permission of the colonial authorities.

1664

Maryland is the first colony to mandate the lifelong servitude of black slaves. Previously, under English law, slaves who became Christians and met other requirements were granted their freedom.

1667

The House of Burgesses in Virginia passes a law stating that Christian conversion does not bring about freedom from slavery. This law encourages slave owners to convert their slaves to Christianity without fear of losing them. In *1670* a countermeasure is passed which allows freedom for those blacks who were Christians before arriving in the colony. This law is a reaction to the moral issue raised by slavery. In *1671* Maryland reacts by declaring the conversion of slaves at any time as irrelevant to their servitude. And in *1682* Virginia repeals its *1670* conversion law, which had limited the importation of slaves.

1672

The English Royal Africa Company is granted a monopoly on the English slave trade which lasts until *1696*. After that date, extensive slave trading is initiated by merchants in the New England colonies.

1688

The Germantown Protest, the first organized demonstration in opposition to slavery and the slave trade, is made by a group of Quakers in Germantown, Pennsylvania.

1700

In Boston, Samuel Sewall (1652–1730) publishes *The Selling of Joseph*, one of America's first antislavery tracts. One of the judges at the Salem witchcraft trials in 1692, Sewall was the only one to later publicly admit his mistake in condemning nineteen people to death.

Virginia rules that slaves are "real estate," restricts the travel of slaves, and calls for strict penalties for miscegenation (marriage or sexual relations between races).

New York enacts a death penalty for runaway slaves caught forty miles north of Albany.

Massachusetts declares marriage between blacks and whites illegal.

1713

A British firm, the South Sea Company, is granted the Asiento, the exclusive right to import black slaves into Spain's American colonies.

1725

The black slave population in the colonies is estimated at seventy-five thousand.

In Williamsburg, Virginia, black slaves are granted the right to organize a separate Baptist church.

1733

Georgia is chartered, the last of the thirteen original colonies. Under the terms of the charter, the importation of slaves is forbidden, but this prohibition is repealed in *1749*.

1739

September A violent slave insurrection takes place in Stono, South Carolina. The slaves kill thirty whites; forty-four blacks die in the aftermath.

1741

March Following a series of crimes and fires in New York, rumors of a black plot to seize power spread through the city. In the hysteria, more than a hundred blacks and poor whites are convicted of conspiracy. Four whites and eighteen blacks are hanged, thirteen blacks are burned alive, and another seventy blacks are banished from the colony.

1754

Quaker clergyman John Woolman (1720–1772), preaching against the evils of slavery, publishes his sermons under the title *Some Considerations on the Keeping of Negroes*. He publishes a second volume in *1762*. Woolman is instrumental in persuading American Quakers to oppose slaveholding.

1775

Black patriots fight in all the early battles of the Revolution including Lexington, Concord, Fort Ticonderoga, and Bunker (Breed's) Hill. However, the Continental Congress, with George Washington's consent, bars slaves and free blacks from the Continental Army. Only when the deposed royal governor of Virginia promises freedom to male slaves who join the British army does Washington reverse his orders and authorize the enlistment of free blacks. Eventually, an estimated five thousand blacks are involved in many of the Revolution's major engagements.

1776

July Thomas Jefferson drafts the Declaration of Independence. Included is a charge that King George III is responsible for the slave trade and has prevented the colonists from outlawing it. This passage is deleted by Congress at the request of Southerners who keep slaves and with the support of Northerners involved in the slave trade.

1777

July Although not yet a state, Vermont abolishes slavery. It will be the first free state when it enters the Union.

1780

The Pennsylvania legislature mandates gradual abolition within the state.

1781

March 1 First proposed in 1776, the Articles of Confederation is ratified by all thirteen states; it calls for a central government subordinate to the states. Previously, the nation was a league of sovereign states, each with a single vote. Congress, made up of delegates chosen by the states, is given certain limited powers, and the states are not bound to carry out any congressional measures.

1783

The Supreme Court of Massachusetts abolishes slavery.

In Virginia, black slaves who served in the Continental Army are granted their freedom by the House of Burgesses.

By year's end the importation of African slaves is banned by all the northern states.

1787

July Still operating under the Articles of Confederation, Congress passes the Northwest Ordinance, establishing a government in the area bounded by the Great Lakes on the north, the Ohio River on the south; and the Mississippi River on the west. This ordinance calls for the creation of new states on an equal level with the original states. Under Jefferson's plan, there will be freedom of religion, the right to public education, and a ban on slavery in these new states.

August In Philadelphia, the form of the new U.S. Constitution is debated. The delegates compromise on three areas concerning slavery:

* Slavery or the slave trade cannot be restricted by law for twenty years.
* Slavery can be taxed.
* Slaves will be counted at three fifths of their total population for the purpose of determining representation in the House of Representatives.

What's the Difference Between a Servant and a Slave?

In 1619 indentured servitude was a widely accepted European tradition. Essentially, it was a contract between a worker and a merchant or farmer. In colonial America, the servant agreed to work for some period of time in exchange for passage to America. In many cases, the agreement was one-sided: convicts, people kidnapped from the street, and young children "apprenticed" to masters by their parents all ended up in America as indentured servants. The word *indenture* comes from the contract between servant and master. The contract was actually torn in two, with each party holding one half. At the end of the term of servitude, the contract would be pieced back together at the indentations. After earning their freedom, the former servants were usually given a piece of land and a small payment.

This arrangement was sanitized in the history books of a generation ago as a tidy system of labor with clean-cut servants happily fulfilling their contracts and gradually working their way toward freedom and a piece of the American dream. The reality was rudely different. Freedom, first of all, demanded survival. And servitude did not generally lend itself to a long career. During the seventeenth century, up to half the servants in the Chesapeake colonies may have died before being freed, usually of disease. Servants often had their terms of indenture arbitrarily extended, sometimes in punishment for sins real or imagined. Often it was the simple whim of a master who knew the law and how to read. Most indentured servants were young and illiterate. Of course, it was harder for women than men: indenture for a young woman often meant sexual servitude as well as long hours and no pay.

But the American colonies were fairly well built on this system, and it took care of colonial labor needs for most of the seventeenth century, when the opportunity to get out of Europe seemed like a good idea if you were poor. In the 1700s, however, the situation changed. Improved economies and political stability in Europe gradually reduced the number of white Europeans who were willing to sell themselves to a master in exchange for passage and a chance for a new life in the New World. As slavery displaced servitude, the number of slaves being imported grew rapidly.

Economics was another reason for the shift to a slave system. On the face of it, slavery seemed to be extremely cost effective. A slave is a slave is a slave. Forever. There was no promise of freedom and a piece of farmland at the end of the rainbow. The only alternatives were escape (unlikely), the granting of purchase of freedom (even more unlikely), or death (inevitable).

Colonial masters had also learned that white servants could easily escape and drift into anonymity. An experiment using Native Americans as slaves also failed. They too could escape easily and return to their people. The Europeans also learned that they died in fearful numbers, usually from diseases to which they had no immunities. But African slaves solved the problem perfectly. They were easily identifiable in a predominantly white society. They were a long way from home. And they could neither speak nor read the language of their masters. More important still, slaves made more slaves. While the child of a servant was born free, a female slave's child was a slave, even if its father was also its master. An African slave woman who could be purchased and then bred was a valuable piece of property, just like any farm animal.

By the late 1600s, the southern plantation system was firmly established, and at the same time American slavery was cemented through a growing number of laws passed by colonial legislatures. The full legalization of slavery, restrictions on the emancipation of slaves, laws forbidding miscegenation, and strict regulations regarding escape all came in quick succession. All of these laws solidified slavery as a legal reign of terror. And the plantations' success in producing cash crops of tobacco, indigo, rice, and sugar for European markets pushed the need for additional slaves to do the labor-intensive work. In fact linked with imported slaves, the southern slave-breeding system was so successful that by the time Jefferson wrote the Declaration of Independence, *one in six Americans was a black slave!* Eighty-four years later, at the beginning of the Civil War, there were nearly four million slaves in America, despite the legal end of the overseas slave trade in 1808.

But the simplistic economic argument for slavery ignored a more complex view of how economies work. In 1776, English philosopher Adam Smith (1723–1790) wrote *The Wealth of Nations*, considered the bible of capitalism. Smith argued that capital is best employed for the production and distribution of wealth when governments stay away from business or practice *laissez-faire* (French for "leave things alone"). Further, Smith believed that slaves were a drag on the economy because they tied up capital that could be more profitably invested elsewhere, a view that was widely accepted by merchants in the American North. In addition, Smith felt that the large plantation system was a wasteful use of land and that slaves cost more to maintain than free laborers.

Here was a real objection to slavery. Sure, it might be wrong. But far worse to the Puritans of New England, it was inefficient! No longer wed to the slave system, Northerners could invest their capital in the ships, factories, railroads, and canals—one reason for the North's Industrial Age, which created the wealth and jobs that lured millions of European working-class immigrants, primarily from Ireland and Germany, to the growing northeastern industrial centers in the eighteenth and early nineteenth centuries.

Adam Smith's ideas didn't have the same appeal in the booming slave economies, however. Holding to their faith in the seemingly obvious economic benefits of slavery, the slaveholders developed an elaborate set of defenses to justify their "peculiar institution." In

Europe, the "savagery" of Africans had been cited for centuries to justify slavery, and slaveholding Americans now parroted that excuse.

But the most fiercely held defense was one that people have been using for a long time to answer almost any question: the biblical justification. First, clergymen provided the white masters with references that seemed to sanction slavery—a perfectly acceptable practice in biblical times—and, more specifically, the enslaving of the black race. The critical passage was Genesis 9:25–27, in which Noah, upset over an indiscretion by his son Ham, cursed all of Ham's descendants for eternity, saying, "a slave of slaves shall he be to his brothers." Ham fathered four sons, who gave rise to the southern tribes of the Earth, including the people of Africa.

Beyond that, a second theological argument claimed that slavery was actually a good thing. White masters converted their slaves to Christianity, bringing these "heathens" hope for salvation. One problem with this argument was that English law had provided that a slave who was baptized a Christian could become free. American slaveholders soon realized that saving souls meant losing bodies, so most colonies passed laws that said Christian baptism was no longer a pathway to freedom. For slaves, Christianity might put them on the stairway to heaven, but it didn't save them from hell on earth.

Civil War Voices

Maryland minister Thomas Bacon's "Sermon to Negro Slaves" (1743).

> Suppose that you were people of some substance, and had something of your own in the world, would you not desire to keep what you had? And that no body should take it from you, without your own consent, or hurt any thing belonging to you? If, then, you love your neighbor as yourself, or would do by others as you could wish they would do by you, you will learn to be honest and just towards mankind, as well as to your masters and mistresses, and not steal, or

take away any thing from any one, without his knowledge
or consent: You will be careful not to hurt any thing belonging to a neighbour, or to do any harm to his goods, his
cattle, or his plantation.

What Was the Triangle Trade?

The economic heart of the slave trade was a smooth system of supply
and demand that came to be called the Triangle Trade. Here, the
Anglo-American genius for commerce settled into a ghastly and murderous brand of efficiency. Operating under the flags of all nations,
the slave ships left those pious ports of Protestant New England with
cargoes of rum along with trinkets, guns, and iron bars, which were
used as currency in Africa. In 1643 the first American slavers had
exchanged shiploads of fish, lumber, and grain—the rich output of
the then-prospering American colonies—for the prized sherry and
Madeira wines from Spain and Portugal. But they also bought Spanish and Portuguese slaves.

Eventually these enterprising New Englanders realized that they
could cut out the middleman and go directly to the source: Africa.
On the first leg of the triangle, the rum, cotton, grain, and other raw
materials and goods produced in the colonies went directly from New
England to Africa in exchange for slaves. On the second and deadliest
leg of the triangle, the captives were taken to the West Indies and
sold to plantation owners. The slavers then returned to the American
ports of Richmond, New York, Newport, and Boston, among others,
on the final leg of the triangle, with sugar, molasses, large profits from
the Caribbean islands, and "seasoned" slaves who had gone through
a period of "breaking in" that lasted as long as three years. During
this time newly arrived Africans were mixed in with veterans. The
newcomer learned his work assignment and any tendencies to commit suicide or escape were dispelled. "Seasoning" doubled his value.
The sugar and molasses were turned into New England rum, which,
along with other colonial products, was sent to Europe and Africa in
return for more slaves.

It wasn't always so geometrically perfect. Ships departed from the triangle route to trade with Europe. But this grim triangular web was the heart of the business. Determining the precise number of Africans who were caught in this deadly triangle is impossible. Various estimates in the past put the number of blacks taken from Africa as high as 24 million. More recent studies, such as Peter Kolchin's *American Slavery* (1993), suggest that 10 to 12 million Africans were actually transported to the Americas. By any measure it was an African Holocaust. And that estimate does not take into account the millions of Africans who did not survive capture or the transatlantic voyage. Taking the conservative estimate of 14 to 20 percent who died in transit—some ships actually lost half their human cargo to disease, starvation, and other grim fates—produces a range of at least another 4 or 5 million Africans who perished before reaching the American slave markets. The vast majority of the survivors ended up working in the Caribbean sugar fields or on Brazilian plantations, but somewhere between 600,000 and 700,000 Africans arrived in the colonies that would eventually become the United States. Not a single American colony—including Rhode Island, which had once forbidden slavery—was free of slaves. Astonishingly, these hundreds of thousands torn from their villages and homes survived degradation and deprivation to become the almost 4 million people held in slavery in 1860, at the eve of the Civil War. (Noting a high birthrate and a lower mortality rate among American slaves than other slaves in the Americas, defenders of American slavery would point to this population increase as a sign that American slavery was humane and healthy for Africans. Historians disagree, pointing out that American crops were easier to work than the sugar crops of the Caribbean and Brazil, there were fewer deadly tropical diseases, and American slaves generally enjoyed a better diet than those in the Caribbean and South America.)

Civil War Voices

From "The Advantages of the African Trade" (1772), by Malachi Postlethwayt, a British economist.

The most approved Judges of the commercial Interests of these Kingdoms have ever been of Opinion, that our West-India and African Trades are the most nationally beneficial of any we carry on. It is also allowed on all Hands, that the Trade to Africa is the Branch which renders our American colonies and Plantations so advantageous to Great-Britain; that Traffic only affording our Planters a constant Supply of Negro Servants for the culture of their Lands....

The Negroe-Trade therefore ... may be justly esteemed an inexhaustible Fund of Wealth and Naval Power to this Nation. And by the Overplus of Negroes ... we have drawn likewise no inconsiderable Quantities of Treasure from the Spaniards.

... Many are prepossessed against this Trade, thinking it a barbarous, inhuman and unlawful Traffic for a Christian Country to Trade in Blacks; to which I would beg leave to observe; that though the odious Appellation of Slaves is annexed to this Trade, it being called by some the Slave-Trade, yet it does not appear from the best Enquiry I have been able to make, that the State of those People is changed for the worse by being Servants to our British Planters in America; they are certainly treated with great Lenity and Humanity: And as the Improvement of the Planter's Estates depends upon due Care being taken of their Healths and Lives, I cannot but think their Condition is much bettered to what it was in their own Country.

... But I never heard it said that the Lives of Negroes in the Servitude of our Planters were less tolerable than those of Colliers and Miners in all Christian Countries.

What Was the Middle Passage?

Of course, the key to the Triangle Trade was a large supply of "merchandise."

In Africa, few Europeans ever actually captured a slave. The work of capturing and enslaving Africans was generally done by other Africans, who sold their prisoners to Arab coastal traders. Who were these unlucky millions? Like slavery through the ages, African slavery began with prisoners of war. Some were tribal royalty, prizes of war, taken by neigboring or enemy tribesmen and sold to Arab traders for rum, codfish, salt, and Spanish money. In some cases they were criminals or debtors. Eventually they were simply the subjects of African kings and local chieftains who were unable to resist the temptation of the European riches and goods placed before them. Linked together in coffles—chains that bound the captives together by the neck, a word derived from the Arabic for *caravan*—the Africans were turned over to the Arab merchants who ran the slave markets of Africa's West Coast.

In *The Africans*, journalist David Lamb described a modern visit to the bleak cells where the African captives once awaited their fate:

> It was here, in the dark, dank slave house that Arab traders bartered and bickered with European shippers, here that the Africans spent the last weeks in their homeland, awaiting a buyer, a boat and, at the end of a long, harrowing voyage, a master.
>
> In the weighing room below the trading office, the men's muscles were examined, the women's breasts measured, the children's teeth checked. Those were the qualities—muscles, breasts, teeth—on which human worth was judged. The slaves were fattened like livestock up to what was considered the ideal shipping weight, 140 pounds, and those who remained sickly or fell victim to pneumonia or tuberculosis were segregated from the rest by an Arab doctor, led

into a corridor whose open door overlooked the ocean and tossed out for the sharks.

The Europeans and Americans brought their trade goods, landing at fortresses along the West African coast where the captives had already been introduced to barbaric tortures meant to break their will to escape. For those who attempted to flee, there was an anklet through which a spike was driven into the foot. Another device used a spike to seal the lips of defiant captives. The masters of the slave ships bartered their material cargo for human cargo, branding the Africans before forcing them into the holds.

On the voyage, the captives were packed as tightly as "books upon the shelf" in the hellhole below the decks, usually with no more than eighteen inches of space above them. One captain recorded, "They had not so much room as a man in his coffin, either in length or breadth. It was impossible for them to turn or shift with any degree of ease."

In this floating coffin, the journey lasted six to ten weeks. The water was impure, the food inedible, if available at all. When a ship ran low on provisions, captives were simply thrown overboard. Those who became weak or sick were also thrown overboard. Many others committed suicide by jumping into the sea rather than submit.

When this dreadful middle passage or journey to the islands of Jamaica or Barbados ended, the survivors were delivered to slave dealers and the vessels were loaded with agricultural products from the Caribbean plantations for the final leg of the voyage north to New England. Many of the merchants and shipowners of New England, those respectable Puritan gentlemen who would foot the bill for the American Revolution, profited handsomely from their slave ships. They invested their profits wisely in factories, mines, and transportation, providing much of the impetus for the great Industrial Revolution that would eventually help separate North and South.

Civil War Voices

From *The Interesting Narrative of the Life of Olaudah Equiano or Gustavus Vassa, the African* (1789), the first published autobiography of an African American and the first of thousands of narratives by former slaves that appeared throughout the eighteenth and nineteenth centuries.

The first object that saluted my eyes when I arrived on the coast was the sea, and a slave ship, which was then riding at anchor, and waiting for its cargo. These filled me with astonishment, that was soon converted into terror, which I am yet at a loss to describe. . . . I was immediately handled and tossed up to see if I was sound, by some of the crew; and I was now persuaded that I had got into a world of bad spirits, and that they were going to kill me. Their complexions too, differing so much from ours, their long hair, and the language they spoke, confirm me in this belief. Indeed such were the horrors of my views and fears at the moment, that if ten thousand worlds had been my own, I would have freely parted with them all to have exchanged my condition with the meanest slave in my own country. When I looked round the ship too, and saw a large furnace or copper boiling and a multitude of black people of every description, chained together, every one of their countenances expressing dejection and sorrow, I no longer doubted of my fate; and, quite overpowered with horror and anguish, I fell motionless on the deck, and fainted. When I recovered a little, I found some black people about me, who I believed were some of those who brought me on board, and had been receiving their pay: They talked to me in order to cheer me, but all in vain. I asked them if we were not to be eaten by those white men with horrible looks, red faces and long hair. . . .

. . . I was soon put down under the decks, and there I

received such a salutation in my nostrils as I had never experienced in my life; so that, with the loathsomeness of the stench, and with my crying together, I became so sick and low that I was not able to eat, nor had I the least desire to taste any thing. I now wished for the last friend, death, to relieve me; but soon, to my grief, two of the white men offered me eatables; and, on my refusing to eat, one of them held me fast by the hands . . . and tied my feet, while the other flogged me severely. I had never experienced any thing of this kind before, and although, not being used to the water, I naturally feared that element the first time I saw it, yet nevertheless, could I have got over the nettings, I would have jumped over the side, but I could not; and besides the crew used to watch us very closely, who were not chained down to the decks, lest we should leap into the water. I have seen some of these poor African prisoners most severely cut for attempting to do so, and hourly whipped for not eating. This indeed was often the case with myself. . . .

. . . At last, when the ship, in which we were, had got in all her cargo, they made ready with many fearful noises, and we were all put under deck, so that we could not see how they managed the vessel.

But this disappointment was the least of my grief. The stench of the hold, while we were on the coast, was so intolerably loathsome, that it was dangerous to remain there for any time, and some of us had been permitted to stay on the deck for the fresh air, but now that the whole ship's cargo was confined together, it became absolutely pestilential. The closeness of the place, and the heat of the climate, added to the number in the ship, being so crowded that each had scarcely room to turn himself, almost suffocated us. This produced copious perspirations, so that the air soon became unfit for respiration, from a variety of loathsome smells, and brought on a sickness among the slaves, of which many died, thus falling victims to the improvident avarice, as I may call it, of their purchasers. This deplorable situation was again aggravated by the galling of the chains, now become insupportable; and the filth of necessary tubs, into which the children often fell, and were almost suffocated. The shrieks of the women, and the groans of the dying, rendered it a scene of horror almost inconceivable. Happily, perhaps, for myself, I was soon reduced so low here that it was thought necessary to keep me almost continually on deck; and from my extreme youth, I was not put in fetters. In this situation I expected every hour to share the fate of my companions, some of whom were almost daily brought upon the deck at the point of death, and I began to hope that death would soon put an end to my miseries.

Equiano survived the trip, the slave market in Barbados, and a colonial Virginia plantation. Subsequently, he was sold to a British naval officer, who thought it clever to rename the slave boy Gustavus Vassa, after a Swedish king. Vassa was taken to Nova Scotia and later to the Mediterranean. From a third master, a Quaker named Robert King, he learned to read and write. Permitted to earn extra money, he purchased his freedom in 1766 and pursued a series of adventures that included a polar voyage. His memoir appeared in 1789, one of the few slave narratives that includes an account of the transatlantic voyage.

What Was the Stono Rebellion?

The dreadful journey finished, the slave ships landed in the West Indies at places like Barbados and Jamaica, where the human cargo was sent to the market to be sold. Those slaves who were sold to the colonies of North America made another voyage, to Atlantic ports like New York, Newport, and Boston, to be sold once again in American slave markets. Though a small fraction went to the northern colonies to become domestics or do factory work, the vast majority ended up in the Deep South to work the rice, tobacco, and, later, cotton plantations. Here was the scene of the greatest myth of the slave period—the image of slaves as docile, somewhat childlike laborers content with their situation. It was a paternalistic image perpetuated by slave owners, who wanted to present slavery as somehow beneficial to the slaves. This myth continued into this century and might be the greatest sin of books and films like *Gone With the Wind*.

Almost from its very beginnings in colonial America, slavery produced violent reactions by slaves, who protested either by working slowly and inefficiently, running away, or rebelling. Although the 1831 rebellion of Nat Turner is the most famous slave revolt—because it became the subject of William Styron's Pulitzer Prize–winning novel *The Confessions of Nat Turner* (1967)—it was by no means unique.

New York was the site of some of the earliest and bloodiest slave uprisings. In Queens County in 1708, an Indian slave and a black woman killed their master and his family of six. In two separate instances in 1712 and 1741, a total of fifty blacks and thirteen whites were killed. But one of the most violent early slave uprisings began on September 9, 1739, in Stono, South Carolina, on the Stono River about twenty miles from Charleston. The Stono Rebellion, the first mass slave revolt, was led by a slave named Jemmy. Apparently encouraged by Spanish missionaries and promised liberation, a group of twenty slaves attacked a store and killed the two storekeepers. Armed with guns and powder, they set off for St. Augustine, Florida. As they marched through the country with their captured weapons, according to one account, they set fire to plantations and cried out, "Liberty." Gathering recruits, Jemmy and his band eventually grew to about a hundred. During a ten-mile march, they killed approximately thirty white people before an armed militia caught them in a field. Forty-four of the rebellious blacks died in the aftermath, and in one account the white planters "cutt off their heads and set them up at every Mile Post they came to." Meant as a warning to other rebellious slaves, the display grimly recalled the famed ancient uprising of Spartacus, who led an army of slaves against the Roman legions. (After their defeat, six thousand followers of Spartacus were crucified and left to rot along a Roman highway.)

The Stono Rebellion was followed by two other uprisings in South Carolina that year. With each successive revolt, the colonies and later the states of the young Republic instituted new and harsher laws to control the slaves. Indeed, a great part of the objection to slavery came not from those who had moral qualms but from those who feared that the rapidly expanding numbers of slaves would eventually lead to more dangerous rebellions.

Documents of the Civil War

Following the Stono Rebellion, the colonial legislature of South Carolina, the Commons House of Assembly, passed An Act to Impose Duties on the Importation of Slaves (1740) out of fear of the growing number of slaves and the likelihood of future rebellions.

> Whereas, the great importation of Negroes from the coast of Africa, who are generally of a barbarous and savage disposition, may hereafter prove of very dangerous consequence to the peace and safety of the Province, and which we have now more reason to be apprehensive of from the late rising in rebellion of a great number of the Negroes lately imported into the Province from the coast of Africa ... and barbarously murdering upwards of twenty persons of His Majesty's faithful subjects of this Province; and whereas, the best way to prevent those fatal mischiefs for the future, will be to establish a method by which the importation of Negroes into the Province should be made a necessary means of introducing a proportionable number of white inhabitants into the same; and whereas, in order to effect this good purpose, it is fit and necessary that a sufficient fund should be appropriated by the laws of this Province for the better settling the frontiers ...

... there shall be imposed and paid by all and every the inhabitants of this Province and other person and persons whomsoever, first purchasing any negro or other slave within the same which hath not been the space of six months within this Province, a certain tax or sum of twelve pounds current money for every such negro and other slave of the height of four feet and two inches and upwards; and for every one under that height, and above three feet two inches, the sum of five pounds like money; and for all under three feet two inches (suckling children excepted), two pounds ten shillings like money.

Civil War Voices

From Phillis Wheatley's "To the Right Honorable William, Earl of Dartmouth" (1773).

> Should you, my lord, while you pursue my song,
> Wonder from whence my love of *Freedom* sprung,
> Whence flow these wishes for the common good,
> By feeling hearts alone best understood,
> I, young in life, by seeming cruel fate
> Was snatch'd from *Afric's* fancy'd happy seat;
> What pangs excruciating must molest,
> What sorrows labour in my parent's breast?
> Steel'd was the soul and by no misery mov'd
> That from a father seiz'd this babe belov'd.
> Such, such my case. And can I then but pray
> Others may never feel tyrannic sway.

Born in Africa, the young woman later called Phillis Wheatley (c. 1753–1794) was captured by slave traders at the age of eight. Sold to the Wheatley family of Boston, she was recognized as a prodigy and, by age thirteen, was writing poetry modeled on such English poets as Alexander Pope and translating from ancient Greek. Taken to England, she astonished the British literary world when her *Poems on Various Subjects: Religious and Moral* was published in London in 1773. Not everyone admired her skills. Jefferson dismissed Wheatley's poetry, perhaps because of its religious nature, writing, "Religion produced a Phillis Wheatley, but it could not produce a poet."

Wheatley, along with other Africans like Olaudah Equiano, gave the lie to another myth of the slave period: that Africans were barbarous and inferior, unable to master complicated tasks, a myth that unthinkably remains alive in modern times in the flickering debate over race and intellect. One who doubted the ability of blacks was Jefferson, who not only knew of Wheatley but was also acquainted with another extraordinary black man of the period, Benjamin Banneker (1731–1806). His grandmother a Welsh convict, his grandfather an African slave, Banneker was born a free black in Maryland, a child of obvious talents. He received an education and established a reputation as a mathematician and astronomer. In 1791 he sent Jefferson, then George Washington's secretary of state, the manuscript of his *Almanac*, along with a letter in which he upbraided Jefferson for his contradictory views.

Suffer me to recall to your mind that time in which the arms of the British crown were exerted, with every powerful effort, in order to reduce you to a state of servitude; look back, I entreat you. . . . You were then impressed with proper ideas of the great violation of liberty, and the free possession of those blessings, to which you were entitled by nature; but sir, how pitiable is it to reflect, that although you were so fully convinced of the benevolence of the Father of Mankind, and of his equal and impartial distribution of these rights and privileges which he hath conferred upon them, that you should at the same time counteract his mercies, in detaining by fraud and violence, so numerous a part of my brethren under groaning captivity and cruel oppression, that you should at the same time be found guilty of that most criminal act, which you professedly detested in others.

Jefferson helped secure Banneker a position surveying the future capital of the United States. But he remained unconvinced that this black man's brilliance offered proof that the races were equal.

Documents of the Civil War

From Thomas Jefferson's Declaration of Independence (1776).

> When in the course of human events, it becomes necessary for one people to dissolve the political bands which have connected them with another, and to assume among the powers of the earth, the separate and equal station to which the Laws of Nature and of Nature's God entitle them, a decent respect to the opinions of mankind requires that they should declare the causes which impel them to the separation.
>
> We hold these truths to be self-evident, that all men are created equal, that they are endowed by their Creator with certain unalienable rights, that among these are life, liberty and the pursuit of happiness.

If Jefferson Believed What He Wrote, Why Did He Keep Slaves?

The cornerstone of American democracy, the clarion cry for generations of independence movements since, Jefferson's Declaration of Independence is familiar to generations of Americans. Less familiar to most Americans is what it didn't say.

On June 7, 1776, Richard Henry Lee (1732–1794; remember the name Lee; it will come up again) rose before the Continental Congress in Philadelphia and proposed, "These united colonies are and ought to be free and independent States." His call was seconded by John Adams of Massachusetts, and a committee was formed to draft a declaration of principles in line with Lee's resolution. Adams, Benjamin Franklin, Roger Sherman, and Robert Livingston—all from the North—were chosen for the task but needed a Southerner for regional balance. Familiar with the writings of a bright, thirty-three-year-old Virginian, Adams suggested Thomas Jefferson. A fiery redhead, Jefferson had written an influential pamphlet claiming that the original settlers of America came as free individuals and had a natural right to choose their rulers. Jefferson not only joined the Committee of Five, but working in Philadelphia's sultry summer heat on a portable desk of his own design, he singlehandedly drafted a document. He first showed it to Adams and Franklin, each of whom recommended some minor word changes. It was then approved by the full committee without further change.

The document was presented on July 2 to the Continental Congress, which, in the nature of all Congresses since, decided to tack on some amendments. Sitting beside Franklin, a humiliated Jefferson fumed as the other delegates picked at his words and ideas for two and a half days. Fortunately for posterity, they left most of the language alone, but they did make some changes.

A substantial part of the Declaration was a long list of outrages committed against Americans by the king of England—this is the section most Americans have never bothered to look at. Some of these accusations were legitimate; others were inflamed rhetoric. But the accusation that caused the most dismay among the delegates was Jefferson's statement that the king was personally responsible for transporting slaves from Africa. In his draft, Jefferson had written:

> . . . He has waged cruel war against human nature itself, violating its most sacred rights of life & liberty in the persons of distant people who have never offended him, captivating & carrying them into slavery in another hemisphere, or to incur miserable death in their transportation thither. This piratical warfare, the opprobrium of *infidel* powers, is the warfare of the *Christian* king of Great Britain. Determined to keep open a market where *MEN*

should be bought and sold, he has prostituted his negative
for suppressing every legislative attempt to prohibit or restrain this execrable commerce, determining to keep open
a market where *MEN* should be bought and sold.

This controversial passage, Jefferson wrote later in his life, "was
struck out in complaisance to South Carolina and Georgia, who had
never attempted to restrain the importation of slaves, and who on
the contrary still wished to continue it. Our Northern brethren also
I believe felt a little tender under those censures, for tho' their people have very few slaves themselves yet they had been pretty
considerable carriers of them to others."

After more wrangling, the amended Declaration was adopted unan-
imously by the Continental Congress on July 4, 1776, two days after
it had already passed Lee's resolution of independence—the day that
by all rights ought to be our national birthday. Disheartened by the
damage he felt had been done to his Declaration, Jefferson withdrew
from the Congress and returned to Virginia, where he served out the
war years as its governor.

But the issue of Jefferson and his objection to the "execrable commerce" gets to the very heart of the American contradiction. Even
more than his fellow Virginians and slaveholders George Washington,
George Mason, the Lees, and Patrick Henry, Jefferson embodied the
enigma of early American slavery. After all, he was the Enlightenment hero who wrote what was and remains one of the most radically
idealistic and democratic documents in history.

His simple and seemingly unconditional statement that "all men
are created equal," a notion we accept today without question, was
then about as radical as proposing to put the king himself in chains
and auction him off. Yet Jefferson kept slaves, profited from them,
and sold them on at least two occasions. A Monticello slave named
Sandy, who had escaped and was returned, was sold to Jefferson's
cousin. Slaves worked on his plantation and in the nail and brick
factories he set up at Shadwell, his other property. During the Revolution, loyal slaves hid Jefferson's silverware when the British came

to Monticello in hot pursuit of the rebel governor. Still, throughout his legal and political career, Jefferson argued against slavery as an institution and whenever possible attacked slavery's foundations. Late in his life he wrote, "Nothing is more certainly written in the book of Fate than that these people are to be free."

The question has confounded historians and biographers. Jefferson was the greatest American example of the Enlightenment or Age of Reason, when European thinkers and writers like Thomas Hobbes and John Locke, René Descartes and Immanuel Kant, had pioneered a new belief that reason—man's mind—could discover the laws underlying all nature. Emerging from centuries of European history dominated by the Roman Catholic Church, these men were saying that simple faith was no longer sufficient. They said, "Dare to know" and "I think, therefore I am." The ideas of the Enlightenment, in which reason and science were elevated along with a new respect for humanity, profoundly influenced Jefferson. So how could a man who represented these ideals maintain a plantation whose economy was based on slave labor? Although Jefferson would write of the natural inferiority of blacks, he still believed they were men, as his bold use of the word *men* in the draft of the Declaration clearly shows. To Jefferson, equal meant in God's eyes, not in mental ability. He was also a man of his times, financially chained to a rigid social and economic system that did not welcome emancipation or Negro equality. Willard Sterne Randall, a recent biographer, may have summed it up well—if short of satisfactorily—when he wrote in *Thomas Jefferson: A Life* (1993):

> All his adult life, Thomas Jefferson seems to have tossed and turned in an agony of ambivalence over the dilemma of slavery and freedom. Repeatedly he sought to have public institutions relieve him of the burden of his conscience while he tried to avoid giving offense to his close-knit family and the slaveowning society of his beloved Virginia. He knew slavery was evil, he called it evil and spoke out against it in a series of public forums, but he would only push so

far—and then he would fall back on a way of life utterly
dependent on slave labor.

Documents of the Civil War

From the Northwest Ordinance, largely drafted by Thomas Jefferson
in 1784 and adopted by the Continental Congress on July 13, 1787:

> There shall be neither slavery nor involuntary servitude in
> said territory.

The Northwest Territory was a large area (more than 265,000
square miles) surrounding the Great Lakes that was ceded to the
United States by Great Britain in 1783. When four existing states laid
claim to parts of this huge, rich territory, the dispute was turned over
to the Continental Congress. It in turn relied on a plan drafted by
Jefferson which set the rules for forming new states out of this land.
(Eventually the states of Ohio, Indiana, Illinois, Michigan, Wisconsin,
and part of Minnesota emerged.) Besides setting the boundaries of
the states and establishing sixty thousand inhabitants as the population required for statehood, the Northwest Ordinance also prohibited slavery. That sanction marked the first federal restriction on
slavery or the slave trade.

Civil War Voices

Oliver Ellsworth, a delegate from Connecticut to the Constitutional
Convention, in 1787: "Slavery in time will not be a speck in our
country."

What Did the Constitution Say About Slavery?

Within months of the passage of the Northwest Ordinance, the Continental Congress was back at work on a much bigger task, drafting
a new Constitution for the United States of America. Just about

everyone agreed that the Articles of Confederation, adopted in 1781, was inadequate to run the new country. To keep the power of the states strong and to avoid having a powerful central government like the one they had just fought, the Founding Fathers had managed to create a toothless beast. Under the Articles of Confederation there was no common currency; individual states were printing their own money. There was no army to protect citizens from Indians, foreign threat, or another popular rebellion. And the last of these is exactly what happened in Massachusetts, "the cradle of liberty," where a group of struggling farmers, veterans of the Revolution, discovered they were not much better off after independence than they had been before. Led in 1786 by a farmer named Daniel Shays, they decided to raise a little hell of their own. Shays's Rebellion, as it was called, was put down without too much bloodshed, but the uproar it caused made America's new ruling class realize they needed a better set of rules to govern this country.

Accordingly, the Constitutional Convention was convened in Philadelphia in the spring and summer of 1787 to devise a new form of government, one that was consistent with the ideals of the Revolution but that would effectively manage a new and growing nation.

The Constitution they debated and drafted was, like all political documents, a compromise. The biggest compromise involved the creation of a legislature. The small states wanted equal representation. The larger states wanted representation based on population. Roger Sherman, a tough, plain-spoken Yankee from Connecticut, proposed a seemingly obvious and elegant solution: The legislature would include two houses. One would be based on population—the House of Representatives. The second would have an equal number of representatives—the Senate.

The other thorny issue was slavery. Many of the delegates wanted to see slavery eradicated, for economic and political reasons as well as moral ones. But since a good many of them had most of their wealth tied up in slaves, the continued existence of slavery was not debatable. Several other convenient compromises took care of these problems—but only for the moment.

Documents of the Civil War

From the Constitution of the United States, drafted in 1787 and ratified in 1789. (Emphasis added.)

We, the people of the United States, in order to form a more perfect Union, establish justice, insure domestic tranquility, provide for the common defense, promote the general welfare, *and secure the blessings of liberty to ourselves and our posterity* do ordain and establish this Constitution for the United States of America.

Article I. Section 2.3: Representatives and direct taxes shall be apportioned among the several states which may be included within this Union, according to their respective numbers, which shall be determined by adding to the whole number of free persons, including *those bound to service* for a term of years, and excluding Indians not taxed, *three-fifths of all other persons*. [This marked a compromise on the issue of counting slaves, who could not vote, in the census that would determine the number of representatives a state would have in Congress. Nonslave states didn't want to count them; slave states did. The compromise meant that slaves were counted as three fifths of a person.]

Article I, Section 9: 1. The migration or *importation of such persons* as any of the States now existing shall think proper to admit, shall not be prohibited by the Congress prior to the year one thousand eight hundred and eight, but a tax or duty may be imposed on such importation, not exceeding ten dollars for each person. [This agreement would cause trouble. Abolitionists pointed to the clause as proof that the Fathers wanted to do away with slavery. Slavers argued that it provided constitutional protection for slavery.]

Article IV, Section 2.3: No person held to service or labor in one State, under the laws thereof, escaping into another, shall in consequence of any law or regulation therein, be discharged from such service or labor, but shall be delivered up on claim of the party to whom such labor may be due. [Another sop to slaveholders, this clause protected slave owners from the loss of runaway slaves. Bolstered by the Fugitive Slave Act of 1793, slave owners had federal law on their side. Escaped slaves had to be returned.]

Fighting for independence didn't guarantee better treatment under a new government, as Daniel Shays and his fellow farmers in Massachusetts had learned. It was also a lesson for the estimated five thousand African Americans who fought in the Revolution. Ironically, they might have been better off fighting for the British, for during the war, the British Army had offered freedom to any slave who came over to fight the American rebels. How many actually did so is a guess. And more than a few who were promised their freedom may have ended up being sold back into slavery in the West Indies. The other irony is that the slaves would have been freed sooner if the British had kept their American colonies, for the British abolished the slave trade in 1807 and West Indian slavery in 1833.

There is no question that black men served conspicuously in the patriot armies. A black soldier named Oliver Cromwell helped row Washington across the Delaware. (He is for some reason absent in the famous painting of the scene. Oops!) Service by blacks in the American navy was even more commonplace, for recruiting for naval service was difficult, so captains became color-blind when a black face was enlisting. But the nation that emerged under the Constitution had little to offer the black people of America, free or slave. While the Constitution never specifically mentioned slaves, it substituted polite legalisms such as "persons held to service." Under the Constitution, slavery could not be touched for twenty years. Slaves would not be able to vote and would be counted in the census as three fifths of a person—a key political victory for the slaveholding states, who would see their political muscle in Congress grow out of proportion to their white population. And slavery could be taxed. It might be wrong, but we still might as well make a few bucks from it.

Elbridge Gerry (1744–1814), a member of the Continental Congress as well as a signer of the Declaration of Independence from Massachusetts, didn't think all this compromising was such a good idea. He prophetically wrote to his wife, "I am exceedingly distressed at the proceedings of the Convention. . . . They will lay the foundation of a Civil War." Later governor of Massachusetts and vice-president in James Madison's second term, Gerry refused to sign the Constitution. Apart from his prescient wisdom, he was immortalized in American politics when the Massachusetts legislature, at his direction, passed a law redistributing the electoral districts to ensure that Gerry's party would retain control. One Federalist complained that the oddly shaped districts Gerry had mapped out looked like salamanders, hence the political term *gerrymander.*

Civil War Voices

Benjamin Franklin, in a fund-raising address for the Pennsylvania Society for Promoting the Abolition of Slavery, and the Relief of Free Negroes Unlawfully Held in Bondage (November 9, 1789).

It is with peculiar satisfaction we assure the friends of humanity, that, in prosecuting the design of our association, our endeavours have proved successful, far beyond our most sanguine expectations.

. . . Slavery is such an atrocious debasement of human nature, that its very extirpation, if not performed with solicitous care, may sometimes open a source of serious evils.

The unhappy man, who has long been treated as a brute animal, too frequently sinks beneath the common standard of the human species. The galling chains that bind his body, do also fetter his intellectual faculties, and impair the social affections of his heart. Accustomed to move like a mere machine, by the will of a master, reflection is suspended; he has not the power of choice; and reason and conscience have but little influence over his conduct, because he is chiefly governed by the passion of fear. He is poor and friendless; perhaps worn out by extreme labour, age, and disease.

Under such circumstances, freedom may often prove a misfortune to himself, and prejudicial to society.

Attention to emancipated black people, it is therefore to be hoped, will become a branch of our national policy; but, as far as we contribute to promote this emancipation, so far that attention is evidently a serious duty incumbent on us, and which we mean to discharge to the best of our judgment and abilities.

To instruct, to advise, to qualify those, who have been restored to freedom, for the exercise and enjoyment of civil liberty, to promote in them habits of industry, to furnish them with employments suited to their age, sex, talents, and other circumstances, and to procure their children an education calculated for their future situation in life; these are the great outlines of the annexed plan, which we have adopted, and which we conceive will essentially promote the public good, and the happiness of these our hitherto too much neglected fellow-creatures. . . .

Signed, by order of the Society,

B. FRANKLIN, *President*

Widely recognized as an author, scientist, diplomat, inventor, philosopher, printer, and the first American to gain international renown, Franklin (1706–1790) had owned slaves for thirty years. As a printer and publisher, he also accepted advertisements for slave sales, and he considered blacks inferior to whites in their "natural capacities."

When Franklin did come to oppose slavery, it seemed to have been only partially out of strict moral reasons. He believed that slavery encouraged idleness, the same reason he opposed relief for the poor. Franklin was, above all, the high priest of honest labor. But in his later years his views changed dramatically, and he became president of the Pennsylvania Society for Promoting the Abolition of Slavery. In the society's platform, he called for a national policy of educating and training freed blacks for gainful employment.

Franklin, along with Jefferson, Washington, and other Founding Fathers, recognized that slavery was wrong. They understood that it hurt the ideals of liberty and democracy for which they had struggled and that it would have to end in America. The only question was one of timing: How long would slavery last? But to many other Americans the issue was not so simple, and the question of slavery ran like a deadly undercurrent beneath every important national question for more than fifty years.

chapter two

"*Fire-bell in the Night*"

We are natives *of this country; we only ask that we be
treated as well as* foreigners. *Not a few of our fathers suffered
and bled to purchase its independence; we ask only to be
treated as well as those who fought against it.*

—PETER WILLIAMS, JR.
PASTOR, ST. PHILIP'S EPISCOPAL CHURCH, NEW YORK

*Can any one of common sense believe the absurdity that a faction of
any state, or a state, has a right to secede and destroy this union and
the liberty of our country with it; or nullify laws of the union;
then indeed is our constitution a rope of sand; under such I would not
live.... The union must be preserved, and it will now be tested,
by the support I get by the people. I will die for the union.*

—PRESIDENT ANDREW JACKSON
NOVEMBER 1832

*I am opposed to slavery, not because it enslaves the black man,
but because it enslaves man.*

—DAVID PAYNE (1839)

✳

* What Did the Industrial Revolution Have to Do with the Civil War?

* What Did the Louisiana Purchase Have to Do with the Civil War?

* Who Were the First Abolitionists?

* What Was the Missouri Compromise?

* What Rights Does a State Have?

* Who Were Gabriel, Denmark, and Nat?

* Why Did the State of Georgia Want to Arrest a Publisher Named William Lloyd Garrison?

* What Was the Gag Rule?

* What Did the War with Mexico Have to Do with the Civil War?

I ronic, isn't it? America's vaunted Industrial Revolution began with a design for a machine that was pilfered from England. This bit of industrial espionage created the huge demand for cotton. And the demand for cotton made healthy and profitable once more a slave system that had seemed headed for a natural death. Those northern cotton mills created the industrial wealth that modernized the North with railroads, canals, and roads and brought hundreds of thousands of Europeans to America. And those immigrants would eventually end up in Union armies.

Born in England, Samuel Slater (1768–1835) was apprenticed at age fourteen to Sir Richard Arkwright, the Rumpelstiltskin of the Industrial Age, who had designed a mechanical spinning frame that turned cotton fiber into yarn. Using first horses, then water, and finally steam to power his mills, Arkwright basically invented the modern factory system. Eager to protect the secret behind its cotton-milling monopoly, however, Great Britain wouldn't allow textile workers with knowledge of Arkwright's invention to leave the country. In the United States, bounties were offered for this information. Slater, having memorized the machine's design, snuck out of England in disguise and brought the secret to America, where he opened the first American spinning mill in Pawtucket, Rhode Island, in 1790. Slater tempered his sin of industrial theft by establishing a Sunday school for his workers in 1796, one of the first such programs in the United States.

What Did the Industrial Revolution Have to Do with the Civil War?

A militant Quaker patriot from Rhode Island, Nathanael Greene (1742–1786) was a self-taught general who became one of George Washington's most trusted lieutenants. (Washington wanted Greene to replace him if he were ever killed or captured.) Greene had been at Boston's Battle of Breed's Hill and said memorably of that costly British victory, "I wish we could sell them another hill at the same price." After taking command of an army in the South, Greene enjoyed significant successes, and following the war he was given three valuable estates confiscated from Tories (Americans who were loyal to the king) in gratitude for his service. It was at one of these, Mulberry Grove on the Savannah River in Georgia, that a young Yale graduate arrived in 1793 to visit the overseer of the plantation, now in the hands of Greene's widow, Catherine, one of Washington's favorite dancing partners.

Planning to work nearby as a tutor, Eli Whitney (1765–1825) was intrigued when a group of planters discussed the difficulty of separating the seeds from the cotton fiber. Working all day, a slave could clean about a pound of cotton. This laborious process kept cotton from becoming a major cash crop for plantation owners at a time when the new mills of New England were demanding more. According to the legend, Whitney put his mind to it and invented a cotton *engine* (shortened to *gin*) that could clean cotton fifty times faster than by hand.

While Whitney clearly produced this machine, he apparently didn't just imagine it out of thin air. According to Richard Shenkman's *Legends Lies & Cherished Myths of American History*, "The cotton gin was invented in Asia and perfected in Santo Domingo in the 1740's—half a century before Whitney produced his gin. The Santo Domingo gin was crude but effective. . . . The Santo Domingo gin, however, did not work on the slippery seeds of American cotton. That was where Whitney came in. His gin was effective on American cotton."

In a way it's appropriate, then, that even though Whitney received a patent in 1794, he never profited greatly from his "invention." So simple to duplicate, the machine was widely pirated. Far beyond legend, though, is the impact of the machine on the future of the Deep South and the course of American history.

Cotton can't grow anywhere. It needs about two hundred consecutive days without frost and about twenty-four inches of rain—conditions that the Deep South could ideally provide. Even though Whitney's machine revolutionized the processing of picked cotton, planting and picking it was still backbreaking, tedious work. Sounds like a perfect job for slaves!

In 1790, the year Slater set up shop in New England, about 3,000 bales of cotton (a bale weighed about five hundred pounds) were produced in the United States. By 1801 the number of bales produced with the spread of the cotton gin had risen to 100,000; after the disruptions of the War of 1812, production reached 400,000 bales by 1820, then surged even further, reaching *4 million* bales just before the Civil War. The leading product of the United States during the first half of the nineteenth century, cotton exports exceeded the value of all other American exports combined. Cotton was king!

To keep the king on top of the hill, plantation owners needed two things: more workers and more land. The slave system had been dying out because the Protestant work ethic was alive in America and favored free—more accurately, cheap—labor. Admittedly, there was also some moral discomfort over the obvious contradiction between the new nation's ideals and the inhumane system of slavery. Fear as well was a big factor in the temporary decline of slavery. Many white Americans were afraid that the slaves, who were being bred so successfully, might eventually overwhelm the white population as they had done in the French Caribbean colony of St. Domingue. In addition, the recognition that the slave trade was scheduled to die in 1808 (as part of the original constitutional compromise) suggested that slavery was headed for extinction, just as Jefferson and Washington had hoped.

But the cotton gin and the demands of the northern factories changed everything. The boom in cotton exports created a drastic need for labor, so overnight a dying vampire came back to life, sustained by human blood. In 1790 America's first official census counted 697,897 slaves. By 1810—two years after the foreign slave trade was banned—their number had grown by *70 percent*, to 1,191,354. During the next fifty years, that number would more than triple to the 3,953,760 slaves counted in the 1860 census. American slaves were capable of reproducing even faster than they could be shipped in from Africa.

The other part of the equation—more land—was easy too. North

America was a pretty big place. And that's where Jefferson comes in again.

What Did the Louisiana Purchase Have to Do with the Civil War?

A little revolution can be a dangerous thing. Like a bad cold, it can be contagious. First the Americans had their rebellion. Then the ideas of liberty and equality caught hold in France, where the head cold became virulent pneumonia, particularly for members of the ruling class. Of course, the guillotine was remarkable for clearing the sinuses. From France, the Revolutionary Flu spread to the colony of St. Domingue, on the Caribbean island Christopher Columbus called Hispaniola. There, slaves inspired by the American and French examples had taken over, made it the second republic in the Western Hemisphere (and the first run by blacks), and eventually renamed it Haiti, its original Arawak name.

Then in 1802 France—now under Napoleon Bonaparte—sent an army to its former colony to put down the slave rebellion and retake the island. France's North American property, the Louisiana Territory, could serve as the breadbasket for Napoleon's European armies. Originally settled by the French in the early eighteenth century, the huge province of Louisiana was passed to Spain in 1763 at the end of the Seven Years' War (a.k.a. the French and Indian War). Under a secret treaty, Louisiana was returned to France in 1800 with the provision that if France ever gave it up, Spain would get the territory back. But France held the territory in name only; to control it meant an armed occupation.

Before he could launch a North American offensive, Napoleon needed an army base, and the former French colony of Hispaniola would make a convenient jumping-off point for North America. But first there was the problem of Toussaint-Louverture, the former carriage driver who had led the slaves in overthrowing the French.

Secretly, President Jefferson began to prepare for a possible war with France, but he wanted to try diplomacy first, so he sent James Monroe to assist Robert Livingston, the American envoy in Paris, in negotiating a deal. Jefferson hoped to purchase parts of either Louisiana or Florida or, at the very least, to guarantee the right of Americans to continue doing business in New Orleans. But the notion of buying this territory was never in Jefferson's mind. Then the French army sent to St. Domingue was wiped out in 1802 by a combination of yellow fever and Toussaint-Louverture's army. With a war against England looming, Napoleon knew that holding on to Louisiana would be a costly drain. France would have to defend the territory against a likely attack from the British, moving south from Canada. Strapped for cash, Napoleon decided to concentrate his energies in Europe, putting aside any dreams for North America. The French offer to the Americans was unexpected and dramatic: all of Louisiana or nothing. Jefferson jumped at the chance.

In other words, Jefferson's greatest presidential accomplishment might not have happened without Toussaint-Louverture and the rebel slaves of Haiti. The price for the entire Louisiana Territory was fixed at $15 million, about four cents an acre. Although Jefferson doubted the constitutionality of his actions, he pressed ahead, and in 1803 the United States more than doubled in size. An area of more than 800,000 square miles was added overnight. It was enough land to create eventually all or parts of Arkansas, Missouri, Iowa, Minnesota, North Dakota, South Dakota, Nebraska, Oklahoma, Kansas, Louisiana, Montana, Wyoming, and Colorado.

Suddenly the question of more land for growing cotton had become academic. Of course, there was a slight problem: the Indians already living there. Jefferson idealistically thought the Indians might be "educated" to give up their traditional way of life and adapt to farming and trading. But the next few presidents, including the Indian fighter Andrew Jackson, had their own agendas for the Indians. It wasn't going to be a pretty picture.

Further, all this new territory to slice, dice, and redecorate in red, white, and blue bunting posed another problem. Were all the new states to be carved out of the Louisiana Territory going to permit slavery when they entered the Union? A lot of people said, "I don't think so."

The Northwest Ordinance, used previously to admit new states, had expressly prohibited slavery; the argument went that the Founding Fathers must have been opposed to the expansion of slavery into any new state. The slaveholders argued that the Constitution talked about ending the slave trade, but it stopped short of restricting slavery itself. And in the three-fifths rule the Constitution seemed to clearly recognize the legitimacy of slavery. Besides, said the slavery supporters, the Constitution reserves certain powers to the states, meaning that the people in each state had the right to choose for themselves. Finally, the Fourth Article in the Bill of Rights protected "property," a category that included slaves. Braced by these legalistic arguments, cotton planters were quick to leave their worn-out land— cotton is hard on soil—and move west to the new lands of the Louisiana Purchase, bringing with them slaves by the hundreds of thousands.

PRELUDE TO THE CIVIL WAR: 1790–1820

1790

December In Pawtucket, Rhode Island, the first American cotton mill begins to turn out yarn using new "Arkwright machines," once a closely held British industrial secret. Operated by children as young as four years old, the machines outproduce adult manual labor, marking the onset of the Industrial Revolution in America and increasing the demand for cotton.

The first official census mandated by Congress, for the purpose of establishing the number of each state's delegates to the House of Representatives, shows a total population of 3,929,625. This includes 697,624 slaves (and 59,557 free blacks).

1791

December 15 The Bill of Rights, the first ten amendments to the Constitution, is ratified.

Following a slave revolt on the island of St. Domingue (modern Haiti) led by François Dominique Toussaint-Louverture (1743–1803), France abolishes slavery on the island in *1794*. In *1801* Toussaint-Louverture liberates the island from the French and becomes its ruler. Captured and later killed by the French, he is succeeded by General Jean-Jacques Dessalines (1758–1806), who proclaims a republic named Haiti in *1804* and declares himself emperor.

1792

April At Kentucky's constitutional convention, clergyman David Rice unsuccessfully attempts to prohibit slavery in the state. (A second attempt at abolition in Kentucky fails in *1799*.)

June 1 Kentucky becomes the fifteenth state, a slave state.

Denmark is the first nation to abolish the slave trade.

1793

February The Fugitive Slave Act is passed, providing legal mechanisms for the recovery of runaway slaves. It is illegal to harbor a slave or prevent his arrest.

October 28 Eli Whitney of Mulberry Grove, Georgia, files an application for a patent on the cotton gin. The patent is granted in *March 1794*, but the machine is so simple to duplicate and construct that it is widely pirated. Cotton production is revolutionized in the Deep South.

1796

June Tennessee becomes the sixteenth state, a slave state.

1800

January In Philadelphia, a group of free blacks present Congress with a petition condemning slavery, the slave trade, and the Fugitive Slave Act. The petition is tabled and dies in committee.

August Attempting to create a free black state in Virginia, slave Gabriel Prosser (c. 1776–1800) leads armed slaves in rebellion. He is caught and executed along with many of his followers.

The 1800 census shows a population of 5.3 million Americans, including 1 million blacks (90 percent of them slaves).

1803

March Ohio admitted as the seventeenth state, a free state under the terms of the Northwest Ordinance.

December After the Louisiana Purchase, the United States takes possession of the territory from the French. More than 800,000 square miles, the land doubles the size of the United States and intensifies the debate over the future of slavery in this new territory.

1804

New Jersey's legislature announces gradual emancipation.

1807

At the urging of President Jefferson. Congress prohibits the importation of any new slaves after *January 1808*.

The British Parliament prohibits British subjects from engaging in the slave trade after *March 1, 1808*. (West Indian slavery is abolished by the British in *1833*.)

1810

The third national census: total population: 7,239,881, a 36 percent increase since 1800; black population: 1,377,808 (free blacks number 186,746).

1812

December Louisiana, formerly the territory of Orleans, becomes the eighteenth state, a slave state. Its capital, New Orleans, becomes America's fifth-largest city.

1814–1815

December 1814–January 1815 The Hartford Convention. Unhappy over the War of 1812 and its effect on New England's fishing and shipping business, a gathering of New England Federalists meets in Hartford, Connecticut, to consider states' rights and possible secession from the Union. The war ends before the convention can press its recommendations. The dying Federalist party doesn't even run a candidate in the presidential election of 1820.

1816

The American Colonization Society is chartered by Congress. Supported by a $100,000 grant from Congress, it begins to organize ship caravans of freed slaves who want to return to Africa. Six years later, the first shipload lands near the mouth of the Mesurado River on Africa's west coast. In *1824* American philanthropist Ralph Randolph Gurley names the colony Liberia and the settlement at Cape Mesurado, Monrovia, in honor of America's President James Monroe. The colony is declared an independent republic in *1847*, although the United States does not recognize it until *1862*.
December Indiana is admitted as the nineteenth state, a free state.

1817

December Mississippi enters as the twentieth state, with slavery allowed.

1818

December Illinois becomes the twenty-first state, a free state.

1819

December Alabama admitted as the twenty-second state, a slave state.

1820

March The Missouri Compromise calls for the admission of Maine as a free state and Missouri as a slave state in 1821, maintaining parity between free and slave states. Under the terms of the Compromise, slavery will be excluded in the future from any Louisiana Purchase territory north of Missouri's southern border.

May Congress makes trading in foreign slaves an act of piracy and includes a death penalty for American citizens caught smuggling slaves.

Maine is admitted as the twenty-third state, with a ban on slavery.

U.S. census. Total population: 9,638,453; black population: 1,771,656 (18.4 percent).

Civil War Voices

Philadelphia Quaker leader Anthony Benezet (1754).

> To live in ease and plenty, by the toil of those, whom vi-
> olence and cruelty have put in our power, is neither con-
> sistent with Christianity nor common justice; and we have
> good reason to believe, draws down the displeasure of
> heaven; it being a melancholy but true reflection, that

where slave keeping prevails, pure religion and sobriety decline; as it evidently tends to harden the heart, and render the soul less susceptible of that holy spirit of love, meekness and charity. . . . How then can we . . . be so inconsistent with ourselves as to purchase such who are prisoners of war, and thereby encourage this anti-Christian practice; and more especially, as many of those poor creatures are stolen away, parents from children and children from parents;

You who by inheritance have slaves born in your families, we beseech you to consider them as souls committed to your trust. . . . Let it be your constant care to watch over them . . . that if you should come to behold their unhappy situation in the same light that many worthy men who are at rest have done . . . and should think it your duty to set them free, they may be more capable to make a proper use of their liberty.

Who Were the First Abolitionists?

Not everyone was happy about the new cotton power, which was growing by leaps and bounds. White workers didn't like it because slavery drove down wages and they didn't want to compete with blacks—slave or free—for jobs in the new western territories. Many in the North didn't like it because all those slaves were giving southern states an advantage in Congress, where the major decisions about tariffs and dividing up all the new land were being made. Every new slave state would get two new senators. And remember the three-fifths rule? The slave states were getting votes in the House of Representatives entirely out of proportion to their voting population.

There were a few people who didn't like slavery because it was wrong. Not many but a few. And they made themselves heard. The Germantown Protest, the first organized demonstration against slavery and the slave trade, was made by Quakers in Germantown, Pennsylvania, in 1688. Slowly and hesitantly, they were joined by a few lonely voices. In 1700 Samuel Sewall, who had been a judge in the

Salem witch trials and the only one to express publicly his regret, published *The Selling of Joseph*, one of America's first antislavery tracts. But his words and the voice of Cotton Mather (1663–1728), the prominent pastor of North Church who established a school for blacks in Boston, had a marginal effect, even in that Puritan city. Although joined by other denominations, the Quakers—or the Society of Friends—were the most persistent group to maintain a steady drumbeat against slavery before the Revolution. Quaker leader John Woolman (1720–1772), preaching against the evils of slavery, published his sermons under the title *Some Considerations on the Keeping of Negroes* (1754). He argued that "free Men, whose minds were properly on their business found a satisfaction in improving, cultivating and providing for their families; but Negroes, laboring to support others who claim them as their Property, and expecting nothing but slavery during Life, had not the like Inducement to be industrious." Woolman urged Quakers not only to free their slaves but to educate them as well.

Woolman's friend Anthony Benezet was even more tireless in his efforts to improve life for blacks. In 1770 he founded a school for slaves and freed blacks in Philadelphia, the city that was home to a vibrant and influential free black community in the early days of the United States. By 1787 no Quakers owned slaves, and many southern Quakers had left their homes to migrate westward rather than remain in a slave society. But the Quakers were on the fringe, a small group of ardent people whose influence was out of proportion to their numbers. While Quaker resistance to slavery had its greatest effect in Pennsylvania and Rhode Island, the Friends had little political or economic impact on the explosion of slavery in the cotton gin era.

Still, these voices in the wilderness made gradual progress, and abolition societies began to spring up in both the North and South. It was in Philadelphia in 1776 that Dr. Benjamin Rush (1745–1813), a signer of the Declaration and surgeon general of the Continental Army, started one of the first antislavery associations. (An unusually progressive thinker, Rush was also among the first to call for humane treatment for the mentally ill. For his book *Medical Inquiries and Observations upon Diseases of the Mind* [1812], he is known as the father of American psychiatry.) It was this group that Benjamin Franklin joined late in his life.

Perhaps the greatest contribution of the Quakers was to help create a young generation of free, educated African Americans who were passionate about ending slavery and who moved into the forefront of the abolition movement. Born free in Philadelphia, James Forten (1766–1842) had enlisted in the Revolutionary navy as a "powder monkey," a young boy who scrambled between the decks on warships, bringing powder for the ship's cannons. After the war, he started a sailmaking shop and prospered in the shipping business, amassing the rather substantial fortune of $100,000, making him the equivalent of a modern millionaire. Among the first to argue that the races were biologically equal, Forten eventually used his wealth to fund the abolition movement.

Two other Philadelphia leaders were Absalom Jones and Richard Allen, who founded the Free African Society. Allen's master had allowed him to earn enough money as a bricklayer to purchase his freedom. When Allen and Jones, another slave who had bought his freedom, were kneeling in prayer one day in 1787, they were yanked from their pew in St. George's Methodist Episcopal Church and told to pray in the gallery because they were black. They walked out. Allen established the African Methodist Episcopal (A.M.E.) Church, the first independent black Protestant denomination in America, and became its first bishop. Allen and Jones became heroes during a yellow fever epidemic in 1793, when they nursed the city's sick as many others fled. In 1800 Jones, Allen, James Forten, and other freemen of Philadelphia petitioned Congress with an attack on slavery and the slave trade. Congress quietly tabled the petition. Harrison Gray Otis of Massachusetts noted, "To encourage a measure of this kind would have an irritating tendency, and must be mischievous to America very soon. It would teach them the art of assembling together, debating, and the like, and would so soon, if encouraged, extend from one end of the Union to the other."

Paul Cuffe, another free black Quaker and a successful shipowner, had a proposal that did interest Congress. The son of an African-born slave father and a Nantucket Indian mother, Cuffe had gone to sea on a Nantucket whaler and eventually owned a fleet of ships. Although he had to pay taxes, he couldn't vote, so he sued in a Massachusetts court, arguing that blacks and Indians had fought in the Revolution and deserved the full rights of citizenship. In 1815 Cuffe began to sponsor the immigration of freed blacks back to Africa, transporting thirty-eight black Americans and bearing the $4,000 cost himself. Cuffe's plan was exactly what many white opponents of slavery thought was the perfect solution to the real problem of abolition: What are we going to do with all those freed slaves?

For all the good intentions of many early white abolitionists, blacks were not especially welcome in the free states of America. Several territories and states, such as Ohio, not only refused to allow slaves, but also had passed laws specifically limiting or excluding any blacks from entering its territory or owning any property. This was the flip side of the abolition question, and in 1816 Cuffe's plan was eagerly adopted by Congress as a solution to "the Negro Problem." Congress authorized $100,000 to form the American Colonization Society, and President Monroe appointed two commissioners to help it establish a settlement. Six years later, a colony came into existence under the group's auspices; in 1824 it was named Liberia by philanthropist Ralph Gurley, an agent for the society.

But the colonization solution presented a big problem. Except for Paul Cuffe and a few other supporters, most African Americans had not been consulted about it. And most of them didn't think it was such a great idea. (Despite the good intentions of the Quakers, Paul Cuffe was buried in a segregated section of a Quaker cemetery.)

A leader among free black abolitionists, Peter Williams, Jr., spoke for many of his brethren when he said, "We are *natives* of this country; we only ask that we be treated as well as *foreigners*. Not a few of our fathers suffered and bled to purchase its independence; we ask only to treated as well as those who fought against it."

While well-meaning whites saw colonization in Africa, the Caribbean, and Central America as the perfect way to deal with freed slaves, it did not suit many free blacks, in particular those who had struggled to gain a place at the American table. Their ancestors had fought in the Revolution. They had worked hard. They no longer considered themselves Africans but Americans. Although colonization was eagerly promoted by many white politicians, including Henry Clay and later Abraham Lincoln, only about eight thousand blacks accepted the offer to return to Africa.

Civil War Voices

Thomas Jefferson, in a letter to John Holmes in response to the passage of the Missouri Compromise (April 22, 1820): "But this momentous question, like a fire-bell in the night awakened and filled me with terror. I considered it the knell of the Union."

What Was the Missouri Compromise?

From the day Jefferson drafted the Declaration through the debates at the Constitutional Convention, slavery was clearly on the American agenda. In their wisdom, the Constitution's framers had delayed a tough decision for twenty years, thinking that slavery would eventually flicker out like a candle. Instead it became a bonfire, and now "the fire-bell" was ringing.

The unavoidable confrontation finally came. By adding the massive real estate deal of the Louisiana Purchase, the free state–slave state debate finally became a national issue in 1818 when the Missouri Territory petitioned for statehood, jeopardizing the equilibrium between free and slave states at eleven each. Once again, the nation tried to compromise.

The chief architect of what came to be known as the Missouri Compromise was Henry Clay (1777–1852), one of the most influential politicians of his day and Lincoln's political idol. Born in Virginia, Clay had moved to Lexington, Kentucky, where he established a law practice and, while still in his twenties, joined the state legislature.

He moved on to the House of Representatives in 1811 and soon became Speaker of the House and one of the leading "War Hawks," who eagerly sought war with England in 1812.

When the bill requesting Missouri statehood was submitted, New York Representative James Tallmadge proposed an amendment forbidding the entry of slaves into Missouri and allowing for the emancipation of the two thousand slaves already there. About the same time, Maine applied for statehood. But the slaveholders in Congress wouldn't accept slaveless Maine into the Union unless Missouri could enter as a slave state. Henry Clay eventually brokered a deal reached on March 3, 1820. Maine would enter as a free state and Missouri as a slave state, but slavery would be prohibited forever from states created from Louisiana Purchase lands north of Missouri's southern border. Since that was the natural boundary of cotton-growing territory, it was accepted by the slaveholders. As a result, the Missouri Compromise regulated the extension of slavery for the next quarter century, in practical terms simply delaying an inevitable showdown.

Civil War Voices

During a heated debate over tariffs and the sale of western lands, Senator Robert Young Hayne of South Carolina had attacked the New England states, questioning the patriotism of its citizens, whom he derided as selfish as he asserted the superiority of the states over the federal government. For two days before a packed gallery, Senator Daniel Webster of Massachusetts replied in one of the most memorable speeches in Senate history (January 1830).

> I have not allowed myself, Sir, to look beyond the Union, to see what might lie hidden in the dark recess behind. I have not coolly weighed the chances of preserving liberty when the bonds that unite us together shall be broken asunder. I have not accustomed myself to hang over the precipice of disunion, to see whether, with my short sight, I can fathom the depth of the abyss below. . . . While the Union

lasts, we have high, exciting, gratifying prospects spread out before us for us and our children. Beyond that I seek not to penetrate the veil. God grant that in my day, at least, that curtain may not rise! God grant that on my vision never may be opened what lies behind! When my eyes shall be turned to behold for the last time the sun in heaven, may I not see him shining on the broken and dishonored fragments of a once glorious Union; on States dissevered, discordant, belligerent; on a land rent with civil feuds, or drenched, it may be, in fraternal blood! Let their last feeble and lingering glance rather behold the gorgeous ensign of the republic, now known and honored throughout the earth, still full high advanced, its arms and trophies streaming in their original luster, not a stripe erased or polluted nor a single star obscured, bearing for its motto no such miserable interrogatory as "What is all this worth?" nor those words of delusion and folly, "Liberty first and Union afterwards"; but everywhere, spread all over in characters of living light, blazing on all its ample folds, as they float over the sea and over the land, and in every wind under the whole heavens that other sentiment, dear to every true American heart— Liberty *and* Union, now and forever, one and inseparable.

Civil War Voices

President Andrew Jackson, in a White House toast on the anniversary of Thomas Jefferson's birth (April 1830): "Our Federal Union it must be preserved!"

To which Vice-President John C. Calhoun responded, "The Union—next to our liberty, the most dear."

What Rights Does a State Have?

In 1994 California's Governor Pete Wilson opposed implementing a federal "motor voter" law, which required the states to register voters at the same time they register their automobiles or renew driver's licenses. Wilson complained that Congress was forcing the states to do something without offering to help pay for it. Sympathetic congressmen agreed, and there was a heated debate over whether Congress can pass legislation that affects the states without providing the funding to carry it out. These are the so-called unfunded mandates that the new Republican Congress opposed so vigorously in 1995. Apart from the fact that Wilson and other Republicans also objected to the motor voter law because they assumed that it would create more new Democrats than Republicans, his objection reopened the debate over states' rights versus federal powers. It is an old fight, one that goes back to the days of the Articles of Confederation.

Early in the nineteenth century, the battle over states' rights nearly brought the nation to civil war. That it did not is important to keep in mind. The main players then were three of the most influential American politicians who were never president: Henry Clay of Missouri Compromise fame; Daniel Webster (1782–1852), the fiery orator from Massachusetts; and the curmudgeonly dean of states' rights theory, John Caldwell Calhoun (1782–1850). Born in South Carolina and educated at Yale, Calhoun practiced law until his marriage to a cousin, whose inheritance allowed him to become the Jeffersonian ideal—a planter-politician. After two years in the state legislature, he began his national political career as a member of the House from South Carolina, serving from 1811 to 1817. A powerful nationalist, along with Henry Clay, Calhoun was an outspoken leader of the War Hawks, who persuaded the House to declare war with England in 1812. He also advocated the federal financing of canals and roads, a modern navy, and a standing army. After serving as secretary of war for eight years under President Monroe, Calhoun had the curious distinction to serve consecutive terms as vice-president to successive presidents. He was first elected vice-president to John Quincy Adams in 1824; four years later he was reelected vice-president when Jackson defeated Adams.

Andrew Jackson was the towering presidential figure of American politics, the only president on a par with the celebrated congressional trio of Clay, Calhoun, and Webster. But Jackson and Calhoun did not like each other. Both were ambitious and shrewd, and they had personal antagonisms as well as political differences. As secretary of war, Calhoun had recommended that Jackson, then fighting Indians in Florida, be summoned to a court-martial. But Calhoun had told Jackson that he supported him. When Jackson later learned the truth, he felt Calhoun had been dishonest. Now they were dueling politically over the issue of federal power.

The core of the argument was a stiff tariff placed on imports in 1828. The recent passionate debate over the North American Free Trade Agreement (NAFTA) might give modern Americans some sense of what the debate over tariffs was all about. But the NAFTA fight was a tea party compared to the vicious regional fighting that took place over these taxes on imported goods. Passed at the insistence of northern merchants, who wanted to protect their products by making European imports more expensive, and reluctantly signed into law by President John Quincy Adams, it was the highest tariff imposed in America and was labeled the Tariff of Abominations. An angry Calhoun anonymously wrote an essay in 1828, "The South Carolina Exposition and Protest." Calling the tariff "unconstitutional, oppressive and unjust," he began to lay the legal groundwork for the right of the states to "nullify" federal laws. In 1831 Calhoun made a clean break with Jackson, publicly issuing an address that made it clear he stood with the "nullifiers." In 1832 he became the first man to resign from the vice-presidency, leaving when he won a seat from South Carolina in the Senate, where he served until his death in 1850.

Nullification, the theoretical right of a state to suspend the operation of a federal law within its boundaries, was not a new idea in America when Calhoun floated it in 1831. It went back to the very beginnings of the republic, after the loose Articles of Confederation had proved ineffective. Although the states definitely feared a tyrannical central government, they had agreed to yield certain powers to the federal government under the Constitution. But the principle of nullification was supported by many of the founders, including James Madison and Thomas Jefferson. It had also been used by slavery opponents as a justification for failing to enforce the fugitive slave laws, which compelled the return of runaway slaves.

On November 24, 1832, a South Carolina state convention issued an Ordinance of Nullification, which declared "null, void, and no law" the high protective tariff. President Jackson was not amused. He wrote to one of his generals, "Can any one of common sense believe the absurdity that a faction of any state, or a state, has a right to secede and destroy this union and the liberty of our country with it; or nullify laws of the union; then indeed is our constitution a rope of sand; under such I would not live.... The union must be preserved, and it will now be tested, by the support I get from the people. I will die for the union."

Jackson, threatening to send fifty thousand troops to enforce the tariff in the port of Charleston, said, "Disunion by armed force is *treason*. Are you really ready to incur this guilt?" The defiant governor and legislature of South Carolina called for ten thousand militiamen to repel any "invasion" by federal troops. America truly seemed to be on the verge of civil war. Faced with the prospect of warfare over the tariff, Calhoun joined with Clay to reconcile the claims of South Carolina with those of the federal government. As a result, the Compromise Tariff of 1833, which gradually reduced tariffs until 1842, was passed, the second of Henry Clay's three compromises to avert war. The South Carolina convention repealed the Ordinance of Nullification, and both sides claimed victory. But the point remains that whenever questions of states' rights came up, they had been solved through compromise. It was only when the issue became attached to the perpetuation of slavery that compromise became impossible.

Civil War Voices

From *The History of Mary Prince, A West Indian Slave, Related by Herself* (1831).

> Our mother, weeping as she went, called me away with the children Hannah and Dinah, and we took the road that led

to Hamble Town, which we reached about four o'clock in the afternoon. We followed my mother to the market-place, where she placed us in a row against a large house, with our backs to the wall and our arms folded across our breasts. I, as the eldest, stood first, Hannah next to me, then Dinah; and our mother stood beside, crying over us. My heart throbbed with grief and terror so violently, that I pressed my hands quite tightly across my breast, but I could not keep it still, and it continued to leap as though it would burst out of my body. But who cared for that? Did one of the many by-standers, who were looking at us so carelessly, think of the pain that wrung the hearts of the negro woman and her young ones? No, no! They were not all bad, I dare say, but slavery hardens the white people's hearts towards the blacks; and many of them were not slow to make their remarks upon us aloud, without regard to our grief—though their light words fell like cayenne on the fresh wounds of our hearts. Oh those white people have small hearts who can only feel for themselves.

. . . The bidding commenced at a few pounds, and gradually rose to fifty-seven, when I was knocked down to the highest bidder; and the people who stood by said that I had fetched a great sum for so young a slave.

I then saw my sisters led forth, and sold to different owners; so that we had not the sad satisfaction of being partners in bondage. When the sale was over, my mother hugged and kissed us, and mourned over us, begging of us to keep up a good heart, and do our duty to our new masters. It was a sad parting; one went one way, one another, and our poor mammy went home with nothing. . . .

. . . The next morning my mistress set about instructing me in my tasks. She taught me to do all sorts of household work; to wash and bake, pick cotton and wool, and wash floors, and cook. And she taught me (how can I ever forget it) more things than these; she caused me to know the exact difference between the smart of the rope, the cart-whip, and the cow-skin, when applied to my naked body by her own cruel hand. And there was scarcely any punishment more dreadful than the blows I received on my face and head from her hard heavy fist. She was a fearful woman, and a savage mistress to her slaves.

... To strip me naked—to hang me up by the wrists and lay my flesh open with the cow-skin, was an ordinary punishment for even a slight offence. My mistress often robbed me too of the hours that belong to sleep. She used to sit up very late, frequently even until morning; and then I had to stand at a bench and wash during the greater part of the night, or pick wool and cotton; and often have I dropped down overcome by sleep and fatigue, till roused from a state of stupor by the whip, and forced to start up to my tasks.

... Poor Hetty, my fellow slave, was very kind to me, and I used to call her my Aunt; but she led a most miserable life, and her death was hastened ... by the dreadful chastisement she received from my master during her pregnancy. It happened as follows. One of the cows had dragged the rope away from the stake to which Hetty had fastened it, and got loose. My master flew into a terrible passion, and ordered the poor creature to be stripped quite naked, notwithstanding her pregnancy, and to be tied up to a tree in the yard. He then flogged her as hard as he could lick, both with the whip and the cow-skin, till she was all over streaming with blood. He rested, and then beat her again and again. Her shrieks were terrible. The consequence was that poor Hetty was brought to bed before her time, and was delivered after severe labour of a dead child. She appeared to recover after her confinement, so far that she was repeatedly flogged by both master and mistress afterwards; but her former strength never returned to her. Ere long her body and limbs swelled to a great size; and she lay on a mat in the kitchen, till the water burst out of her body and she died. All the slaves said that death was a good thing for poor Hetty; but I cried very much for her death. The manner of it filled me with horror. I could not bear to think about it; yet it was always present to my mind for many a day.

. . . For five years I remained at his house, and almost daily received the same harsh treatment. At length he put me on board a sloop, and to my great joy sent me away to Turk's Island. I was not permitted to see my mother or father, or poor sisters and brothers, to say good-bye, though going away to a strange land, and might never see them again. Oh the Buckra [white] people who keep slaves think that black people are like cattle, without natural affection. But my heart tells me it is far otherwise.

Not quite the same picture Margaret Mitchell gave us of happy slaves doing their jobs around the Big House!

Mary Prince's account of her life on Barbados and Antigua appeared in England in 1831, the first published narrative of a female slave. Sold to a series of owners of only slightly differing levels of sadism, Mary was eventually taken to England, where she gained her freedom. Her story was published by a British abolitionist group, and two years later slavery in the British West Indies was ended.

Her tales of beatings, hard labor, and cruelty would have been equally common among American slaves, although many slaveowners took care not to inflict such grievous wounds. Many would no more beat a slave than they would a prized horse or cow. After all, it was their own valuable property they would be damaging or destroying. But it was also no small wonder that some slaves struck back.

Who Were Gabriel, Denmark, and Nat?

A few years after the Revolution, Jefferson observed, "A little rebellion now and then is a good thing. . . . God forbid that we should ever be twenty years without such a rebellion. . . . The tree of Liberty must be refreshed from time to time with the blood of patriots and tyrants." In advocating "a little rebellion," he probably did not have in mind the likes of Gabriel, Denmark Vesey, and Nat Turner, three men who may have never heard Jefferson's words. These three black men represented white America's worst nightmare come true.

America in 1800 was turbulent and politically unsettled. The nation had settled into two fractious parties, the Federalists (led by John Adams and Alexander Hamilton) and Jefferson's Republicans. Under the election procedures of the period, Jefferson had become Adams's vice-president in 1796, and they had endured a stormy four years together. With the presidential election of 1800, the differences between these men and their parties were not simply polite arguments over policy but serious confrontations over the direction of the new nation that spilled over into nasty personal attacks on each candidate.

At just this time, a young slave living on a plantation near Richmond got caught up in all those ideas of freedom and democracy that the white gentlemen planters had been talking about since 1776. Born in 1776, ironically, Gabriel belonged to Thomas Henry Prosser, a successful tobacco planter whose father owned nearly sixty slaves. Thomas Prosser and Gabriel had grown up together as boys—until master went one way and slave another. Besides the plantation near Richmond that Prosser had inherited, he owned a tavern and an auction and real estate business. Following his father, Gabriel had become a blacksmith, a skilled tradesman who was permitted to hire himself out and even keep a portion of his earnings. In fact, he was hired out so often, it was assumed he was a free black. He had another distinction unusual for slaves: He could read, although who taught him is a mystery. (Douglas Egerton, the author of *Gabriel's Rebellion*, suggests that as many as 5 percent of the slaves could read.) Standing over six feet tall and shaped by years of smithing, Gabriel was physically imposing. The limited freedom he enjoyed, along with his literacy and physical prowess, made him a dangerous black man. Gabriel and other blacks in Richmond were well aware of the events in St. Domingue and inspired by the exploits of Toussaint-Louverture and his slave army. Using his freedom to move about the Richmond vicinity, Gabriel, with his two brothers, recruited a small army of slaves and free blacks willing to revolt against their masters. They even had a banner inscribed with a slogan that must have held a familiar (and embarrassing) ring to Virginians of that era: "Death or liberty."

Gabriel's secret army, carrying scythes that had been turned into swords, prepared to march on Richmond. Although Gabriel estimated that as many as 600 men would join the uprising, approximately 150 men, including at least 2 white men, were actually awaiting the order to begin. Late in August, the word went out on the slave grapevine that the day was to be Saturday, August 30, 1800. The target was Richmond, then the home of 5,700 residents, half of whom were black; with another 4,600 slaves on nearby plantations, whites were actually in the minority in the area. Gabriel planned to take the guns stored in the Richmond armory and perhaps even capture the governor of Virginia, James Monroe, a friend and neighbor of Thomas Jefferson's and a future president. The only whites to be spared were Quakers, French, and others known to oppose slavery.

Confounded by a torrential rainstorm and then betrayed by two slaves from a neighboring plantation, the rebellion fizzled. An armed militia quickly put down the uprising, and although Gabriel escaped briefly, he was soon captured and brought to trial with most of his followers. After seventeen of the slaves were hanged, Jefferson counseled Monroe to stop the hangings. The revolutionaries of 1776 must have been a little uncomfortable when one of the rebels told the court, "I have nothing more to offer than what General Washington would have had to offer had he been taken by the British and put to trial. I have adventured my life in endeavouring to obtain the liberty of my countrymen, and am a willing sacrifice in their cause."

Ultimately, Gabriel, his brothers, and twenty-four other followers went to the gallows. Two years later a second rebellion, the Easter Plot, was planned by one of Gabriel's followers and was similarly put down by Monroe. The immediate outcome of these two conspiracies was simple. Having determined that slavery would continue, the slaveowners realized that they had become too lenient toward their slaves. As one Virginian commented, "It is a pity, but slavery and Tyranny must go together." Virginia and other slave states enacted new codes that restricted slave meetings on Sundays after work was finished. More important, slave literacy was going to end. A little learning for slaves was clearly a dangerous thing. Finally, Virginia ruled that all free slaves—those who had purchased or been given their freedom—had to leave the state within a year of their emancipation.

But the desire for freedom didn't end there. The same year that Gabriel tried to lead his army to freedom, Denmark Vesey purchased his freedom from a ship's captain. A skilled carpenter, he was able to save several thousand dollars, not a small amount for a free black man in Charleston. Vesey was a leader in the city's black church and could have easily returned to Africa, but like other free blacks, he rejected that option because it meant leaving too many fellow slaves in chains. When Vesey's church was seen as too outspoken, it was closed by Charleston authorities. (So much for the First Amendment.)

Having visited the black republic on St. Domingue and well aware of Gabriel's conspiracy of 1800, Vesey began to plot another insurrection, much along the lines of Gabriel's plan, in 1822. His band would also attack the armory and seize weapons, burn the city, and kill as many whites as possible. The men hoped to seize ships and eventually make their way to Haiti. As the plot built toward its culmination, it too was betrayed by a slave. By June Vesey and other conspirators had been arrested, and more than thirty of them were executed on July 2, 1822.

But the "little rebellions" didn't end. Once more slaves would lash out in a violent frenzy that threw the slaveholding territory of America into utter panic. Born in 1800, the year of Gabriel's rebellion, Nat Turner may well have grown up hearing of these exploits, for Gabriel had acquired mythic status on the plantations of Virginia. (During the Civil War, black troops made up a verse about him that was sung to the tune of "The Battle Hymn of the Republic.") Raised on Samuel Turner's small plantation in Virginia by his African-born mother, Nat learned to read from his master's son. His father escaped and did not return, and he ran away once himself but returned voluntarily after thirty days. Over the years, Turner became fanatically religious and served as a preacher for the slaves of his area, who began to call him the Prophet, and people claimed that he had healing powers. By 1825 he had visions of a second coming of Christ, and a later vision told him to "slay my enemies with their own weapons." A solar eclipse in 1831 was God's sign to Nat Turner that the time had come. He said, "I heard a loud voice in the heavens, and the Spirit instantly appeared to me and said . . . I should arise and prepare myself, and slay my enemies with their own weapons . . . for the time was fast approaching when the first should be last and the last should be first."

On the night of August 21, Turner and seven fellow slaves used hatchets and axes to murder the Travis family (who now owned him) while they slept. Then they set out on a campaign that terrorized the Virginia countryside. Picking up some sixty slave recruits as they traveled from one plantation to another, the group moved through Southampton County toward the county seat of Jerusalem, where they planned to capture the armory but apparently had no further goal. For forty-eight hours Turner's followers rampaged, but the alarm had been spread. The group was disorganized, with a few of them drunk, and the rebellion fell apart. Meeting a well-armed militia, they were killed, captured, or dispersed outside Jerusalem. Turner managed to escape with twenty followers. After another battle, he and four others escaped once more and hid out for six weeks before being captured. Turner was hanged along with sixteen of his followers on November 11, 1831. This rebellion also spawned a wave of terror for all blacks in the area, as state and federal troops swept through, killing as many as two hundred blacks in reprisal.

The slaveowners pointed at the radical northern abolitionists and the impact they were having on slaves. In the wake of the Turner rebellion, harsh new laws were passed forbidding slaves to read and write and eliminating the few freedoms that blacks enjoyed at the time. Postal officials were also instructed to seize abolitionist materials in the mail.

PRELUDE TO THE CIVIL WAR: 1821–1835

1821

January Moses Austin receives permission from Mexico to move three hundred American families from Missouri into the Mexican territory of Texas. This grant will be taken over by Stephen F. Austin after his father's death, and the first Americans move to Texas in *1825*.

August Missouri is admitted as the twenty-fourth state, a slave state, bringing temporary parity between free and slave states at twelve each. Congress admits Missouri on the condition that the state constitution will not try to limit the rights of citizens, specifically free blacks.

The Genius of Universal Emancipation, one of the earliest abolition journals, is founded by Quaker Benjamin Lundy in Mount Pleasant, Ohio.

1822

May The Denmark Vesey conspiracy. In one of the most elaborate slave plots, slaves led by Denmark Vesey, a freed slave, plan a revolt in Charleston, South Carolina. The plot is betrayed, and thirty-seven blacks are hanged.

1827

The State of New York abolishes slavery.

1828

May 19 The "Tariff of Abominations" is signed into law. Passed to promote American industry, it is viewed as harmful to the southern economy while greatly benefiting the northern states.

The Tariff of 1828 is denounced as unconstitutional, oppressive, and unjust by the South Carolina legislature. An anonymous essay written by Calhoun argues that single states can nullify federal law.

1829

August Mexico rejects President Andrew Jackson's offer to buy the territory of Texas, by now home to thousands of Americans.
September Mexico outlaws slavery but exempts Texas from the law a month later.

1830

January 12–27 In a series of Senate debates, Robert Hayne of South Carolina and Daniel Webster of Massachusetts debate the origin of the Constitution and states' rights versus federal power. Hayne, speaking for most southern and western senators, supports state sovereignty and nullification. Webster argues that the states derive their power from the Constitution and that the national government is the final authority.

Mexico passes a law forbidding further American colonization of Texas and prohibits the importation of more slaves into the territory.

1831

January Radical abolitionist William Lloyd Garrison publishes the first issue of his journal, the *Liberator*.
August The Nat Turner rebellion in Virginia. After his capture, Turner is hanged in *November*. In the wake of this uprising, harsher slave laws are passed, abolitionist writings are censored, and southerners organize militias to suppress future uprisings.

1832

January In the Virginia legislature a grandson of Thomas Jefferson, Thomas Jefferson Randolph, presents his grandfather's plan for gradual emancipation.

July Congress passes a new Tariff Act, which is again seen as benefiting the northern states at the expense of the South.

November In response to the Tariff Act, South Carolina adopts a law to nullify the Tariff Acts of 1828 and 1832. The legislature calls for military preparations and secession if necessary.

December Following South Carolina's actions, President Andrew Jackson reinforces the federal forts off Charleston and warns that no state can secede from the Union.

The New England Anti-Slavery Society is organized at the African Baptist Church in Boston.

1833

January South Carolina's legislature defiantly calls the president "King Andrew" and prepares a volunteer unit to repel any "invasion." Jackson responds with a "force bill," enabling him to enforce the Tariff Acts.

February As Calhoun argues against the force bill, Henry Clay works out a new compromise tariff act aimed at gradually reducing the tariffs seen as onerous in the South. Both the force bill and the Compromise Tariff pass and are signed into law by Jackson, defusing the confrontation over nullification.

June In Connecticut, Prudence Crandall attempts to admit black girls to her private school, but the state legislature quickly passes a law against educating blacks. Crandall is jailed, and the school is closed.

August Great Britain outlaws slavery in its colonies.

December The Female Anti-Slavery Society is organized in Philadelphia by Lucretia Cottin Mott (1793–1880), who becomes its first president.

The American Anti-Slavery Society is organized, also in Philadelphia, with the backing of wealthy New Yorkers Arthur and Lewis Tappan and prominent abolitionist minister Theodore Weld.

About this time, abolitionist voices emerge elsewhere. Elijah P. Lovejoy begins the publication of his antislavery newspaper, the *Observer*, in St. Louis, and noted writer John Greenleaf Whittier publishes an abolitionist tract, *Justice and Expediency*.

Ohio's Oberlin College, the first coeducational college in America, opens; it is the first college to admit blacks.

1834

January After presenting a petition to the Mexican government requesting independence for Texas, Stephen Austin is arrested and jailed for eight months in Mexico City.

July On Independence Day, a mob breaks up an antislavery meeting in New York City because blacks and whites are mingling. Eight days of rioting follow.

August England abolishes slavery.

October In Philadelphia, the center of abolitionist societies, proslavery rioters go on a rampage and destroy forty houses belonging to blacks.

1835

July Chief Justice John Marshall dies. President Jackson nominates Virginian Roger B. Taney, a former slaveholder, to replace him. Due to vigorous opposition by northern senators, the nomination is not confirmed until *March 1836*.

July In Charleston, South Carolina, a mob burns abolitionist literature impounded by a local post office. Antiabolitionist laws are passed throughout the South. In Georgia, one of these calls for the death penalty for anyone who publishes material that incites a slave insurrection.

August In Canaan, New Hampshire, the Noyes Academy is destroyed by a white mob after it enrolls fourteen black students.

October In Boston, a proslavery mob disrupts an address by English abolitionist George Thompson. The mob parades radical abolitionist William Lloyd Garrison through the streets with a noose around his neck.

December Partly in response to the growing violence of the anti-abolitionists, President Jackson recommends laws to prevent the circulation of abolitionist materials by mail.

December Mexican President Santa Anna extends the law against slavery to include Texas. But American settlers there announce they will secede from Mexico.

Unitarian leader William Ellery Channing publishes his abolition tract *Slavery*.

Civil War Voices

Alexis de Tocqueville, from *Democracy in America* (1831).

> . . . I do not think that the white and black races will ever be brought anywhere to live on a footing of equality. But I think that the matter will be still harder in the United States than anywhere else. It can happen that a man will rise above prejudices of religion, country, and race, and if that man is a king, he can bring about astonishing transformations in society; but it is not possible for a whole people to rise, as it were, above itself.
>
> . . . I confess that in considering the South I see only two alternatives for the white people living there: to free the Negroes and to mingle with them or to remain isolated from them and keep them as long as possible in slavery. Any intermediate measures seem to me likely to terminate shortly, in the most horrible of civil wars, and perhaps in the extermination of one or the other of the two races.
>
> . . . If freedom is refused to the Negroes in the South, in the end they will seize it themselves.

Why Did the State of Georgia Want to Arrest a Publisher Named William Lloyd Garrison?

In 1831, the year that Nat Turner led his doomed band, the State of Georgia offered a reward for the arrest and conviction of William Lloyd Garrison. A slight, ascetic, owl-faced man, Garrison had never been to Georgia and had broken no laws there, but he was considered very dangerous throughout the slaveholding South. He received hundreds of abusive letters, many of which threatened him with assassination. What had this printer done to deserve hate mail and the reward of arrest?

Born in Newburyport, Massachusetts, in 1805, Garrison was indentured at age fourteen to the owner of the *Newburyport Herald*. He became an expert printer and began to write anonymously, attempting to arouse Northerners from their apathy on the question of slavery.

In 1829 Garrison began to work with antislavery agitator Benjamin Lundy (1789–1839), who started *The Genius of Universal Emancipation*, a monthly periodical, in 1821. In 1827 Lundy, who believed in gradual emancipation, had described a prosperous Baltimore slave trader as "a monster in human shape." (The slaver knocked him down in the street. Lundy sued and won a dollar, the judge ruling that the assault had been provoked.) Impressed by Garrison's writings, Lundy had walked from Baltimore to Vermont to recruit the Northerner. Working in Baltimore, then a center of the domestic slave trade, Garrison began to denounce this trade in uncompromising terms. When he wrote an attack on two New Englanders whose ship was carrying slaves from Maryland to New Orleans, he was sued for libel by the slave traders, fined fifty dollars, and jailed when he refused to pay. From his cell, he continued to attack the slave trade until he was released after forty-nine days.

Garrison returned to New England and launched the *Liberator* on January 1, 1831. James Forten, the wealthy Philadelphian freedman, was one of his strongest financial supporters. Later Garrison formed the Massachusetts Anti-Slavery Society and the New England Anti-Slavery Society.

Viewed as fanatics by the general public, Garrison and his fellow abolitionists were relatively few in number—only about 160,000 in the years from 1833 to 1840. Most were educated church people from a middle-class New England Puritan and Quaker heritage. Support from the working and upper classes was minimal, and even in the northern states, abolitionists were often attacked by mobs. Garrison himself was paraded through the streets of Boston with a noose around his neck in 1835.

Convinced that slavery had to be abolished by moral force, Garrison appealed through the *Liberator*, especially to the clergy, for the practical application of Christianity in demanding freedom for the slaves. But few listened, and Garrison grew more militant and less compromising. Believing that the slavery clauses of the Constitution were immoral and that stronger action was needed, he took increasingly radical views that opened a rift in the abolition movement. In 1839 the society split over tactics and policy into two main groups. The moderates, or gradualists, included James Birney, Arthur and Lewis Tappan, and Theodore Weld, who believed that emancipation could be achieved legally through religious and political pressure. In 1840 the Tappans founded the American and Foreign Anti-Slavery Society. The "Garrisonians" wanted immediate action and were far more radical in their approach.

A further cause of dissension was Garrison's advocacy of equal rights for women in general and especially within the abolitionist movement. When women abolitionists were excluded from an international anti-slavery convention held in London in 1840, Garrison walked out with them. The radicals, led by Garrison and others, including the Quaker feminist Lucretia Mott and her husband, James, refused to remain in a party committed to gradual and legal emancipation, and with Garrison, they formed the American Anti-Slavery Society. In 1840 Garrison publicly burned a copy of the Constitution and denounced it as "a covenant with Death and an agreement with Hell." Instead, he chose as his motto, "No union with slaveholders." True to his pacifist beliefs, Garrison advocated the peaceful separation of the free states from the slave states, a view shared by Horace Greeley, who started the *New York Tribune* in 1841.

Garrison traveled widely as a lecturer. It was on an April night in 1839 that he may have had his greatest impact. Speaking in New Bedford, Massachusetts, he captivated one young man in the audience with his words and impassioned delivery. A recently escaped slave, this young man decided that he too would speak out against slavery. His given name was Frederick Augustus Washington Bailey, but having made his escape, he had changed it to Frederick Douglass.

What Was the Gag Rule?

You flip on C-Span to catch the House of Representatives at work. There they are, legislating. One politician rises to say that the budget deficit is a big problem but that to discuss it on the House floor would be inflammatory. This representative introduces a rule that prohibits the House or any committee of the House from discussing the deficit and petitions from the public would be ignored. Preposterous idea? Well, it happened. Only the issue wasn't the deficit but slavery.

From 1836 to 1844, the House opened each session by adopting a number of procedural rules designed to exclude from consideration by the House, or by its committees, petitions asking for the abolition of slavery in the District of Columbia, where the federal government held the power to legislate. Known as gag rules, they were supported by southern congressmen and by many northern representatives who regarded the petitions as a dangerous threat to the Union.

For all those years one congressman fought to have this so-called gag rule lifted. He was a former president, John Quincy Adams, then a member of the House from Massachusetts. Adams contended that the Bill of Rights protected the right of petition and that the refusal of Congress to consider petitions was unconstitutional. At the beginning of each session, therefore, when the House adopted its rules of procedure, he moved to strike out the offending gag rule.

Born in Braintree (now Quincy), Massachusetts, John Quincy Adams (1767–1848) was the son of John and Abigail Adams. At age twelve, he had traveled to Europe with his father and served as his secretary during the peace negotiations that ended the American Revolution in 1783. Graduating from Harvard College, he opened a law office and later became minister to Russia. He then served as the principal American negotiator of the Treaty of Ghent (1814), ending the War of 1812. In 1815 he became minister to Great Britain and continued to ease tensions between the two nations after the war.

In 1817, Adams became James Monroe's secretary of state, a long and harmonious association during which he helped formulate the Monroe Doctrine, which asserts that Europe should remain out of the Americas. In the 1824 election, Adams was involved in a bitter political contest in which none of the four candidates obtained a majority in the electoral college. With eighty-four votes, Adams ran behind Andrew Jackson but ahead of William Crawford and Henry Clay. The vote went to the House of Representatives, where Clay, the Speaker, threw his significant support to Adams. Adams won and later named Clay his secretary of state, prompting Jackson's charge of a "corrupt bargain." This bitter political feud meant that Adams's presidency was marred by constant fighting with Jackson's allies. Committed to a protective tariff much favored in his native New England. Adams signed the Tariff of Abominations in 1828; later that year he was overwhelmingly defeated by Jackson.

Two years later Adams returned to the House and led the anti-slavery forces in Congress. A vigorous speaker known as Old Man Eloquent, he finally succeeded on December 3, 1844, when the gag rule was abolished. On February 21, 1848, Adams suffered a stroke on the House floor and died two days later without regaining consciousness.

Civil War Voices

Abolitionist James Birney (September 1835): "The antagonist principles of liberty and slavery have been roused into action and one or the other must be victorious. There will be no cessation of strife until slavery shall be exterminated or liberty destroyed."

A slaveholder from Kentucky who converted to abolitionism, James Gillespie Birney (1792–1857) became, like many religious converts, a zealot for his cause. Educated at the College of New Jersey (now Princeton), he gave up a prosperous law practice in Alabama to become an agent of the American Colonization Society, which supported political actions to end slavery and resettle freed slaves in Africa or the Caribbean. Having freed his inherited slaves, Birney began to publish an abolitionist newspaper, the *Philanthropist*, in Cincinnati and became executive director of Garrison's American Anti-Slavery Society. In the presidential elections of 1840 and 1844, he was the nominee of the Liberty party, which advocated the abolition of slavery by moral persuasion and political action. His third-party campaign in 1844 was particularly significant, because he drew enough popular votes to allow the Democratic candidate, James K. Polk, to win the election, defeating the Whig candidate, Henry Clay. An adroit politician and master of the art of compromise, Clay was a slaveholder who was dedicated to the Union and looked for an end to slavery. Ironically, he might have been the strong president who would have found a political solution to the issue.

What Did the War with Mexico Have to Do with the Civil War?

Although the Liberty party's James Birney may well have altered the course of the election of 1844, most Americans weren't thinking about slavery when they went to the polls. The big issue was America's expansion under "Manifest Destiny," a term coined by a newspaperman (John L. O'Sullivan in the *United States Magazine and Democratic Review*) who stated the widely held belief that America should stretch from the Atlantic to the Pacific. There were even those who believed that Mexico should be part of America. The stormy relations between the two countries had been simmering since the

1820s, when the first Americans were invited by the Mexican government, eager for development, to farm the sparsely settled borderland of Texas, then a Mexican province. Led by Stephen Austin (1793–1836), they brought their slaves with them.

In 1833 the colonists met and asked Mexico for the right to govern themselves. When Austin told them to declare independence, he was jailed in Mexico City and held until 1835. By then there were twenty thousand Americans in Texas and four thousand slaves. When General Antonio López de Santa Anna became Mexico's military dictator in 1835, the Texans revolted. In February 1836 Santa Anna led four thousand Mexicans against 189 Texans barricaded inside an old Franciscan mission in San Antonio, the Alamo. When the Texans finally fell, only six survived, and the general ordered them killed; all the bodies were then doused in oil and burned. Santa Anna allowed a woman, her child, and a single slave to warn other Texans of what would happen to them. A second massacre of Texans followed at Goliad. But "Remember the Alamo" became the rallying cry at the Battle of San Jacinto, fought on April 21. With only eight hundred men, Texas commander Sam Houston attacked a larger force, routing the Mexicans in an eighteen-minute battle. Capturing Santa Anna, Houston basically ended Texas's war for independence. The Republic of Texas was established but was never recognized by Mexico, who wanted the territory returned. For the next decade, American politicians would battle over admitting Texas into the Union as a slave state.

Most Texans and most Americans wanted Texas to join the Union, and President Jackson recognized the independence of Texas in 1836, on his last day in office. But Mexico refused to recognize the republic and disputed the boundary of the Rio Grande. Nine years later, in November 1845, President James K. Polk, attempting to settle with Mexico, offered to purchase Texas and California, but the Mexicans refused to negotiate with Polk's ambassador, James Slidell. After this failure, a U.S. Army under General Zachary Taylor advanced to the mouth of the Rio Grande, which Texas claimed as its southern boundary. Mexico claimed that the boundary was the Nueces River, northeast of the Rio Grande, and considered the advance of Taylor's troops an act of aggression. In April 1846 the Mexicans sent troops across the Rio Grande.

After a border incident in which an American soldier died under questionable circumstances, President Polk had sufficient pretext to tell Congress that war existed. Congress happily went along, with only a few dissenting voices coming from several Easterners, scattered Whigs, and a rather unusual writer named Henry David Thoreau, who went to jail rather than pay taxes in support of the war and wrote about it in an 1849 essay called "Civil Disobedience."

For Polk, the biggest complication was that his two senior military commanders, Zachary Taylor and Winfield Scott, were both Whigs, the party formed in opposition to Andrew Jackson's power. All three men knew that battlefield heroics could lead to the White House. While greatly outnumbered by Mexican troops, the Americans were generally better organized and trained and held a marked advantage in artillery, probably the decisive factor in their victory. A relatively brief war, the conflict with Mexico was costly. In *So Far from God*, a history of the Mexican War, John S. D. Eisenhower, son of the general and president, wrote, "Of the 104,556 men who served in the army . . . 13,768 men died, the highest death rate of any war in our history."

With victory, America had acquired another massive new territory, including California and Texas. Of course, this meant more land into which slaves could be brought. This issue would divide the country into increasingly hostile camps over the next twelve years. That the aftermath of the Mexican War would be complicated by slavery became clear immediately. At the conclusion of the war, Polk requested $2 million from Congress for reparations to the Mexican government for the territory annexed by the United States. But David Wilmot (1814–1868), a Democrat from Pennsylvania, attached an amendment to the House bill appropriating this money: The Wilmot Proviso moved to exclude slavery from the acquired territory; it was approved by the House on August 8, 1846. The Senate adjourned without considering the measure. After a second approval by the House on February 1, 1847, the bill was rewritten by the Senate to exclude the amendment. The differences of opinion on slavery, presumably settled by the Missouri Compromise, were back squarely on the American agenda. The Wilmot Proviso simply turned the heat up a few notches.

Perhaps one of the saddest retributions for the sins of the Mexican War was that, thirteen years later, former comrades-in-arms would use their hard-won experience against one another on the battlefields of the Civil War. Almost every significant Civil War commander served in the Mexican War, including Confederate President Jefferson Davis, Robert E. Lee, Thomas Jackson, Ulysses S. Grant, William T. Sherman, George B. McClellan, George Meade, James Longstreet, George Pickett, Pierre Beauregard, and Joseph E. Johnston. Some of these Mexican veterans, like Lewis Armistead and Winfield Hancock, had become close friends but went their separate ways in 1861, only to face each other on the fateful third day at Gettysburg in 1863.

Civil War Voices

From *Narrative of the Life of Frederick Douglass, an American Slave* (1845).

> I have no accurate knowledge of my age, never having seen any authentic record containing it. By far, the larger part of the slaves know as little of their ages as horses know of theirs, and it is the wish of most masters within my knowledge to keep their slaves thus ignorant. I do not remember to have ever met a slave who could tell of his birthday. . . .
>
> My father was a white man. He was admitted to be such by all I ever heard speak of my parentage. The opinion was also whispered that my master was my father; but of the correctness of this opinion, I know nothing; the means of knowing was withheld from me. My mother and I were

separated when I was but an infant—before I knew her as my mother. It is a common custom in the part of Maryland from which I ran away, to part children from their mothers at a very early age. Frequently, before the child had reached its twelfth month, its mother is taken off from it, and hired out on some farm a considerable distance off, and the child is placed under the care of an old woman, too old for field labor. For what this separation is done, I do not know unless it be to hinder the development of the child's affection towards its mother, and to blunt and destroy the natural affection of the mother for the child. This is the inevitable result. . . .

The whisper that my master was my father, may or may not be true; and, true or false, it is of but little consequence to my purpose whilst the fact remains, in all its glaring odiousness, that slaveholders have ordained, and by law established, that the children of slave women shall in all cases follow the condition of their mothers, and this is done too obviously to administer to their own lusts, and make a gratification of their wicked desires profitable as well as pleasurable; for by this cunning arrangement, the slaveholder, in cases not a few, sustains to his slaves the double relation of master and father.

I know of such cases; and it is worthy to remark that such slaves invariably suffer greater hardships, and have more to contend with, than others. They are, in the first place, a constant offence to their mistress. She is ever disposed to find fault with them; they can seldom do any thing to please her; she is never better pleased than when she sees them under the lash, especially when she suspects her husband of showing his mulatto children favors which he withholds from his black slaves. The master is frequently compelled to sell this class of his slaves, out of deference to the feelings of his white wife; and, cruel as the deed may strike any one to be, for a man to sell his own children to human flesh-mongers, it is often the dictate of humanity for him to do so; for, unless he does this, he must not only whip them himself, but must stand by and see one white son tie up his brother, of but a few shades darker complexion than himself, and ply the gory lash to his naked back; and if he lisp one word of disapproval, it is set down to his parental partiality, and only makes a bad matter worse, both for himself and the slave whom he would protect and defend.

Born Frederick Augustus Washington Bailey in Maryland, he was the son of a slave mother. His father's identity was a mystery, though certainly it was one of his masters. After a farmhouse childhood on Chesapeake Bay, young Fred was sent to Baltimore as a companion to a white boy. There his mistress illegally taught him to read until her husband forced her to stop, and he later worked on the docks, learning to write and continuing his reading. Sent back to the country, he was rented by a cruel farmer who beat him often until he faced the man down at risk to his life. In 1838 he escaped from Baltimore in the guise of a free sailor and, using forged papers, made his way to New York. He eventually reached the whaling port of New Bedford, Massachusetts, where he was joined by his wife. Renouncing his slave name, he took the name Frederick Douglass.

Having heard William Lloyd Garrison lecture, Douglass himself rose before a Nantucket abolition convention in 1841 and delivered an unrehearsed speech that marked him as a natural speaker—eloquent, commanding, dramatic. Garrison immediately hired him to work as a speaker and agent for the Massachusetts Anti-Slavery Society. Beaten and stoned by a mob in Indiana, Douglass eventually wrote his narrative, which became an international success. As his fame grew, he went to England for fear of being captured and returned to slavery. Friends later raised the money that allowed him to purchase his freedom.

Returning to the United States, Douglass established an abolitionist newspaper, *The North Star*, in Rochester, New York, although he and Garrison later parted ways over tactics and personality traits.

(Neither man was especially easy to get along with.) Without question, Frederick Douglass became one of the most influential men of his times; he was eventually powerful enough to advise and even contradict President Lincoln during the war years.

PRELUDE TO THE CIVIL WAR: 1836–1848

1836

January Congress is presented with abolitionist petitions aimed at outlawing slavery in Washington, D.C. Senator John C. Calhoun rejects them as "foul slanders" on the South.

February A Mexican army of six thousand led by General Santa Anna is raised to defend Mexico's claim on Texas. Four thousand Mexican troops lay siege to the Alamo, overwhelming the garrison and massacring its defenders.

March Texas adopts a constitution that formally legalizes slavery.

April The Texans defeat Santa Anna, ratify their constitution, and elect Sam Houston as president. An envoy sent to Washington requests annexation or recognition as an independent republic. Slavery is again at issue in the following congressional debate over the annexation or admission of Texas.

May The House passes a "gag" resolution, designed to table any petitions relating to slavery, a rule that continues through *1844.*

June Arkansas, a slave state, becomes the twenty-fifth state.

1837

January Michigan, a free state, becomes the twenty-sixth state.

March President Jackson recognizes Texas as an independent republic on his last day in office.

August Texas formally petitions the United States for annexation. The request is denied because Congress wants to avoid the issue of slavery there.

November Abolitionist publisher Elijah Lovejoy, whose presses have twice been destroyed by proslavery rioters, is murdered by a mob in Alton, Illinois.

December The gag rule is temporarily lifted, allowing northern congressmen to present an antislavery petition. Southerners present a countermeasure that will protect slavery or recommend dissolving the Union. The House responds with a stricter gag rule.

1838

February Former President John Quincy Adams, now a representative in the House, introduces 350 petitions opposed to slavery, but they are all tabled under the gag rule.

May Philadelphia's Pennsylvania Hall, the site of antislavery meetings, is burned to the ground.

1839

January Henry Clay condemns abolitionists as he bids for southern support in a campaign for the presidency. He claims that there is no constitutional right to interfere with slavery where it already exists.

The *Amistad* rebellion. Slaves capture a Spanish slave ship near Cuba. They are later captured in Long Island Sound, near Connecticut. In *March 1841* the Supreme Court orders the release of the men, who have been defended by John Quincy Adams, saying they are free persons. Thirty-five *Amistad* survivors return to Africa.

1840

December William Henry Harrison defeats Martin Van Buren to become the ninth president. He dies one month after his inauguration and is replaced by Vice-President John Tyler, formerly the governor and a senator from Virginia and a strong advocate of states' rights.

* Edgar Allan Poe publishes his first collection of stories, *Tales of the Grotesque and Arabesque*. Though he later becomes an editor and critic of some note, Poe never achieves any financial success from his writing.

1841

November A group of slaves being transported from Virginia to New Orleans mutinies and captures the transport ship *Creole*. After sailing to the British port of Nassau, all are freed except those accused of murder. (In 1855 a British court awards the United States $119,330 in damages over the case.)

* Ralph Waldo Emerson publishes his first series of *Essays*. In Philadelphia, Edgar Allan Poe publishes "The Murders in the Rue Morgue," the first American detective story.
* *The Deerslayer*, the last of James Fenimore Cooper's "Leatherstocking Tales," is published.

1842

George Latimer, said to be a runaway slave, is captured in Boston. But the city's abolitionists force the authorities to allow Latimer to buy his freedom from his Virginia owner. This is one of a growing number of incidents in which the rights of runaways are defended by antislavery forces in the northern states.
March The Supreme Court overturns a Pennsylvania law forbidding the seizure of fugitive slaves. But the ruling also makes the enforcement of fugitive slave laws a federal responsibility.

1843

August The Liberty party holds another convention and denounces the extension of slave territory. James Birney is again nominated for president.

1844

June The Texas annexation treaty, negotiated by John C. Calhoun and submitted to the Senate by President Tyler, is rejected, because antislavery forces convince a majority that admitting a slave state will simply lead to another North-South confrontation.

December Tyler asks the Congress to annex Texas. The next day, Democrat James K. Polk defeats Henry Clay for president. Polk is an aggressive expansionist who favors annexing Texas and acquiring other territories in Oregon and California.

John Quincy Adams successfully urges the House to revoke the gag rule that prohibits discussion of antislavery petitions.

The Baptist Church, divided over the question of slavery, splits into northern and southern conventions.

* The first telegraphic message, "What hath God wrought?," is sent from the Supreme Court room in Washington, D.C., to Baltimore by Samuel F. B. Morse.
* Amos Bronson Alcott moves his family to a utopian community he founds called Fruitlands. His daughter, Louisa May Alcott, will later write about the family and their life in *Little Women*.

1845

February At the request of President-elect Polk, the House and Senate adopt a joint resolution annexing Texas.

March Texas breaks off negotiations with Mexico. Mexico severs diplomatic relations with the United States. In his inaugural address, Polk makes it clear that he considers Texas a part of the United States.

Florida, a slave state, is admitted as the twenty-seventh state.

May Polk sends troops under General Zachary Taylor to the southwestern border of Texas to guard against an "invasion" by Mexico. Texas is still recognized internationally as part of Mexico.

November Polk offers to buy Texas, New Mexico, and California from Mexico.

December Texas is admitted as the twenty-eighth state.

Former slave Frederick Douglass publishes his autobiography, *Narrative of the Life of Frederick Douglass.*

* The U.S. Naval Academy opens at Annapolis, Maryland.
* John C. Frémont publishes *The Report of the Exploring Expedition to the Rocky Mountains in the Year 1842 and to Oregon and Northern California in the Years 1843–44.* The book sparks enormous interest in the West.
* The potato blight strikes Ireland. About 1.5 million Irish emigrate to America in the next few years.

1846

January Polk orders Zachary Taylor to move his troops, an "Army of Observation," near the Rio Grande. Polk's thirty-five hundred men are equal to about half the U.S. Army at the time.

March Taylor moves his troops to the Rio Grande's left bank, recognized as Mexican territory.

April A small troop of Mexicans fire on American soldiers blockading a Mexican town. Taylor reports to Polk that "hostilities may now be considered as commenced." This serves as the pretext for war that Polk has been awaiting.

May Following an attack on Fort Texas, Taylor pursues the Mexicans and at the Battle of Resaca de la Palma on May 9 defeats a much larger force in the first concentrated action of the war with Mexico.

On *May 13,* Congress approves a declaration of war. From the outset, the country is divided mostly along North-South lines, with the South favoring war and the expansion of slavery it will mean and the North opposing it for the same reason.

May–June Polk orders a naval blockade of Mexican ports on the Pacific and Gulf of Mexico.

The Senate ratifies a treaty with Great Britain over the Oregon Territory.

Americans declare a Republic of California, breaking away from Mexican rule.

August The Wilmot Proviso. After the war with Mexico begins, Polk asks Congress for funding to purchase territory from Mexico. Pennsylvania Representative David Wilmot amends the appropriations bill to include the phrase, taken from the Northwest Ordinance, that "neither slavery nor involuntary servitude shall ever exist in any part of" the new territories that might be acquired from Mexico. Divided sharply along regional lines, the House passes the bill, but the Senate doesn't act on it.

December Iowa, a free state, becomes the twenty-ninth state.

* The first organized American baseball game takes place in Hoboken, New Jersey, in June.
* Congress establishes the Smithsonian Institution, funded with the $550,000 willed in 1829 by James Smithson, the illegitimate son of the Duke of Northumberland.
* Elias Howe patents the first sewing machine. Despite having acquired great wealth, Howe will enlist in the Union army during the Civil War.

1847

January General Winfield Scott takes command of the Gulf expedition and orders nine thousand men from Taylor's force to attack Vera Cruz, a port city on the Gulf of Mexico.

February The Battle of Buena Vista, Taylor, vastly outnumbered, defeats Mexico's Santa Anna, who retreats to Mexico City.

Congress passes the Mexico appropriations without the Wilmot Proviso, leaving the question of slavery unresolved. During the debate, Senator Calhoun introduces resolutions laying out the southern position. Among them:

* Congress has no right to limit extant or prospective states in matters pertaining to slavery;

* Slaves are property, and Congress is obligated to protect them as such.

March Scott's army lands near Vera Cruz, one of the most heavily defended cities in the Western Hemisphere. Scott captures the city on *March 29.*

June Peace negotiations begin. The American delegation is led by Nicholas P. Trist.

September 8–14 Scott takes the strategic town of Chapultepec and marches victoriously into Mexico City.

December Elected to the House of Representatives from Illinois, Abraham Lincoln takes his seat in Congress. His first speech from the floor attacks the Mexican War.

The concept of popular sovereignty, introduced by New York Senator Dickinson, would allow local legislatures to permit or outlaw slavery in the territories. The resolution is adopted by influential politicians such as Lewis Cass, as it allows Congress to sidestep the slavery issue.

* A party led by George and Jacob Donner, traveling west through the Sierra Nevadas of California, is trapped by snow. Many of the party resort to cannibalism to survive.
* Stephen Foster's "O, Susanna" is performed for the first time and is Foster's first major success.

1848

February The United States and Mexico sign the Treaty of Guadalupe Hidalgo, ending the war. The United States receives more than 500,000 square miles of land—a territory that will yield the future states of California, Nevada, Utah, Arizona, most of New Mexico, and parts of Wyoming and Colorado. Texas is also conceded to the United States, with the boundary fixed at the Rio Grande. The United States agrees to pay $15 million for the land and reparations of $3.25 million. The Senate ratifies the treaty and provides the funding but without approving the Wilmot Proviso.

May General Lewis Cass of Michigan is nominated for the presidency by the Democrats after Polk keeps his pledge not to run for reelection. The party platform criticizes any attempt to bring the slavery question before Congress.

May 29 Wisconsin, a free state, becomes the thirtieth state.

August In Buffalo, an antislavery coalition forms the Free-Soil party and nominates former President Martin Van Buren. The party opposes slavery and upholds the Wilmot Proviso. Its slogan is "Free soil, free speech, free labor, and free men."

August 14 Polk signs a bill organizing the Oregon Territory without slavery. Southerners agree, with the tacit understanding that other new territory will be open to slaveholders.

November 7 General Zachary Taylor, a slaveholder, is elected president. Van Buren's Free-Soil candidacy wins nearly 300,000 popular votes, helping to assure Taylor's victory by drawing votes from Democratic candidate Lewis Cass, an advocate of popular sovereignty.

* The Associated Press is organized in New York to speed newsgathering for newspaper publishers.
* Gold discovered in California sparks the great Gold Rush of 1849.

chapter three

"*The Edge of the Precipice*"

*I implore gentlemen... whether from the South or the North...
to pause at the edge of the precipice, before the fearful and dangerous
leap be taken into the yawning abyss below, from which none
who ever take it shall return in safety.*

—SENATOR HENRY CLAY
FEBRUARY 5, 1850

*I hate [slavery] because it deprives the republican example of its
just influence in the world—enables the enemies of free
institutions, with plausibility, to taunt us as hypocrites—causes
the real friends of freedom to doubt our sincerity.*

—ABRAHAM LINCOLN
OCTOBER 16, 1854

I... am now quite certain that the crimes of this *guilty land* will
never be purged *away* but with Blood.

—JOHN BROWN
DECEMBER 1859

✳

* Where Did the Underground Railroad Run?

* What Was the Compromise of 1850?

* What Was the Fugitive Slave Act?

* Who Was Uncle Tom?

* Who Were the Know-Nothings, the Free-Soilers, and the First Republicans?

* What Are Beecher's Bibles and Border Ruffians?

* How Did a Slave Named Dred Scott Change History?

* What Did Lincoln and Douglas Debate?

* What Happened at Harpers Ferry?

* Who Ran Against Lincoln?

It would be a decade of bad presidents and broken promises, the fateful, fearful period that finally tipped the scales toward war. But in 1850 there was still a wide range of opinion about American slavery's future. Standing at two extremes were the uncompromising foes and supporters of slavery. Between them ranged those who desperately sought compromise and the great majority of Americans, who wished the problem would just go away. In these years of growing animosity, one question screamed out. Could the voices of moderation drown out the vitriolic tirades of violent hostility?

As the words flew in the editorial pages of increasingly partisan newspapers, in state capitals and in Washington, the propaganda war escalated. Then the rhetoric reached a boiling point. Swords replaced speeches. And in this fateful decade the words became flesh and blood.

But for the nearly four million in chains, all the words—the newspaper editorials, the politicians' debates, the abolitionists' sermons— were empty. For slaves, there were few choices: serve for life or attempt to escape.

Where Did the Underground Railroad Run?

In his first autobiography, Frederick Douglass deliberately left out many of the details of his 1838 escape from Baltimore. He feared that describing his route and methods and the assistance he had received along the way would only threaten the safety and continued effectiveness of those who had helped him. He also did not want to alert slaveowners, who could shut down one of the routes that escaping slaves might use.

Ten years after Douglass fled Maryland, a Georgia couple made an even more extraordinary escape. Ellen, a twenty-two-year-old slave, was the daughter of her owner, James Smith of Macon, Georgia; Ellen's mother was one of Smith's slaves. At the age of eleven, Ellen was "given" as a wedding present to Smith's legitimate daughter—Ellen's white half sister—and became her house servant. Sounds like Cinderella, but it was no fairy tale. When Ellen met a slave carpenter named William Craft and the two fell in love, they planned to run away. Since the light-skinned Ellen could easily pass as white, William believed they might escape in the guise of master and slave. But no reputable southern girl of the day would travel alone with a black man, so Ellen had to disguise herself doubly. With her arm in a sling so she wouldn't be asked to write—she was illiterate—and her head wrapped in bandages to disguise her face, Ellen called herself William Johnson, a sick young white *man* on the way to Philadelphia for treatment, with William acting the part of his (her) slave.

Traveling by steamship and train, they made their way to Philadelphia and freedom in December 1848, narrowly averting detection with only the luckiest of breaks along the way. In Philadelphia they were welcomed by black abolitionists, who took them to the home of white abolitionist Barley Ivens. Afraid that these white people would only try to return them to slavery, the Crafts initially refused their help. But they eventually made their way to Boston and became popular speakers on the abolitionist lecture circuit, just as Douglass had done. When slavecatchers were dispatched from Georgia to capture the Crafts, the couple was taken to England by abolitionist Henry Ingersoll Bowditch, a son of Nathaniel Bowditch, whose book on celestial navigation was a standard for sailors for nearly a century.

Both escapes—by the Crafts and Douglass—made international celebrities of these former slaves. The Crafts made their escape relying mostly on their own wits until they reached Philadelphia. There, like Douglass, they received help. The last phase of the Crafts' escape and Douglass's route brought them in touch with the Underground Railroad, a legendary chapter in American history that has been both mythologized and misunderstood.

The first myth concerning the Underground Railroad is that it was the creation of benign whites who oversaw its efficient, friendly operation. The second is the number of slaves who actually used it. Called by some the Liberty Line, the Underground Railroad was a loose national network of mostly black abolitionists who illegally helped fugitive slaves reach safety in the free states or Canada. Begun in the 1780s under the auspices of Quakers and other church groups, the network gradually grew into a widespread chain of safe houses along routes to the free states and Canada (although some escapes from the Deep South went through Mexico, where slavery was also illegal). There was never any actual railroad, but the system gradually came to be called the Underground Railroad, and the safe houses became known as stations or depots. Along the route, black fugitives were concealed, provisioned, and guided from one stop to the next, often at great risk to the "conductors" and "stationmasters."

Although routes ran all the way north from the Deep South, most of those who escaped successfully were, like Frederick Douglass, from the upper South. Also like Douglass, those most able to escape were generally unattached young men. Some were highly skilled, although few were literate. Flights by families or couples, like the Crafts, were rare. The slaves were often initially contacted by a conductor posing as a peddler or mapmaker.

Traveling by night, many of the fugitives relied on rumors of safe houses and little more than the North Star for guidance. Usually they sought isolated stations (farms) or vigilance committees in towns a night's journey apart, where sympathetic free blacks could effectively conceal them. By day, they hid in barns and caves, fed and cared for by the stationmaster. Eventually other conductors met them at such border points between the free and slave states as Cincinnati, Ohio, and Wilmington, Delaware, until they made their way to the Great Lakes ports of Detroit, Sandusky, Erie, and Buffalo—all terminals for the final leg of the journey into Canada and freedom.

Other routes went through upstate New York and New England. Like Douglass, not all fugitives went all the way to Canada, choosing to remain in the anonymity and relative safety of large cities like New York or Philadelphia, where sympathetic vigilance committees helped them find work and housing.

Besides Douglass, who sheltered as many as eleven fugitives at a time in his Rochester home, other notable stationmasters were the nation's most outspoken abolitionist, William Lloyd Garrison, famed police detective Allan Pinkerton (whose name resurfaces in Civil War history), and abolitionist and feminist Susan B. Anthony, one of many women who adopted emancipation and women's rights as equal causes. Cincinnati lawyer Salmon Chase, a founder of the Liberty and Free-Soil parties and later President Lincoln's secretary of the treasury, frequently defended fugitives at no cost. The man often identified as the "president" of the Underground Railroad was a Quaker from North Carolina, Levi Coffin; he was said to have ferried more than three thousand slaves through Indiana and Cincinnati.

The risks involved in aiding slaves were not small. After having his printing presses thrown into the river three times, Elijah Lovejoy, a prominent publisher and stationmaster, was killed in 1837 by an Alton, Illinois, mob who set fire to his presses and then shot him. Another well-known stationmaster, Jonathan Walker, was branded with the letters "SS" (for slave stealer) for assisting fugitives. In 1850 a Maryland couple suspected of aiding runaways was tarred and feathered, then driven out of the community. And for black operatives, there was the double fear of being kidnapped and sold back into slavery. Black bookshop owner David Ruggles, a stationmaster in New York, had his shop burned and was constantly on guard for slavecatchers, who would have taken him south.

Besides the physical danger, there were legal dangers. In Lexington, Kentucky, Calvin Fairbanks was jailed on two occasions, serving a total of seventeen years in prison for assisting runaways in reaching Ohio and the safe houses of Levi Coffin; he remained in prison until 1864, well into the Civil War. A Maryland minister, Charles T. Torrey, who aided hundreds of runaways, was sentenced to six years of hard labor in 1844. He died in a Maryland penitentiary in May 1846, becoming, along with Lovejoy, one of the first martyrs of the abolitionist cause.

Perhaps the most famous conductor was the legendary Harriet Tubman, called Moses by her people. Born Araminta, a slave in rural Dorchester County, Maryland, Harriet worked as a field hand and plantation house servant. After marrying a free black man named John Tubman, she was still a slave when she learned that she and her children might be sold. Harriet decided to escape. Walking by night, she made it to Philadelphia, where she met William Still, a central figure in the black Underground Railroad, and was given work as a hotel domestic.

Deeply religious, Tubman soon became the Railroad's best-known conductor, making nineteen trips back into slave territory. She carried laudanum, an opiate, to quiet crying babies and a gun for dealing with slavecatchers. Occasionally she used that gun to threaten skittish passengers who wanted to get off the "train." "Dead niggers tell no tales. You go or die," she supposedly warned them.

Tubman personally helped more than three hundred runaways gain their freedom, including her own parents and six of her ten siblings, taking them through Pennsylvania to her home in upstate New York and finally across a suspended bridge at Niagara Falls and into Canada. Her notoriety grew to such an extent that a reward of $40,000—a huge amount of money—was offered for her capture.

While Douglass and Tubman are the best known of the prominent blacks in the Underground, it could not have existed without the active support of large numbers of free blacks. When Douglass arrived in New York, he sought out the home of David Ruggles, a free black from Norwich, Connecticut, who helped hundreds of slaves move through New York. In Philadelphia, Tubman had been assisted by William Still, one of the most effective and significant figures in the Underground and author of a monumental memoir, *The Underground Railroad* (1872). It was to Still's office that a slave named Henry Brown was shipped in a wooden box from Richmond, his twenty-hour trip earning him the nickname of "Box" Brown. Samuel Smith, the man who had shipped Box Brown, tried this method again with less success. The two boxed slaves were caught, and Smith was jailed. Still recalled another encounter with a fugitive named Peter Freeman, who asked if Still knew two former slaves, Levin and Sidney. Still realized immediately that the man was asking about Still's own parents and that Freeman was actually an older brother he had never known.

The daughter of James and Charlotte Forten, pioneering black abolitionists in Philadelphia, Margaretta Forten founded the integrated Philadelphia Female Anti-Slavery Society. Another Forten daughter Harriet, was married to Robert Purvis, a young man of mixed background who inherited a fortune from his white father, a cotton broker, and devoted these funds to the abolitionist movement and assisting runaways. One of his notions was the Free Produce Movement, which used no food or other items produced with slave labor.

But when the first histories of the Underground Railroad were written after the war, these black names were omitted or overlooked. Despite the Civil War, few white historians were willing to recognize the achievements of blacks. The worthy efforts of Quakers like Coffin and other white abolitionists were elevated, and the role of blacks as conductors and organizers was seriously diminished. An even worse historical miscarriage was the presentation of the actual fugitives. Paternalistic historians presented them as passive and helpless. Of course, even to attempt such an escape required enormous courage. Usually illiterate, with no maps, often guided only by whispered plantation rumors of safe houses, and with the threat of dogs and slavecatchers behind them at every step, these fugitives had to be resourceful and fearless.

A second aspect of the Underground Railroad mythology is more difficult to assess. The network was not a massive, organized bureaucracy with its own record-keeping division. Since their actions were illegal in the eyes of the federal government, operatives rarely documented their work. In an unfortunate effort to underscore the importance of the Underground Railroad, early historians may have exaggerated the number of slaves who actually escaped. It was once thought that more than sixty thousand slaves gained their freedom through the Railroad, but that estimate is probably high. By 1860 Southerners owned a total of about four million slaves. The number successfully escaping—only about a thousand to twenty-five hundred a year between 1830 and 1850—was virtually insignificant in terms of making an impact on the slave economy. More recent suggestions have reduced the number to forty thousand slaves, who escaped over the course of several decades, but there is no way of knowing with any certainty how many slaves made it safely to freedom either on the Railroad or on their own.

The deliberate inflation of the number of slaves who may have been helped by the Railroad does nothing to undercut the Railroad's significance. To minimize its role by lowering the number of slaves it did help is a disservice. It would be the equivalent of saying the French and Danish resistance movements against the Nazis didn't accomplish much because they didn't save many Jews from the Holocaust. For every fugitive slave assisted, the Railroad's significance was clear.

Perhaps far more important than the number of safe arrivals was the impact of the Underground Railroad. The publicity given to freed slaves and their accounts of the inhumanity of the slave system were major factors in the transformation of public opinion, where indifference and racism once prevailed. By the mid-1850s there was a growing antipathy toward slavery, and it produced dramatic confrontations over the return of fugitives who had been captured and were being forced back into shackles under federal law. In Ohio and Boston, in particular, popular outrage at these slave laws led to large-scale incidents of civil disobedience.

However, this publicity was double-edged. The open defiance of the federal fugitive slave laws by abolitionists infuriated slaveowners. The antagonism over the open assistance to fugitives and the celebrity status acquired by such former slaves as Frederick Douglass, Harriet Tubman, and the Crafts were crucial in fanning the flames of hatred and mistrust between slave and free regions, making any compromise over slavery difficult if not impossible.

Songs of the Civil War

"Follow the Drinking Gourd"

The riverbank will make a very good road
The dead trees show you the way
Left foot, peg foot, traveling on
Follow the drinking gourd

Chorus
Follow the drinking gourd
Follow the drinking gourd
For the old man is a-waiting for to take you to freedom
If you follow the drinking gourd

The river ends between two hills
Follow the drinking gourd
There's another river on the other side
Follow the drinking gourd
Chorus

Where the great big river meets the little river
Follow the drinking gourd
The old man is a-waiting for to take you to freedom
If you follow the drinking gourd.
Chorus

Frederick Douglass chose *The North Star* as a symbol of black freedom. This traditional song was sung by slaves and provided a sort of
road map to freedom. The "drinking gourd" was the Little Dipper
constellation, in which the North Star, at the top of the gourd's "han-
dle," was used by the slaves as a beacon in attempting to escape.
The "big river" was the Tennessee River, the "little river," the
Ohio. The "old man" of the song was a legendary character known
as Peg Leg Joe, a white sailor who supposedly went around planta-

tions teaching this song. It was his peg leg that made the "left foot, peg foot" impression that blazed the trail to freedom.

Civil War Voices

Senator Henry Clay of Kentucky in a speech to the Senate (February 5–6, 1850).

> In my opinion there is no right on the part of any one or more of the states to secede from the Union. War and the dissolution of the Union are identical and inevitable, in my opinion. There can be a dissolution of the Union only by consent or by war. Consent no one can anticipate, from any existing state of things, is likely to be given; and war is the only alternative by which a dissolution could be accomplished. . . . And such a war as it would be, following the dissolution of the Union! Sir, we may search the pages of history, and none so ferocious, so bloody, so implacable, so exterminating . . . would rage with such violence. . . .
>
> I implore gentlemen, I adjure them, whether from the South or the North . . . to pause at the edge of the precipice, before the fearful and dangerous leap be taken into the yawning abyss below, from which none who ever take it shall return in safety.

Senator John C. Calhoun of South Carolina (March 4, 1850). Read to the Senate, this speech was the ailing Calhoun's last public statement before his death.

> I have, Senators, believed from the first that the agitation of the subject of slavery would, if not prevented by some timely and effective measure, end in disunion. Entertaining this opinion, I have, on all proper occasions, endeavored to call the attention of both the two great parties which divide the country to adopt some measure to prevent so great a

disaster, but without success. The agitation has been permitted to proceed, with almost no attempt to resist it, until it has reached a point when it can no longer be disguised or denied that the Union is in danger. You have thus had forced upon you the greatest and the gravest question that can ever come under your consideration—How can the Union be preserved?

... What has endangered the Union?

To this question there can be but one answer,—the immediate cause is the almost universal discontent which pervades all the States composing the Southern section of the Union. [This discontent] commenced with the agitation of the slavery question and has been increasing ever since.... What has caused this widely diffused and almost universal discontent?

One of the causes is, undoubtedly, to be traced to long-continued agitation of the slave question on the part of the North, and the many aggressions which they have made on the rights of the South during the time....

As then, the North has the absolute control over the Government, it is manifest, that on all questions between it and the South, where there is a diversity of interests, the interests of the latter will be sacrificed to the former, however oppressive the effects may be; as the South possesses no means by which it can resist, through the action of the Government. But if there was no question of vital importance to the South, in reference to which there was a diversity of views between the two sections, this state of things might be endured, without the hazard of destruction to the South. But this is not the fact. There is a question of vital importance to the Southern section, in reference to which the views and feelings of the two sections are as opposite and hostile as they can possibly be.

I refer to the relation between the two races in the Southern section, which constitutes a vital portion of her social organization. Every portion of the North entertains views and feelings more or less hostile to it. . . . On the contrary, the Southern sections regards the relation as one which cannot be destroyed without subjecting the two races to the greatest calamity, and the section to poverty, desolation, and wretchedness; and accordingly they feel bound, by every consideration of interest and safety, to defend it.

What Was the Compromise of 1850?

Compromise has always been one of the necessary evils of American history. In Congress and the White House, political compromises have almost always been the answer to reconciling differences and easing the passage of legislation. It may not sound like a good way to run a country, but it has worked for the most part, as long as principles aren't being compromised. Few politicians have mastered this art of the deal better than Henry Clay, the congressman, senator, secretary of state, and presidential candidate from Kentucky.

Twice in his extraordinary political life, with the Missouri Compromise and the Compromise Tariff, Clay had found ways to settle sharp differences before they spilled over into disunion. In 1850 a white-haired, seventy-three-year-old Clay faced the most difficult task yet. Challenged by the ongoing question of whether slavery would be allowed in new states formed out of the land acquired from Mexico, Clay had to find a way to craft another compromise.

His effort would climax in the last great clash among the congressional Big Three: Clay, Calhoun, and Webster. In February 1850, Clay rose to speak and, over the course of two days, offered his solution. (See excerpt, pages 101–103) Pleading for tolerance and moderation, he proposed five acts, the Compromise Measures of 1850. Two of them represented concessions by slaveholders: the first authorized the abolition of the slave trade in the District of Columbia, and the second admitted California as a free state. By the terms of the third measure, the territory east of California won from Mexico was divided into the territories of New Mexico and Utah, and they were opened to settlement by both slaveholders and antislavery settlers. Using the notion of popular sovereignty to determine the future of slavery, this measure effectively undermined the Missouri Compromise of 1820. The fourth measure provided that Texas, already a slave state, be awarded $10 million in settling claims to adjoining territory, further strengthening the political clout of slaveowners. The fifth and last bill was the Fugitive Slave Law of 1850, a major concession to slaveholders; it provided for the return of runaway slaves to their masters and the vigorous prosecution of anyone involved in assisting runaways.

Too ill to hear Clay, the aging Calhoun was given the details of his speech. Wrapped in a cloak, he returned to the Senate floor a few weeks later and was assisted to his seat. The old lion of states' rights was feverish and too ill to speak, so his speech was read for him. Dissatisfied with Clay's compromises, Calhoun threatened secession at any cost. (See excerpt, pages 101–103.) A month later, he was dead.

Webster then offered one of the most famous speeches in Senate history to a crowd that overflowed the chamber and galleries. Long Clay's political foe, Webster, the "Godlike Daniel," now joined Clay in an effort to save the Union: "I wish to speak today, not as a Massachusetts man, nor as a northern man, but as an American. . . . I speak today for the preservation of the Union. Hear me for my cause." In a speech that left listeners in tears, Webster threw his full weight alongside Clay, going on to say "Peaceable secession is an utter impossibility."

While Webster's support was crucial, the real key to the passage of the Compromise of 1850 was the death of President Zachary Taylor from cholera. A slaveowner himself and a Mexican War hero, Taylor had rejected the notion of slave state secession. His death opened the way for Vice-President Millard Fillmore, a New Yorker who was far more conciliatory toward the slavery forces, to sign the legislation. (A few years ago, a historian suggested that Taylor might have been poisoned by the forces who favored slavery and the Compromise; in 1994 his body was actually exhumed for a sophisticated medical examination. The tests revealed that Taylor had indeed died of natural causes.)

Civil War Voices

Abolitionist and women's rights pioneer Sojourner Truth attended a women's rights convention in Ohio. In response to a man's comment that women were weak and inferior to men, baring her arms, she spoke (1851).

> Ain't I a woman? Look at me. Look at my arm. I have ploughed, and planted, and gathered into barns, and no man could head me! And ain't I a woman? I could work as much and eat as much as a man—when I could get it—and bear the lash as well! And ain't I a woman? I have borne thirteen children, and seen most all sold off to slavery, and when I cried out with my mother's grief, none but Jesus heard me! And ain't I a woman?

What Was the Fugitive Slave Act?

The bone Henry Clay tossed to slaveowners known as the Fugitive Slave Act was, without question, the most controversial of his compromises. Derided by outraged abolitionists as "the Man-Stealing Law" or the "Bloodhound Bill," it was stronger than the law already on the books, the 1793 runaway act. Under the Fugitive Slave Act, aid to escaping slaves in the form of food, shelter, or other assistance was a federal crime, punishable by a $1,000 fine and six months in prison. Constitutional guarantees such as a trial by jury were simply ignored. Under a bounty system, the law created the office of commissioners, who determined whether a black person was in fact a runaway. They were paid $10 for every slave they returned to slavery but only $5 for every one they determined to be free. Guess what they usually decided?

The act commanded citizens "to aid and assist in the prompt and efficient execution of this law, whenever their services may be required." It created a wave of forcible reenslavements in which slavecatchers had free rein to swoop down on entire black families and accuse them of being runaways.

But it also created new opponents to slavery where none had existed before. Many Americans had been content to see slavery go on as it had, but the idea that the federal government could now compel what amounted to kidnapping was repugnant. Typical was the reaction of Ralph Waldo Emerson, who called it "a filthy enactment," adding, "I will not obey it, by God!" And in Emerson's Boston, the issue came to a head when a fugitive named Frederick Wilkins, also known as Shadrach, was captured in the coffeehouse where he had worked for a year. Before he could be tried and returned, a throng of blacks rescued him and whisked him off to Canada. Among others, Henry Clay damned this black protest and wondered whether "the government of white men is to be yielded to a government of blacks." President Fillmore threatened to send federal troops to compel Bostonians to comply with the law, raising ominous memories of the days when the quartering of British troops in Boston had created the animosity that helped start a revolution.

But to get the Compromise of 1850 through Congress, proslavery politicians demanded this act as a major concession. Following its passage, many of those who detested slavery, like Emerson, considered the law a violation of moral law, and many more began to support the fugitives' efforts. Some, like Frederick Douglass, were getting tired of words. When the two major political parties, Whigs and Democrats, accepted the 1850 Fugitive Slave Act as part of their presidential platforms in 1852, Douglass thundered,

> The only way to make the Fugitive Slave Law a dead letter is to make half a dozen or more dead kidnappers. A half dozen more dead kidnappers carried down South would cool the ardor of Southern gentlemen, and keep their rapacity in check. . . . A black man may be carried away without any reference to a jury. It is only necessary to claim him, and that some villain should swear to his identity. There is more protection there for a horse, for a donkey, or

anything, rather than a colored man—who is, therefore, justified in the eye of God in maintaining his right with his arm.

Documents of the Civil War

From Harriet Beecher Stowe's *Uncle Tom's Cabin* (1851).

"Mas'r, if you mean to kill me, kill me; but, as to raising my hand agin anyone here, I never shall—I'll die first."

Tom spoke in a mild voice, but with a decision that could not be mistaken. Legree shook with anger; his greenish eyes glared fiercely, and his very whiskers seemed to curl with passion; but like some ferocious beast, that plays with its victim before he devours it, he kept back his strong impulse to proceed to immediate violence, and broke out into bitter railery.

"Well, here's a pious dog, at last, let down among us sinners!—a saint, a gentleman, and no less, to talk to us sinners about our sins! Powerful holy crittur, he must be! Here, you rascal, you make believe to be so pious—didn't you never hear, out of yer Bible, 'Servants obey your masters?' An't I yer master? Didn't I pay down twelve hundred dollars, cash, for all there is of inside yer old cussed black shell? An't yer mine, now, body and soul?" he said, giving Tom a violent kick with his heavy boot; "tell me!"

In the very depth of physical suffering, bowed by brutal oppression, this question shot a gleam of joy and triumph through Tom's soul. He suddenly stretched himself up, and looking earnestly to heaven, while the tears and blood that flowed down his face mingled, he exclaimed:

"No! no! no! my soul an't yours, mas'r! You haven't bought it—ye can't buy it! It's been bought and paid for by he that is able to keep it; no matter, no matter, you can't harm me!"

"I can't!" said Legree, with a sneer, "we'll see—we'll
see! Here, Sambo, Quimbo, give this dog such a breakin'
in as he won't get over this month!"

Who Was Uncle Tom?

One of the most immediate and far-reaching results of the Fugitive
Slave Act was the reaction of a New England writer named Harriet
Beecher Stowe (1811–1896). Born in Litchfield, Connecticut, home
of the Litchfield Female Academy, one of the first women's schools
in America, Harriet Beecher was the daughter of a liberal Presbyte-
rian minister, Lyman Beecher, who became one of the most eloquent
preachers of his day. When her father left New England to head Lane
Seminary in Cincinnati, she went with him and in time married one
of his students, Calvin Ellis Stowe, also an ardent abolitionist. One
of Harriet's brothers, Henry Ward Beecher (1813–1887), became pastor of the Plymouth Church of the Pilgrims in Brooklyn and emerged
as one of the most famous orators and preachers in American history.
An outspoken abolitionist, he attracted huge crowds to services and
revival meetings.

While living in Ohio, Harriet Beecher Stowe began writing children's stories and texts, including a successful schoolbook, *Primary
Geography for Children* (1833), which earned her a considerable amount
of money, especially for a young woman in nineteenth-century America. In 1839 she began contributing to *Godey's Lady's Book*, a popular
magazine that supported an early generation of American writers, and
in 1843 a collection of her stories, *The Mayflower; or Sketches of Scenes
and Characters Among the Descendants of the Pilgrims*, was published.
She was well on her way to being one of the best-known writers in
America.

The Stowes moved to Maine, where Calvin Stowe had taken a
position at Bowdoin College. Following the passage of the Fugitive
Slave Act in 1850, Harriet Stowe and the other Beechers elevated
their rage against slavery to a new level. When she condemned the

act in the *National Era,* an abolitionist journal, Stowe received a congratulatory letter from one of her sisters, who urged her "to write something that would make this whole nation feel what an accursed thing slavery is."

Drawing on newspaper accounts of fugitive slaves and her experience with blacks during her stay in Cincinnati—where the Beechers had hired slaves from their white owners as domestic help—Stowe began a series of literary sketches about slaves that ran in the *National Era* beginning in 1851. She was also drawing on the deep emotional resonance of the recent loss of her own infant son; the loss of children became a major theme in her book. In serial form, the stories attracted little attention outside the literary and abolitionist world. But in March 1852 they were published as a book, *Uncle Tom's Cabin, or Life Among the Lowly,* and its astonishing success was unprecedented in American publishing history. Ten thousand copies sold in the first week and three hundred thousand by the end of the first year. Stowe netted an astonishing $10,000 in royalties in the first three months. Within five years, five hundred thousand copies of the book were sold in the United States, and it was eventually translated into twenty languages.

The book had an immediate impact on the slavery debate. For the first time, the issue had been removed from dry political discussion and humanized in the stories of the Christ-like Uncle Tom, a pious slave who is torn from his family in Kentucky and sold into the Deep South, eventually to be martyred; the sadistic overseer Simon Legree; Eva, the angelic sick child whom Tom had cared for; and Eliza, the slave mother desperately attempting to reach freedom while clutching her baby by crossing the icy Ohio River separating slave state Kentucky from free Ohio. The book instantly galvanized antislavery sentiments and once more hardened the lines between free and slave.

Harriet Beecher Stowe's Uncle Tom was a character of great moral strength and nobility. Unfortunately, as time went by, Tom became a caricature in minstrel-style shows, reduced to a mindless, shuffling lackey of white people. The real character of Uncle Tom—heroic, unbending, willing to die for his people—couldn't have been further from that image.

A hated symbol of northern abolitionism (she once received the ear of a slave in a parcel), Stowe was vilified by slavery's defenders for her portrait of slavery and was compelled to write a second book, *A Key to Uncle Tom's Cabin*, in which she documented the sources of her material. Invited to the White House during the Civil War, she was supposedly greeted by Lincoln: "So you're the little woman who wrote the book that started this great war."

Civil War Voices

George Fitzhugh (1854).

The earliest civilization of which history gives account is that of Egypt. The Negro was always in contact with that civilization. For four thousand years he has had opportunities of becoming civilized. Like the wild horse, he must be caught, tamed and domesticated. . . .

. . . The worst feature of modern civilization, which is the civilization of free society, remains to be exposed. Whilst laborsaving processes have probably lessened by one half, in the last century, the amount of work needed for comfortable support, the free laborer is compelled by capital and competition to work more than he ever did before, and is less comfortable. . . . These reflections might seem, at first view, to have little connection with Negro slavery; but it is well for us of the South not to be deceived by the tinsel glare and glitter of free society. . . .

The Southerner is the Negro's friend, his only friend. Let no intermeddling abolitionist, no refined philosophy, dissolve this friendship.

A somewhat eccentric Virginian, Fitzhugh not only advocated slavery for blacks but wanted to return the nation to a form of medieval feudal system in which working-class whites would once again be made serfs. He railed against the northern factories and merchants, claiming that the slave system was better and fairer. While his views belonged to the lunatic fringe, abolitionists and politicians opposed to slavery pointed to Fitzhugh as the voice of the "Slave Power" of the South.

MILESTONES IN THE CIVIL WAR: 1849–1854

1849

December With the Gold Rush in full fever, California applies for admission as a free state, which will leave slave states in the minority.

* Henry David Thoreau publishes "Civil Disobedience" (originally "Resistance to Civil Government"), an essay that grew out of his refusal to pay taxes in support of the Mexican War.
* The Spiritualism movement, begun in Rochester, New York, sweeps the country. Séances become very popular throughout America, though Spiritualist leader Margaret Fox later confessed that she was a fraud.

1850

July President Zachary Taylor, who opposes the Clay compromise, dies and is replaced by Vice-President Millard Fillmore, who favors the compromise.
September Congress adopts the five measures known as the Compromise of 1850. Fillmore signs all five bills into law:

* California is admitted as the thirty-first state, a free state.
* The territories of New Mexico and Utah are organized without restrictions on slavery.

* The boundaries of Texas are set, also without restrictions on slavery.
* The slave trade, but not slavery itself, is abolished in the District of Columbia after *January 1851*.
* The Fugitive Slave Act provides stricter federal enforcement to the Fugitive Slave Act of 1793.

November–December At conventions throughout the South, the right to secede is upheld if the Compromise of 1850 is not observed strictly by the northern states.

The U.S. Census shows: total population, 23,191,876; black population 3,638,808 (15.7 percent).

* Nathaniel Hawthorne publishes *The Scarlet Letter.* The then-daring tale of adultery becomes an instant best-seller, with all four thousand copies in the first printing selling in four days.
* Herman Melville's novel *White-Jacket* exposes the harsh, often inhumane treatment of American sailors.

1851

February An accused runaway, Shadrach, is rescued from a Boston jail by an angry black mob. Fillmore insists that the law must be obeyed by Northerners, but several other notable fugitive rescues are made and there is strong resistance to the law.

June Harriet Beecher Stowe's novel *Uncle Tom's Cabin* begins to appear as a serial in the *National Era*. It is published in book form in Boston in *March 1852*.

December Pro-union candidates win congressional elections throughout the South, indicating that the Compromise of 1850 has eased slaveowners' concerns.

* The first issue of *The New York Daily Times*, the newspaper that will become *The New York Times*, is published in September.
* Nathaniel Hawthorne publishes *The House of the Seven Gables* and Herman Melville publishes *Moby-Dick*.
* Stephen Foster's popular song "Old Folks at Home" is published.

* The American branch of the Young Men's Christian Association (YMCA) is begun with a chapter in Boston.

1852

June After forty-eight ballots, the Democrats nominate New Hampshire's Franklin Pierce (1804–1869) for president. Long an opponent of abolitionists, Pierce had sponsored the gag rule restricting the presentation of antislavery petitions to Congress. The party platform supports the Compromise of 1850 as the best solution to the slavery issue. The Whigs meet in Baltimore and nominate Mexican War hero General Winfield Scott, Pierce's commander in Mexico. They also accept the Compromise of 1850.

August In Pittsburgh, the Free-Soil party nominates John P. Hale for president. Its platform condemns slavery and the Compromise of 1850 and supports free homesteads for settlers and easy entry into the U.S. for immigrants.

October In a four-hour speech, Massachusetts Senator Charles Sumner attacks the Fugitive Slave Act and submits a resolution calling to overturn it.

November Franklin Pierce is easily elected president. The Free-Soil candidate receives only 156,000 votes, suggesting that the abolition movement is in decline following the Compromise of 1850, but several Free-Soilers are elected to Congress.

1853

March President Pierce is inaugurated, promising compliance with the Compromise of 1850 and the peaceful acquisition of new territory. His Cabinet includes War Secretary Jefferson Davis, a hero of the Mexican War and a former senator from South Carolina.

* Gail Borden introduces a machine for making evaporated milk. Intended to provide milk for city-dwelling immigrant children, evaporated milk had huge sales during the Civil War because of large Army purchases.

1854

May The Kansas-Nebraska Act is passed, overturning the Missouri Compromise and opening the territory north of the old Missouri line to slavery. Both sides of the issue begin to send settlers into the Kansas-Nebraska Territory to influence the future of these territories.

In Boston, an antislavery mob attempts to rescue fugitive slave Anthony Burns from jail. When Burns is later shipped to Virginia, fifty thousand people protest his return to slavery.

July The Wisconsin Supreme Court rules that the Fugitive Slave Act is unconstitutional and frees a man convicted of rescuing a runaway.

* Henry David Thoreau publishes *Walden*.

Who Were the Know-Nothings, the Free-Soilers, and the First Republicans?

Faced with presidential elections in recent years, most of us today would agree that a lot of candidates are "know-nothings." The fact is, there once was a real Know-Nothing candidate and a Know-Nothing party, which stood on the brink of becoming a major political force in America.

Before the 1850s, party lines weren't as clearly drawn as they are today. More people aligned themselves with a particular politician than with a party, so it was more common to be a Clay man or a Calhoun man than a Whig or Democrat, at that time America's two principal parties. The Democrats traced their roots to Thomas Jefferson's Democratic-Republicans and became plain Democrats under Andrew Jackson. The Whig party had grown purely out of resentment and fear of Jackson's enormous popularity and power. (The Whigs' predecessors had been the Federalists, the party of John Adams and Alexander Hamilton.) As slavery proved increasingly divisive, both parties were splitting along geographic as well as philosophical lines. The question of slavery and its expansion into the new territories and states was destroying any semblance of party unity. Into the vacuum created by the internal weakening of the two major parties, offshoot parties were springing up around the country and siphoning power and numbers away from them.

Two of the most influential of these offshoots before the war were the abolitionist Free-Soil party and the American or Know-Nothing party. The Free-Soil party came together in 1848, absorbing what was left of the Liberty party, the first antislavery political party. Formed in 1839 in upstate New York by a group of abolitionists, the Liberty party had nominated James Birney, who drew only 7,000 votes in the election of 1840. Four years later, Birney drew more than 62,000 votes. A vast increase over his 1840 showing, it was still a relatively small figure. But Birney's campaign drew enough votes away from Whig Henry Clay to ensure the election of James K. Polk, a proslavery Democrat. In the midterm election of 1846, Liberty candidates polled 75,000 votes and elected several congressmen. Looking to broaden its base, the party merged in 1848 with foes of slavery from the weakened Whig party and a radical group of New York Democrats, known as "barnburners"—extremists "who would burn down the barn to get rid of the rats"—to form the Free-Soil party in Buffalo, New York.

Calling for a halt to the extension of slavery rather than abolition, the Free-Soilers nominated former President Martin Van Buren and Charles Francis Adams—the son of John Quincy Adams, the sixth president—on a platform that called for "free soil, free speech, free labor and free men." The new party won no states but polled 291,263 votes and especially turned the election in New York, playing a decisive role in the election of Zachary Taylor, "Old Rough and Ready," the Mexican War hero and last Whig president. The Free-Soilers also elected two senators and fourteen members of the House that year, adding more congressional seats in 1852.

Around 1849, about the same time that the Free-Soilers were building a political base, a small secret society was being organized. Its members were brought together by their opposition to foreigners, especially the growing numbers of Roman Catholics in America. (It may be difficult for modern Americans to realize the great suspicion and fear of Catholics that once existed in mostly Protestant America and that continued well into this century. It was maintained until the election of John F. Kennedy in 1960; even then, Kennedy's religion was a significant issue in the campaign.) Called Know-Nothings, the group aimed to prevent foreign-born citizens from holding political office. Their name derived from their secretiveness. When asked about the party, members would usually answer that they didn't know anything about it—hence Know-Nothings.

Declaring in its platform that no immigrant be granted American citizenship until he had lived in the United States for twenty-one years, the party aimed its prejudice at Irish Catholics. Between 1825 and 1855, more than 5 million foreigners had entered the United States as immigrants; most were Roman Catholics, primarily from Germany and Ireland. The power of the Know-Nothings came from a widespread fear of the surging political power of these immigrants. This was not the first time a rabid anti-immigrant fever swept the country, and it certainly wouldn't be the last. In 1854 the party dropped its cloak of secrecy and adopted the name American party; its candidates won governorships in Massachusetts and Delaware. For a time it looked as if the American party might become the dominant second party in America, replacing the worn-out Whigs. But the issue of slavery also split the Know-Nothings. By the presidential election of 1856, when former President Fillmore was the Know-Nothing candidate with the slogan "Americans must rule America," many Know-Nothings were leaving for another new party that was quickly emerging as the challenger to the Democrats on the national scene.

Meeting at a schoolhouse in Ripon, Wisconsin, in February 1854, a diverse coalition of antislavery politicians, former members of the Whig, Free-Soil, and Know-Nothing parties along with disaffected northern Democrats, organized a new party opposed to the further extension of slavery. At a second mass meeting in July in Jackson, Michigan, the group officially took on the name Republicans. With a base of western farmers and northern businessmen, the Republicans drew antislavery forces from the other splinter parties, growing rapidly in size and influence. Less a party of outright abolitionists, the Republicans were the party of the free white workingman; the majority of them were opposed not to the existence of slavery but to its expansion into the western territories. Generally, Republicans also stood for a strong central government, high tariffs to encourage and protect American manufacturers, federally funded improvements—public works projects, or infrastructure in modern parlance—such as railroads and canals and other measures deemed repugnant by many Southerners.

After only two years, the Republicans met in Philadelphia on June 17, 1856, and were able to field their first presidential candidate, the western explorer John C. Frémont, a popular hero who was married to Jesse Benton, the daughter of the powerful senator from Missouri Thomas Hart Benton. Running on the slogan "Free Soil, Free Speech, Free Men, Frémont," the Republicans carried all eleven northern states in the election but received less than 1 percent of the southern vote. For the first time, a purely northern major political party had positioned itself squarely against the supporters of slavery. The distinctly regional division that the country had been moving toward was now firmly established.

What Are Beecher's Bibles and Border Ruffians?

When did the Civil War really begin? When did Americans start killing fellow Americans? Did it start at Fort Sumter in 1861? Or was it back when those first Africans were unloaded in Virginia? Maybe the war started when the Founding Fathers sat down to thrash out the Constitution and left so many troublesome questions. Or perhaps it began when Nat Turner and his fellow slaves took up arms and killed their American masters in a doomed quest for liberty. Then again, maybe it started in Alton, Illinois, when a mob murdered Elijah Lovejoy in 1837 because of his abolitionist agitation. But if a civil war means groups of citizens of the same nation taking up arms against one another, then it's fair to say that the Civil War really started in the territory of Kansas in 1854, long before the first shots were fired at Fort Sumter.

In the broad sweep of history, three years is not a long time. But that is all it took to undo whatever peace had been reached with Clay's Compromise of 1850. The two major parties, the Whigs and Democrats, had accepted the Compromise as a solution to the slavery issue, and the heat was lowered. But the pot was soon going to boil over in the western frontier of Kansas.

Looking to win a cross-country railroad route that would go through his own state instead of taking a southern route, Illinois Senator Stephen Douglas (1813–1861) wanted to divide a single large territory in the nation's center into two, the Kansas and Nebraska territories. Known as "the Little Giant" at just over five feet tall, the Vermont-born Douglas had made a fortune in the new American industrial empire of railroading; through marriage, he also owned a plantation and 140 slaves in Mississippi.

In the Senate, Douglas's role on the committee that oversaw the territories gave him tremendous power. Although he had sided with Clay in crafting the Compromise of 1850, Douglas was now looking for southern votes to open up the railroad route through Illinois. With ambition outweighing any political philosophy or idealism, Douglas was a bundle of contradictions, eager to please whichever group he was addressing. His eye cocked toward a White House run, he proposed to settle the question of slavery in the western territories by popular sovereignty, or allowing the settlers to vote on whether they wanted slavery, in exchange for southern support for his railroad plan and his presidential ambitions.

Douglas's proposal, the Kansas-Nebraska Act, shattered the letter and spirit of the 1820 Missouri Compromise; after its passage, the debate over the future of slavery became more violent than ever. The law split the Democratic party along North-South lines, completed the death throes of the Whig party, and opened the way for the emergence of the Republicans. Among the Whigs who became Republicans was Illinois lawyer Abraham Lincoln, a one-term congressman who acknowledged the legal right to keep slaves in existing slave states while supporting the idea of gradual emancipation. But Lincoln straddled no fences when it came to the Kansas-Nebraska

Act, condemning it and targeting as a dangerous enemy of democracy Douglas, an old political foe who had briefly courted a young Mary Todd—later Lincoln's wife.

The most immediate and deadly reaction to Douglas's act was that it opened the floodgates for both sides to pour men, money, and guns into the territory to influence the vote on slavery. After the act was proposed, Salmon Chase of Ohio, an abolitionist lawyer and politician, branded Douglas's bill as part of a "Slave Power" conspiracy to take control of the federal government.

Abolitionist-minded industrialists formed the Emigrant Aid Society in Massachusetts to encourage antislavery settlers to relocate in Kansas. Boxes of Sharp's breechloading rifles, known as Beecher's Bibles, were sent by Henry Ward Beecher's Plymouth Church in Brooklyn to arm these settlers. At the same time, proslavery factions flooded into the territory. Several thousand slaveowners, known as Border Ruffians because they came from neighboring Missouri, crossed into the Kansas Territory to influence the voting. They were inflamed by the likes of Missouri Senator David Achison, a rabid promoter of slavery who took a leave from the Senate to lead the Border Ruffians. The temper of the times was clear from Achison's threats: "There are eleven hundred men coming over from Platte County to vote and if that ain't enough we can send five thousand—enough to kill every God-damned abolitionist in the Territory."

Atchison's Missouri men succeeded in electing a proslavery legislature in March 1855 which was recognized by the appointed territorial governor. But at an August meeting in Lawrence, Kansas—named for Amos Lawrence of Massachusetts, the wealthy financial backer of the Emigrant Aid Society—antislavery supporters claimed this election was fraudulent. They called for a constitutional convention banning slavery while excluding all blacks from the Kansas Territory. By October, separate elections had named a proslavery and antislavery representative in Congress for Kansas.

Through the winter of 1855–1856, the Kansas controversy went back and forth between Free-Soilers and slaveholders. Finally the powder keg exploded into a bloody and brutal civil war. Violent clashes became commonplace as lynching, murder, and burning replaced popular sovereignty. And federal authorities, including President Franklin Pierce, were helpless to stop the escalating bloodshed. The area around Lawrence was heavily settled by Free-Soilers, who harbored abolitionists, fugitive slaves, and newspaper editors indicted for treason by the proslavery territorial government. On May 21, 1856, armed with five cannons, an eight-hundred-man army of Border Ruffians, calling itself a posse, swept into Lawrence, killing one man, wrecking the newspaper offices, and burning a hotel and the Free-Soil governor's home.

It was the signal for the beginning of all-out guerrilla warfare. Four days later, the fanatical abolitionist John Brown and four of his sons dragged five proslavery settlers—completely innocent of the Lawrence attack—from their homes along Pottawatomie Creek and, in front of their families, shot and hacked them to death with broadswords, weapons once favored by medieval knights. Brown and his sons evaded capture and were never indicted or punished for the massacre. Four years later, Brown would emerge for the attack on Harpers Ferry that gave him such a posthumous notoriety (see page 130). His actions in Kansas simply provoked more retaliation. More than two hundred people died in the border wars fought over "Bleeding Kansas."

Civil War Voices

From "The Crime Against Kansas," a speech by Massachusetts Senator Charles Sumner (May 19–20, 1856): "[Senator Butler] has chosen a mistress to whom he has made his vows, and who, though ugly to others, is always lovely to him; though polluted in the sight of the world, is chaste in his sight. I mean the harlot Slavery."

At the exact moment that the violence was rising to murderous heights in Kansas, the normally decorous halls of Congress also became the scene of bloodshed. Though not deadly, it still profoundly transformed the political landscape.

On May 19–20, 1856, Charles Sumner (1811–1874), a firm abolitionist, made a speech called "The Crime Against Kansas," in which he attacked slavery's supporters in the Senate. Personally singled out for attack was Senator Andrew Butler of South Carolina, whom Sumner called "the Don Quixote of Slavery." Two days later Sumner was at his desk on the Senate floor when South Carolina Representative Preston Brooks, Butler's nephew, approached him. Saying that Sumner had libeled his uncle, Brooks began to beat him with a bamboo cane until blood ran down his face and he was on the floor, nearly unconscious. Sumner withdrew from the Senate and convalesced in Europe for nearly three years, although he was reelected by the Massachusetts legislature during this absence.

The caning of Sumner stunned America. Southerners hailed Brooks as a hero, and he was inundated with canes to replace the one he had broken on Sumner; some were inscribed, "Hit him again." The *Richmond Whig* regretted that "Mr. Brooks did not employ a horsewhip or a cowhide upon his slanderous back, instead of a cane." The House failed to expel Brooks for his assault, although he resigned his seat. Later reelected, his only punishment was a $300 fine.

How Did a Slave Named Dred Scott Change History?

During the Kansas debate and the ensuing violence, James Buchanan (1791–1868) had been out of the country as Franklin Pierce's ambassador to England. That made him a safe candidate for the presidency; there was no blood on his hands. Neither Senator Stephen Douglas nor President Pierce could expect the Democratic party nomination because of their roles in the Kansas-Nebraska controversy. After his 1856 victory over Frémont, Buchanan gave an inaugural speech stating that the slavery question "belongs to the Supreme Court of the United States, before whom it is now pending, and will, it is understood, be speedily and finally settled."

This was a terminal case of wishful thinking. The case to which Buchanan referred had been brought on behalf of Dred Scott, the slave of John Emerson, an army doctor from Missouri. Scott had been taken to different army posts in the United States and the western territories, and having spent two years in the free territory of Minnesota, he claimed he was no longer a slave under the Missouri Compromise. Backed by abolitionists, Scott sued for his freedom in 1846 after failing to purchase his and his wife's freedom from Emerson's widow.

Thus began an eleven-year legal battle that changed American history. New York abolitionist John F. A. Sanford bought Dred Scott in order to take the case to court. (Sanford's name was misspelled as the other party in *Dred Scott* v. *Sandford*.) The case went through a number of lower courts, all but one ruling against Scott, until 1855, when it reached the Supreme Court. The case was argued by Montgomery Blair (1813–1883), the son of a prominent Democratic politician and later Lincoln's postmaster general. The Chief Justice, Roger B. Taney (pronounced *Taw*-ney), was an eighty-year-old Marylander, Andrew Jackson's appointee to fill the vacancy left by the death of the legendary John Marshall. Taney (1777–1864) had freed his own slaves in 1818 but believed passionately that slavery was necessary. Coming right after Buchanan's inaugural, the 7–2 vote (seven Democratic justices were opposed by two Republican justices) was a blow to the hopes of antislavery factions. Although nine different opinions were issued, Taney's was the most far-reaching.

First, Taney said that no black man, free or slave, was a U.S. citizen; therefore a black man had no right to sue in federal court. He could have stopped there without further comment, but he seemed bent on the complete destruction of every piece of antislavery legislation ever passed, going back to the Northwest Ordinance. Taney, speaking for the court, then ruled that Congress never had the right to ban slavery in territories because the Constitution protected people from being deprived of life, liberty, or property. According to Taney, slaves, like cows or goats, were property and could be taken anywhere under U.S. jurisdiction.

Slaveholders rejoiced and felt relief and vindication. One newspaper reported happily, "The Southern opinion upon the subject of Southern slavery . . . is now the supreme law of the land." Suddenly, the issue was not only nonextension, but the removal of all laws limiting or prohibiting slavery. Bills to reopen the African slave trade were offered in Congress.

Abolitionists, and moderate "nonextensionists" like Abraham Lincoln, were outraged. Journalist and poet William Cullen Bryant (1794–1878), a vigorous abolitionist and Republican party organizer, wrote that slavery was now "a Federal institution. . . . Hereafter, wherever our . . . flag floats, it is the flag of slavery."

Ironically, the decision actually aided the antislavery cause. The new Republican party gained strength when many fence-sitters joined the ranks of those more passionately opposed to slavery. And the decision also widened the divisions over slavery within the Democratic party. The Dred Scott decision went a long way toward building the foundation for a Republican victory in 1860.

Dred Scott and his family were freed by his abolitionist owner after the ruling. Scott died the following year.

Civil War Voices

Abraham Lincoln, on accepting the nomination for U.S. senator at the convention of the Illinois Republican party in Springfield (June 1858).

We are now in the fifth year, since a policy was initiated, with the avowed object, and confident promise, of putting an end to slavery agitation.

Under the operation of that policy, that agitation has not only, not ceased, but has constantly augmented.

In my opinion, it will not cease, until a crisis shall have been reached and passed—

"A house divided against itself cannot stand."

I believe this government cannot endure, permanently half slave and half free.

I do not expect the Union to be dissolved—I do not expect the house to fall—but I do expect it will cease to be divided.

It will become <u>all</u> one thing or <u>all</u> the other.

Either the <u>opponents</u> of slavery will arrest the further spread of it, and place it where the public mind shall rest in the belief that it is in course of ultimate extinction; or its <u>advocates</u> will push it forward, till it shall become alike lawful in <u>all</u> the states, <u>old</u> as well as <u>new</u>—<u>North</u> as well as <u>South.</u>

What Did Lincoln and Douglas Debate?

The most famous standoff in American history—until John F. Kennedy and Richard M. Nixon faced each other on television in 1960—was a series of seven debates between Senator Stephen Douglas of Illinois, one of the most famous men of his time, and his Republican challenger, Abraham Lincoln. Their confrontation was nothing like the televised debates of modern presidential politics, in which the candidates are tossed softballs by a panel of reporters. This was two men going at it like prizefighters, slugging away, questioning each other, throwing insults in the heat and cold, sun and rain, of frontier Illinois. It was not high-minded and polite. It was often crude. It was nasty. And it enthralled the thousands of spectators who flocked to each of the open-air spectacles.

Douglas was one of the most influential men in the Senate since the passing of Clay, Calhoun, and Webster. He had battled his way up through the young state's rough-and-tumble political world to stand at the edge of a run for the presidency. His controversial involvement in the Kansas-Nebraska controversy had prevented him from taking the Democratic nomination in 1856. But he was preparing for a run in 1860, and his way was now blocked by an old Springfield foe, Abraham Lincoln (1809–1865). When the two stood side by side, reporter Carl Schurz observed, "No more striking contrast could have been imagined than between these two men as they appeared on the platform. By the side of Lincoln's tall, lank, ungainly form, Douglas stood almost like a dwarf. . . . As I looked at him, I detested him deeply." (Schurz was not disinterested. He helped found the

Republican party and later became a close friend and political ally of Lincoln's.)

Few people in American history have been more idolized, more mythologized, and perhaps less understood than Douglas's opponent, Abraham Lincoln. From his storied log cabin beginnings to his assassination and martyrdom, Lincoln seemed to embody the American myth of a man who could pull himself up by his bootstraps and go on to become anything. That part of the Lincoln story was no myth.

Born in Kentucky to dirt-poor pioneer farmers, Thomas and Nancy Hanks Lincoln, young Abraham had moved as a child to Indiana and then to Illinois. When he was nine, his mother died, in October 1818, the victim of an epidemic of "milk sickness," a disease that came from drinking the milk of cows grazing on a poisonous root which often struck pioneers. His father later married Sarah Johnston, who was an affectionate and much-loved stepmother to Lincoln. His early years were punctuated by occasional schooling and hard work; his father's moves paid off when the family began to prosper in Illinois. At twenty-one, Lincoln set off on his first venture alone, a flatboat journey down the Mississippi, taking a cargo to New Orleans. Along the way, he and his partner were attacked one night by seven slaves from a nearby plantation but fought them off. It was not Lincoln's first encounter with slaves. Hatred of slavery was one reason that Thomas Lincoln had taken his family out of Kentucky. While not a pious churchgoer, the elder Lincoln had already formed the strong opinion that slavery was immoral.

An attempt at shopkeeping in the town of New Salem, a town Lincoln hoped would grow with commerce on the Sangamon River, went nowhere. When this venture failed and his partner died, he was left with what he called "the National Debt." Taking surveying work and a job as postmaster—his first political appointment—he gradually paid off the debt, building a local reputation for honesty, bawdy jokes, a sharp wit, and impressive physical strength. Ambitious and strong-willed, he was drawn to the law as the road to financial success and began to read law books; he also read the Constitution and the Declaration of Independence. Immersing himself in state politics, he was elected to the legislature and then gained admission to the Illinois bar in 1836. His first law partner, John T. Stuart, was also politically ambitious. In a run for the House of Representatives, Stuart and his opponent got into a street fight in which Stuart was actually bitten. His adversary was Stephen Douglas.

A loyal Clay man, Lincoln was elected to Congress in 1846, serving a single term most notable for his opposition to the Mexican War and his proposal for the gradual abolition of slavery in Washington, D.C. By the standards of his day, Lincoln was progressive if not liberal in his views. He balked at the beliefs of the Know-Nothings, saying,

> I am not a Know-Nothing. How could I be? How can anyone who abhors the oppression of negroes, be in favor of degrading classes of white people? Our progress in degeneracy appears to me to be pretty rapid. As a nation, we began by declaring that "all men are created equal." We now practically read it "all men are created equal, except negroes." When the Know-Nothing get control, it will read, "all men are created equal, except negroes, and foreigners, and catholics." When it comes to this I should prefer emigrating to some country where they make no presence of loving liberty—to Russia, for instance, where despotism can be taken pure, and without the base alloy of hypocracy.

Convinced of slavery's immorality, Lincoln was not, however, an abolitionist. In the strange world of mid-nineteenth-century politics, when slavery was the one issue on which most politicians rose or fell, Lincoln walked a careful line. Believing that slavery was wrong, he also held that it was legal under the Constitution. A pragmatic and ambitious "nonextensionist," opposed to allowing slavery to spread beyond its current boundaries, Lincoln thought it would gradually die out, echoing the same vain hope that Jefferson and Washington had voiced some seventy years earlier.

By modern standards, much of what Lincoln thought and said

about blacks would be called racist. He did not think blacks equal to whites in intellect or ability. He opposed the idea that blacks should vote, serve on juries, or intermarry with whites. For the sake of winning a state election, he was willing to keep emancipated blacks out of Illinois.

Earlier in his political life, Lincoln believed that all blacks should be removed from the United States and resettled in some other country: "My first impulse would be to free all the slaves, and send them to Liberia, to their own native land. But a moment's reflection would convince me, that whatever high hope (as I think there is) there may be in this, in the long run, its sudden execution would be impossible." Still, he did believe fiercely that when the Declaration said "all men are created equal," it included Negroes, and Lincoln took that to mean that blacks should be given the opportunity to labor for wages as white men did.

Little known outside Illinois, Lincoln began to attract the attention of national Republicans with his speeches outlining what would become staple Republican principles: preservation of the Union; non-extension of slavery; a strong federal government committed to making improvements that would bolster the national economy. But when he was nominated to run for the Senate seat of the famous Democrat Douglas, there was only one issue between the two men: the future of slavery in America. And Douglas knew he was in for a fight. "I shall have my hands full," he said. "He is the strong man of the party—full of wit, facts, dates—and the best stump speaker, with his droll ways and dry jokes, in the West. He is as honest as he is shrewd."

Civil War Voices

Stephen Douglas, in a debate with Abraham Lincoln (1857).

Mr. Lincoln tried to avoid the main issue by attacking the truth of my proposition, that our fathers made this government divided into free and slave states, recognizing the right of each to decide all its local questions for itself. Did they not thus make it? It is true they did not establish slavery in any of the States, or abolish it in any of the territories; but finding thirteen States, twelve of which were slave and one free, they agreed to form a government uniting them together, as they stood divided into free and slave States, and to guarantee forever to each State the right to do as it pleased on the slavery question. Having thus made the government, and conferred this right upon each State forever, I assert that this government can exist as they made it, divided into free and slave States, if any one State chooses to retain slavery. He says that he looks forward to a time when slavery shall be abolished everywhere. I look forward to a time when each State shall be allowed to do as it pleases. If it chooses to keep slavery forever, it is not my business, but its own; if it chooses to abolish slavery, it is its own business—not mine. I care more for the great principle of self-government, the right of the people to rule, than I do for all the Negroes in Christendom.

In the debates that followed from August to mid-October 1858, Douglas quickly took the offensive, trying to paint Lincoln as a fanatical abolitionist, a "black Republican" who wanted to put freed slaves on an equal footing with whites. He capitalized on what was then a common fear for whites, the specter of black men marrying or sexually assaulting white women. It was a typical slaveholder's tactic, and it played to a very basic fear among white Americans. Douglas conjured up images of tens of thousands of free slaves sweeping into Illinois territory, taking jobs and women from white men. When abolitionist Frederick Douglass visited Illinois during the debates, he was driven through the town of Freeport by a local abolitionist family. Senator Douglas used the image to castigate Lincoln and his party: "If you, Black Republicans, think the negro ought to be on a social equality with your wives and daughters, and ride in a carriage with your wife, whilst you drive the team, you have a perfect right to do so."

Lincoln responded to this "counterfeit logic" as he had before in public: "Because I do not want a black woman for a slave I must necessarily want her for a wife. I need not have her for either, I can just leave her alone." He had also noted that if the mixing of races were the true concern, slavery had done much more to foster that situation. Of more than 405,000 mulattoes in the United States, nearly 350,000 of them lived in the South. Lincoln argued that Republicans, by excluding slavery from the territories, would curtail this "amalgamation" of the races.

The Hollywood image of Lincoln (Raymond Massey, Henry Fonda) is of a deep-voiced and elegant speaker. But in the report of Carl Schurz, Lincoln's voice "was not musical, rather high-keyed, and apt to turn into a shrill treble in moments of excitement. . . . His gesture was awkward. He swung his long arms sometimes in a very ungraceful manner. Now and then he would, to give particular emphasis to a point, bend his knees and body with a sudden downward jerk, and then shoot up again with a vehemence that raised him to his tip-toes."

After the debates, Lincoln lost the election to Douglas. (Lincoln was not officially a candidate. At the time, senators were elected by state legislatures, not by direct vote. Douglas won the seat on the basis of more Democratic state legislators' being elected, even though statewide the Republican candidates received more votes than the Democrats.) But Lincoln's star was clearly on the rise in the western sky. Lincoln began to collect press clippings detailing his campaign and defeat. This scrapbook would become a playbook for the future. Far from being viewed as a radical like Seward of New York or Chase of Ohio, Lincoln had a moderate voice; this, along with his roots in the West—a growing political power—would place him center stage for a shot at the Republican nomination in 1860.

Civil War Voices

Abraham Lincoln, in a speech at Edwardsville, Illinois (September 11, 1858).

> When ... you have succeeded in dehumanizing the Negro; when you have put him down and made it forever impossible for him to be but as the beasts of the field; when you have extinguished his soul and placed him where the ray of hope is blown out in the darkness like that which broods over the spirits of the damned, are you quite sure that the demon you have roused will not turn and rend you?

What Happened at Harpers Ferry?

The "demon" Lincoln warned about was not a Negro but a fanatic who burned with the furies of an Old Testament prophet. If John Brown could have gotten hold of a pickup truck and some high explosives in 1859, a tragic disaster of Oklahoma City proportions might have happened a lot sooner in American history. In 1995 Americans watched in horror as they saw the work of a fanatic in Oklahoma City. Imagine if CNN had been around to chronicle the siege and last stand of a band of fanatical abolitionists bent on nothing short of armed insurrection against the federal government. Then imagine if some of the country's most prominent writers and politicians had decided that these terrorists were actually martyrs to a noble cause. That scenario offers a modern context for the events of late 1859, which ratcheted up the tension between the pro- and antislavery camps.

Late in August 1859, Buchanan's secretary of war, John B. Floyd, received an anonymous letter reporting "the existence of a secret association, having its object the liberation of the slaves of the South by a general insurrection." The letter warned that "an armory in Maryland" was the object of a conspiracy. Secretary Floyd, who was later accused of secretly sending federal guns and ammunition to armories in the South, dismissed the notion of such an uprising as impossible and buried the letter in a desk drawer.

The letter also identified the leader of this supposed uprising as John Brown. Born to a poor family in Torrington, Connecticut, on May 9, 1800, Brown was taken to Ohio at age five. He grew up in Hudson, about thirty-five miles south of Cleveland, receiving little formal schooling. Daily Bible reading and prayers were part of his devout religious family life, not at all unusual for that time. The family was also intensely antislavery. During the Mexican War, his father supplied cattle to American troops in Michigan, and it was then that Brown encountered a slave for the first time. Approximately his own age, this boy was "badly clothed, poorly fed . . . & beaten . . . with Iron Shovels or any other thing that came first to hand."

Returning home swearing "eternal war with Slavery," Brown decided on a life in the ministry. He went back east, and in Litchfield, the birthplace of Harriet Beecher Stowe and a center of abolitionist sentiment, studied theology without much success. Illness forced his return to Ohio. At twenty he married Dianthe Lusk, then nineteen, and they had seven children before she died in 1832. Within a year Brown remarried. With his second wife, Mary Ann Day, he had thirteen more children. Drifting from state to state in an effort to feed and clothe his twenty children, Brown held a succession of jobs— tanner, surveyor, farmer, postmaster. Unsuccessful in every venture, however, he was always a step away from bankruptcy and starvation.

When five of his sons—Owen, Jason, Frederick, Salmon, and John, Jr.—went to Kansas in 1855, they appealed to him for help in fighting the slave factions in the territory. Taking another son, Oliver, and his son-in-law Henry Thompson with him, Brown went to Kansas, the violent epicenter of the slavery issue. Forming a militia he called the Liberty Guards, he quickly earned a reputation as a ruthless leader, capable of murder. When proslavery forces attacked and burned the town of Lawrence in May 1856, Brown had proven his willingness to kill when he struck back quickly, slaughtering five settlers who had not participated in the attack on Lawrence.

With the clandestine support of northern abolitionists who came to be called the Secret Six, Brown started to plan an uprising he thought would liberate the slaves and create a black nation in the Appalachian Mountains of Virginia. His mission was to use the weapons to arm the slaves in Virginia and to march throughout the South with this army of freed slaves, inciting other slaves to join them in a great uprising that would put an end to slavery once and for all. His backers included abolitionist Gerrit Smith, a wealthy New Yorker whose pacifist views were altered by the killings of abolitionists in Kansas; Unitarian preacher Thomas Higginson; Theodore Parker, a prominent Unitarian; linseed oil manufacturer George L. Stearns, who had shipped guns into Kansas; Franklin Sanborn, a schoolteacher and student of Ralph Waldo Emerson's; and Samuel Gridley Howe, a wealthy physician, whose wife, Julia Ward Howe, would eventually eclipse her husband's notoriety when she wrote "The Battle Hymn of the Republic." All six were prominent abolitionists and had worked on vigilance committees that fought the return of runaways under the Fugitive Slave Act. They got to know John Brown during the Kansas violence, and in 1858 Brown returned east to tell them of his plan, which also included a role for Frederick Douglass.

Douglass and Brown had first met in 1848, when Brown invited the now-famous former slave to visit his home in Springfield, Massachusetts. Douglass recalled that Brown's eyes were "full of light and fire." Unlike many upper-class abolitionists who hated slavery but wanted to remain at arm's length from blacks, Brown seemed to be genuinely concerned with blacks, even if his notions of himself as a Moses for blacks made Douglass uncomfortable. When the two met, Brown was already beginning to formulate the outlines of his plan for a free black state in the Appalachians.

Meeting secretly in the fall of 1859, Brown appealed to Douglass to join his band. "I want you for a special purpose," Brown told the former slave. "When I strike, the bees will begin to swarm, and I want you to help hive them." Douglass warned Brown that he was heading for a "steel trap," realizing that his plan, which involved arming freed slaves with pikes—poles with sharp metal points—was suicidal in Virginia, where memories of Nat Turner were still alive.

With a band of twenty-two men—black and white—Brown descended on the border town of Harpers Ferry on the night of October 16, 1859. Set in the mountains of Virginia at a meeting of the Shenandoah and Potomac rivers (in what is now West Virginia), the town held a federal arsenal and armory. Initially successful in capturing the armory and the nearby Hall's Rifle Works, Brown set up defensive positions. Two of his black "soldiers" were sent to scour the countryside for slaves to join the insurrection. Several prominent citizens of Harpers Ferry, including a great-grandnephew of George Washington's, were taken as hostages.

If it weren't all so tragic it would have been comical. Ironically, the first civilian killed by Brown's men was a free black man. Brown's plan fell apart immediately. He had expected thousands of slaves to "swarm" under his leadership, but none came. Instead, the people of Harpers Ferry surrounded the arsenal and fired on the raiders, killing two of Brown's sons. By the afternoon of October 17, Brown had barricaded the remainder of his band and the hostages in the fire engine house next to the armory. A company of marines soon arrived under the command of Lieutenant Colonel Robert E. Lee, accompanied by his student at West Point, cavalry officer Lieutenant James Ewell Brown "Jeb" Stuart. Both men were officially on leave and had been pressed into service, Lee wearing civilian clothes. Stuart had been in the offices of War Secretary Floyd when the first reports of the trouble at Harpers Ferry were received. In Washington, the rumors flew that thousands of men were involved in a slave uprising.

On the morning of October 18, Lee sent Stuart with a white flag to demand Brown's surrender. (Curiously, Stuart and Brown had already crossed paths in Kansas, where Stuart, commanding federal troops attempting to keep order during the Border Wars, rode into Brown's encampment.) When surrender was refused, Stuart waved his broad-brimmed cavalry hat as a signal for the marines to charge. Quickly battering down a door but holding fire to protect the hostages, the marines stormed into the engine house and bayoneted two of the raiders. It was over in three minutes. Wounded by a sword,

Brown and his four remaining men were captured. Four civilians, including the mayor of Harpers Ferry, and one marine had been killed during the two-day raid. The raiders were charged with murder, treason, and inciting insurrection. The trial began ten days after the raid. The verdict was certain: John Brown and his men would hang.

Civil War Voices

John Brown at his sentencing, following his conviction for treason, rebellion, and murder (November 2, 1859).

> I have, may it please the Court, a few words to say. In the first place, I deny everything but what I have all along admitted, the design on my part to free the slaves. It is unjust that I should suffer such a penalty. Had I interfered in the manner which I admit, and which I admit has been fairly proved (for I admire the truthfulness and candor of the greater portion of the witnesses who have testified in this case), had I so interfered in behalf of the rich, the powerful, the intelligent, the so-called great, or in behalf of any of their friends—either father, mother, brother, sister, wife, or children, at any of that class, and suffered and sacrificed what I have in this interference, it would have been all right; and every man in this court would have deemed it worthy of reward rather than punishment.
>
> This court acknowledges, as I suppose, the validity of the law of God. I see a book kissed here which I suppose to be the Bible, or at least the New Testament. That teaches me that all things whatsoever I would that men should do to me, I should do even so to them. It teaches me further, to "remember them that are in bonds, as bound with them." I endeavored to act up to that instruction. I say, I am yet too young to understand that God is any respecter of persons. I believe that to have interfered as I have done—as I

have always freely admitted I have done—in behalf of His despised poor, was not wrong, but right. Now, if it is deemed necessary that I should forfeit my life for the furtherance of the ends of justice, and mingle my blood further with the blood of my children and with the blood of millions in this slave country whose rights are disregarded by wicked, cruel, and unjust enactments—I submit; so let it be done.

John Brown was executed on December 2, 1859. Witnesses to the hanging included a group of Virginia Military Institute students and their instructor, Thomas Jackson, a West Point graduate and Mexican War veteran; Virginia secessionist Edmund Ruffin; and an actor named John Wilkes Booth.

The note Brown left contained a grim prophecy: "I, John Brown, am now quite <u>certain</u> that the crimes of this <u>guilty land</u> will never be purged <u>away</u> but with Blood. I had <u>as I now think vainly</u> flattered myself that without <u>very much</u> bloodshed, it might be done."

Civil War Voices

Henry David Thoreau, "A Plea for Captain John Brown" (1859).

I wish I could say that Brown was the representative of the North. He was a superior man. He did not value his bodily life in comparison with ideal things. He did not recognize unjust human laws; but resisted them as he was bid. For once we are lifted out of the trivialness and dust of politics into the region of truth and manhood, no man in America has ever stood up so persistently and effectively for the dignity of human nature, knowing himself for a man, and the equal of any and all governments. . . .

. . . These men, in teaching us how to die, have at the same time taught us how to live. If this man's acts and words do not create a revival, it will be the severest possible

satire on the acts and words that do. It is the best news that
America has ever heard. It has already quickened the feeble
pulse of the North, and infused more and more generous
blood into her veins and heart than any number of years of
what is called commercial and political prosperity could.
How many a man who has lately contemplated suicide now
has something to live for!

Thoreau's militant tone on the subject of John Brown was ironically at odds with everything else that the prophet of civil disobedience previously stood for. But his passionate reaction to Brown's
execution was typical of that of many abolitionists, who viewed
Brown as a martyr or even a saint. He was even memorialized in a
song,"John Brown's Body Lies A-Moldering in the Grave." (It later
provided the tune for Julia Ward Howe's "Battle Hymn of the Republic.")

This attitude served only to harden the battle lines between abolitionist and slaveholder. The supporters of slavery could point to
Brown as proof that, not only was slavery threatened, but the lives
and freedom of Southerners were under attack by abolitionist forces.
When the involvement of the Secret Six became apparent, there
were fears of a massive abolitionist conspiracy, and Frederick Douglass was thought to be part of it. The very real fear of slave rebellions,
in a part of the country that still remembered the Nat Turner rebellion vividly, was another nail in the coffin of moderation.

Who Ran Against Lincoln?

A few months after the raid on Harpers Ferry came the most momentous election in American history.

But the outcome was probably decided before any votes were cast.
The Democratic convention of April 1860, held in Charleston, South
Carolina, had adjourned without a nomination after fifty of the southern delegates walked out when the party refused to adopt a platform
that guaranteed the constitutional rights of slaveowners. The Dem-

ocrats reconvened in June—this time in Baltimore—to try again. Again unable to reach a compromise, the southern delegates walked out, but this time most of the delegates from the Upper South went with them. The remaining delegates nominated Senator Stephen Douglas as the party's candidate on a platform of popular sovereignty, a policy now discredited in the North and unacceptable in the South. The 110 southern Democrats who had walked out gathered at another location in Baltimore and chose to nominate a separate candidate, Kentuckian John C. Breckinridge, on a platform that called for the U.S. government to "protect the rights of persons and property in the Territories and wherever else its Constitutional authority extended." President Buchanan and former presidents Tyler and Pierce all supported Breckinridge.

By this time the Republicans had already held their convention in Chicago, in May. Denouncing Brown's raid, opposing the extension of slavery, and upholding the Union, they also supported tariffs and a transcontinental railroad to hold their strength among northern business and western farming interests. One contender for the nomination was New York's William H. Seward (1801–1872). A former governor of New York and U.S. senator (as Whig and Republican), Seward was a well-known national politician who had made as many enemies as friends and who was viewed as too radical on the abolition issue. Other contenders were Salmon Chase of Ohio and Edward Cameron, a Pennsylvania power broker with a reputation for corruption. Lincoln was not initially seen as a front-runner, but he was the second choice of many delegates. His views on slavery were far more moderate than those of Seward; he also had broad appeal in the West, which was growing in political power as settlers surged into the open territory. Lincoln took the nomination on the third ballot; the Pennsylvania delegation switched to him on the second ballot—Cameron may have been promised a Cabinet post and, indeed, became Lincoln's first war secretary—and some critical Ohio votes clinched the nomination on the third.

Also in May a small faction of former Whigs had met in Baltimore as the Constitutional Union party and offered John Bell of Tennessee as a compromise candidate. But they took no stand on the issues other than to pledge to uphold the Constitution, and they were widely ridiculed as the "Do-Nothing" and "Old Man's" party.

The election of 1860 ultimately came down to two races: Lincoln—who wasn't even on the ballot in the southern states—against Douglas in the North and Bell versus Breckinridge in the South. Featuring the first nationwide presidential campaign tour, by Douglas, the race was vicious and filled with propaganda and prophecies of doom. But the deep split in the Democratic ranks guaranteed Lincoln's election.

Lincoln won approximately 40 percent of the popular vote, but the combined totals of his three opponents were greater by a million votes. Lincoln had crafted a victory by carrying the electoral votes of the big states in the North and West. The combined votes of the three unionist candidates (Lincoln, Douglas, and Bell) showed that a clear majority of Americans wanted to preserve the Union, but such electoral hair-splitting didn't matter in the slaveholding South.

Civil War Voices

Editorial in the *New Orleans Daily Crescent* (November 13, 1860), an indication of the reception Lincoln's election received in the South.

> The history of the Abolition or Black Republican party of the North is a history of repeated injuries and usurpations, all having in direct object the establishment of absolute tyranny over the slaveholding States. And all without the smallest warrant, excuse or justification. We have appealed to their generosity, justice and patriotism, but all without avail. . . . They have robbed us of our property, they have murdered our citizens while endeavoring to reclaim that property by lawful means, they have set at naught the decrees of the Supreme Court, they have invaded our States and killed our citizens, they have declared their unalterable

determination to exclude us altogether from the Territories, they have nullified the laws of Congress, and finally they have capped the mighty pyramid of unfraternal enormities by electing Abraham Lincoln to the Chief Magistracy, on a platform and by a system which indicated nothing but the subjugation of the South and the complete ruin of her social, political and industrial institutions.

Lincoln's election was particularly galling in South Carolina. On December 20, 1865, a special convention met in Charleston, and South Carolina became the first state to leave the Union.

Documents of the Civil War

From South Carolina's "Declaration of the Causes of Secession" (December 24, 1860).

We, therefore, the people of South Carolina, by our delegates in Convention assembled, appealing to the Supreme Judge of the world for the rectitude of our intentions, have solemnly declared that the Union heretofore existing between this State and the other States of North America is dissolved. . . .

PRELUDE TO THE CIVIL WAR: 1855–1860

1855

March In the elections in Kansas, thousands of Border Ruffians from Missouri come to influence the vote. A proslavery legislature is elected and recognized by the territorial governor. This body meets in *July* and passes a series of strict proslavery laws. Antislavery legislators are expelled from office.

August Meeting in Lawrence, Kansas, antislavery forces protest the election and call for their own constitutional convention. In *September* they declare the Kansas legislature illegal and request admission as a free state.

October Separate elections in Kansas produce proslavery and antislavery representatives in Congress. Free-Soil Kansans adopt a constitution that bans slavery but also excludes all blacks from the territory.

* *Leaves of Grass*, by Walt Whitman, is published.

1856

January 15 Free-Soil Kansans elect their own governor. President Pierce condemns this act because the federal government had already recognized the proslavery legislature.

February 22 The Know-Nothing party, renamed the American party, nominates former President Millard Fillmore for president.

March 4 Congressional Republicans support antislavery legislation in Kansas. But Stephen Douglas introduces a bill allowing Kansas statehood only after another constitutional convention there.

May 19 After a speech in which he attacks supporters of slavery, Senator Charles Sumner is beaten on the Senate floor by Preston Brooks, a congressman from South Carolina and a nephew of Senator Andrew Butler.

May 21 One man dies when proslavery forces attack Lawrence, Kansas, a center of the antislavery movement.

June The Democrats nominate James Buchanan for president, rejecting both Stephen Douglas and Franklin Pierce because they are associated with the problem in Kansas. The new Republican party nominates John Frémont on a platform admitting Kansas as a free state.

July 3 The House and Senate fail to agree on the admission of Kansas as a state.

September Nearly defunct, the Whig party nominates the same candidates selected by the American party, with Millard Fillmore their choice for president.

November 4 Buchanan is elected president, with the voting running along sectional lines. Buchanan wins all the slave states and four free states; Frémont wins eleven free states; Fillmore takes one free state.

* Commodore Matthew Calbraith Perry publishes *Narrative of the Expedition of an American Squadron to the China Sea and Japan Performed in the Years 1852, 1853, and 1854.*
* The Western Union Telegraph Company is established.

1857

January 15 Expressing the radical idea of disassociating with slavery, William Lloyd Garrison addresses a Disunion Convention. Its slogan: "No Union with slaveholders."

March 6 The Dred Scott decision. The Supreme Court rules that Scott is still a slave and has no rights to sue in federal court. The Court also rules that the Missouri Compromise of 1820 is unconstitutional and that Congress has no right to deprive citizens of their property, including slaves, anywhere in the United States.

October The Free State party wins a majority in Kansas territorial elections. However, an amendment to the state constitution allows slaves to remain in Kansas. The proslavery constitution, with President Buchanan's support, is approved in *December*, but Congress remains at an impasse over admitting Kansas.

* Hinton Rowan Helper publishes *The Impending Crisis of the South: How to Meet It.* This book argues that slavery is impoverishing many southern whites. It is banned in the South, but in *1860* Republicans distribute one hundred thousand copies of a condensed edition during Lincoln's presidential campaign.
* The first issues of *The Atlantic Monthly*, edited by James Russell Lowell, and *Harper's* edited by George William Curtis, are published.

* A financial panic sweeps the nation. During the next two years, more than twelve thousand businesses fail, largely because of overspeculation in the railroad and real estate businesses.
* The first New Orleans mardi gras is organized.

1858

May Minnesota, a free state, becomes the thirty-second state.

June 16 In Springfield, Illinois, Lincoln is selected by the Republicans to challenge for his Senate seat the incumbent Stephen Douglas, a Democrat.

August 21–October 15 The Lincoln-Douglas debates. In a series of seven debates, Lincoln confronts Senator Douglas. Lincoln takes an antislavery stand; Douglas defends the American right to decide the issue by popular sovereignty, even though the Dred Scott decision has negated that idea. Douglas wins the election but is set back in his national aspirations while Lincoln achieves new national prominence for his performance in the debates.

* In New York, work is begun on St. Patrick's Cathedral, and the design of Central Park is begun by Frederick Law Olmsted.
* George Pullman begins to design sleeping cars for railroads. Later adding dining cars and chair cars, the Pullman Company will become one of the nation's largest businesses. (Abraham Lincoln's son, Robert, later heads the company.)
* Pioneering photographer Mathew Brady sets up studios in New York and Washington.

1859

February An Arkansas law requires free blacks to choose between leaving the state and enslavement.

Oregon, a free state, becomes the thirty-third state.

March The Supreme Court rules that a state court may not release federal prisoners in the case of a man jailed for rescuing an escaped slave. The decision upholds the Fugitive Slave Act of 1850.

May 12 Meeting in Vicksburg, Mississippi, the Southern Commercial Convention states that all laws prohibiting the African slave trade should be repealed.

October John Brown's raid. At Harpers Ferry, radical abolitionist John Brown leads a plot to seize a federal arsenal and armory in order to set up a state for freed blacks. Within twenty-four hours his band is captured; Brown is hanged on *December 2.*

* Oil is struck in Pennsylvania, and the first oil well is drilled, opening an era that will transform American industry.
* A general store in New York opens and later becomes the Great Atlantic and Pacific Tea Company (A&P).
* *Our Nig: or Sketches from the Life of a Free Black* is published by Harriet E. Wilson. It is recognized as the first novel by a black to appear in the United States.

1860

February Senator Jefferson Davis introduces a bill saying that the federal government cannot prohibit slavery but must safeguard the rights of slaveholders.

May The Republicans nominate Abraham Lincoln to run for president.

June Following a walkout by southern delegates, the Democrats choose Stephen Douglas to run for president. The southern Democrats who left the convention then select John C. Breckinridge of Kentucky on a platform calling for the protection of slavery. During the campaign that follows, Southerners make it clear that they will secede if Lincoln is elected.

November 6 Abraham Lincoln is elected president.

December 3 In his final address, President Buchanan says that the states have no right to secede but that the federal government can do nothing to prevent such an action.

December 20 A state convention in South Carolina, the most outspoken state, votes to secede from the Union.

December 26 Major Robert Anderson, commanding the federal forts in the harbor of Charleston, South Carolina, moves his forces into the stronger of the two forts, Fort Sumter.

December 27 South Carolina troops seize Fort Moultrie and take over the federal arsenal at Charleston.

December 31 Buchanan announces that Fort Sumter will be defended against attack and orders *The Star of the West* to sail there with supplies.

* The legendary Pony Express service opens and thrives briefly until the transcontinental telegraph replaces the famed riders in *1861*.

* The first major labor strike occurs in Lynn, Massachusetts, where thousands of shoe factory workers, many of them women, strike for higher wages. Although their union is not recognized, the workers' demands are met.

Civil War Voices

Louisiana Senator Judah Benjamin's farewell to the Senate (New Year's Eve, 1860).

> We desire, we beseech you, let this parting be in peace....
> Indulge in no vain delusion that duty or conscience, interest
> or honor, imposes upon you the necessity of invading our
> States or shedding the blood of our people. You have not
> the possible justification for it.
>
> What may be the fate of this horrible contest, no man
> can tell ... but this much, I will say: the fortunes of war
> may be adverse to our arms, you may carry desolation into
> our peaceful land, and with torch and fire you may set our
> cities in flame ... you may, under the protection of your
> advancing armies, give shelter to the furious fanatics who
> desire, and profess to desire, nothing more than to add all
> the horrors of a servile insurrection to the calamities of civil

war; you may do all this—and more too, if more there be—
but you can never subjugate us; you never can convert the
free sons of the soil into vassals, paying tribute to your
power, and you never, never can degrade them to the level
of an inferior and servile race. Never! Never!

Born to British Jews living on St. Croix in the British West Indies,
Judah Benjamin (1811–1884) was taken to Charleston, South Carolina, as a child and entered Yale College at age fourteen. Before
graduating, he moved at seventeen to New Orleans, where he studied law and was admitted to the bar in 1834. He became a law partner
of John Slidell's, a future ally in the Senate and Confederacy. Benjamin wrote a digest of Supreme Court decisions that became a classic in legal literature. Elected to the Senate in 1852 and again in
1858, he resigned in February 1861 with Louisiana's secession.

Appointed attorney general in the Confederate Cabinet, Benjamin
was considered the most brilliant of the men surrounding Jefferson
Davis. Although he married a Catholic and his child was raised as a
Catholic, he was also mistrusted and hated by many in the Confederacy because he was a Jew. He would be doubly hated in the Union
because he was a Confederate Jew. Some of the wartime attacks on
Benjamin were virulently anti-Semitic, neither an unusual nor controversial sentiment in mid-nineteenth-century America.

chapter four

1861
"In Dixie Land, I'll Take My Stand"

You cannot transform the negro into anything one-tenth as useful or as good as what slavery enables them to be.

—JEFFERSON DAVIS
FEBRUARY 1861

It is safe to assert that no government proper ever had a provision in its organic law for its own termination.

—ABRAHAM LINCOLN'S FIRST INAUGURAL ADDRESS
MARCH 4, 1861

A reckless and unprincipled tyrant has invaded your soil. Abraham Lincoln, regardless of all moral, legal, and constitutional restraints, has thrown his Abolitionist hosts among you, who are murdering and imprisoning your citizens, confiscating and destroying your property, and committing other acts of violence and outrage, too shocking and revolting to humanity to be enumerated.

—PIERRE BEAUREGARD
JUNE 1, 1861

✳

* Why Did Secession Come?

* Who Elected Jefferson Davis President?

* Why Did Lincoln Sneak into Washington for
His Inauguration?

* Did Lincoln Deliberately Provoke the Shots Fired
at Fort Sumter?

* What Was the Anaconda Plan?

* Who Were the "Plug Uglies"?

* Why Did Robert E. Lee Resign from the U.S. Army?

* What Is Habeas Corpus and What Did Lincoln
Do with It?

* What Happened at Manassas?

* Who Was "Little Mac"?

* What Was the *Trent* Affair?

* What Was a "Radical Republican"?

* Was It Really "the Brothers' War"?

The nation held its breath. The days between Abraham Lincoln's election in November 1860 and inauguration in March 1861 were momentous times, dark and tense. Before Lincoln entered office, the first states of the Confederacy had left the Union. Eleven federal arsenals and forts in the South had been seized by state militias. In the White House, lame duck President James Buchanan wanted to leave office with the country at peace. But South Carolina, the first and most hawkish of the seceded states, was moving toward a war footing.

In the back rooms of Congress and at a Peace Conference presided over by former President John Tyler—who would become the only former president to join the Confederacy—well-meaning politicians on both sides desperately tried to avert bloodshed. Among them was a Mississippi senator who advised the governor of South Carolina to proceed cautiously, especially with regard to Fort Sumter, an unfinished and obscure federal fortress in Charleston Harbor. "The little garrison in its present position presses on nothing but a point of pride," said the senator. But he was also a realist. "We are probably soon to be involved in that fiercest of human strife, a civil war."

That cautious senator was Jefferson Davis.

Civil War Voices

Mary A. Ward, relating the mood in her state of Georgia following secession in postwar testimony before Congress:

> The day that Georgia was declared out of the Union was a day of the wildest excitement in Rome [Georgia]. There was no order or prearrangement about it at all, but the people met each other and shook hands and exchanged congratulations over it and manifested the utmost enthusiasm. Of course, a great many of the older and wiser heads looked on with a great deal of foreboding at these rejoicings and evidence of delight, but the general feeling was one of excitement and joy.
>
> Then we began preparing our soldiers for the war. The

ladies were all summoned to public places, to halls and lecture rooms, and sometimes to churches, and everybody who had sewing machines was invited to send them; they were never demanded because the mere suggestion was all-sufficient. The sewing machines were sent to these places and ladies that were known to be experts in cutting out garments were engaged in that part of the work, and every lady in town was turned into a seamstress and worked as hard as anybody could work; and the ladies not only worked themselves but they brought colored seamstresses to these places, and these halls and public places would be filled with busy women all day long.

But even while we were doing all these things in this enthusiastic manner, of course, there was a great deal of the pathetic manifested in connection with this enthusiasm, because we knew that the war meant separation of our soldiers from their friends and families and the possibility of their not coming back. Still, while we spoke of these things, we really did not think that there was going to be actual war. . . . We got our soldiers ready for the field, and the Governor of Georgia called out the troops. . . . The young men carried dress suits with them and any quantity of fine linen. . . .

Every soldier, nearly, had a servant with him, and a whole lot of spoons and forks, so as to live comfortably and elegantly in camp, and finally to make a splurge in Washington when they should arrive there, which they expected would be very soon indeed. That is really the way they went off; and their sweethearts gave them embroidered slippers and pincushions and needle-books, and all sorts of such et ceteras, and they finally got off.

Why Did Secession Come?

The census of 1860 is quite revealing. On paper, the Confederacy had no chance. Twenty-three states remained in the Union (more would be added during the war). The population of these states was about 22 million, 4 million of them men of combat age. There were 100,000 factories employing 1.1 million workers. The Union possessed more than 20,000 miles of railroad—more than the rest of the world's railroads combined—and 96 percent of the country's railroad equipment. Union banks held 81 percent of the nation's bank deposits and $56 million in gold.

The eleven Confederate states held a population of approximately 9 million, which included nearly 4 million slaves. There were only 20,000 factories in the Confederate states employing about 100,000 workers. With only 9,000 miles of track, the Confederacy's rail lines were insignificant compared to the Union's. More important as the war progressed, the Confederacy lacked the factories to produce new rails and equipment to replace those captured or destroyed in what would become the first war to move by train. Nor could it come remotely close to the Union in the production of weapons. As in World War II, when the U.S. production capacity simply buried the other warring powers, the Union was capable of vastly outproducing the Confederacy. In addition, the Union states thoroughly outproduced the Confederate states in every category of agricultural production, with the exception of cotton. But the most important agricultural product of the South would basically go unsold through the war years, once the Union began effectively to blockade cotton shipments out of Confederate ports.

In other words, the Union and Confederate states were not "one nation under God." They were two nations, divided by politics, economy, and culture—all of those differences generated by the institution of slavery. The America of the Union states was racing toward the twentieth century, with banks, booming factories, railroads, canals, and steamship lines. Its population was mushrooming with the influx of immigrants escaping the famine and political turmoil of Europe, particularly in Ireland and Germany.

The southern states of the Confederacy were, in many respects, standing still in time. They remained stuck in the agricultural, slave-based economy of Jefferson's time, when the gentlemen planters of

Virginia helped create the nation. Most of its cash wealth was tied up in slaves, and its chief product was the cotton produced by those slaves. In other words, the Confederacy was an easy economic hostage to slavery and the cotton system.

Two countries, two cultures, two ideologies. Despite the common ground of language, religion, race, and heritage—America was primarily a white Anglo-Saxon Protestant nation—the people of the day saw more differences between themselves. The simplest explanation for the war was that many in the Confederacy saw themselves being steamrollered by a northern economic machine that threatened every aspect of their way of life, hence this seemingly irrational contradiction in people proclaiming a fight for "liberty" by defending the enslavement of someone else.

Why did this war come? There was a widely shared feeling among many in the Confederacy that their liberty and way of life were being overpowered by northern political, industrial, and banking powers. That defensive posture was whipped into flames with a race-baiting hysteria that politicians of the future Confederacy—the slaveowning class, with the most to lose from emancipation—used to spread fear. There was a fear of blacks taking control and the certain sexual ravishing of the flower of southern womanhood if the slaves were allowed to go free. (That's where Mark Twain got it right about Sir Walter Scott's being to blame for the Civil War.) The potent mix of economic fear and racially tinged emotions coalesced over the question of slavery. Even those in the Confederacy who held no slaves thought the powers of the North had no right to tell them how to live their lives. With their political power diminishing in Congress as the free states were growing in number and population, the men who made the Confederacy turned to the one weapon they thought they had a right to use—secession.

Human nature and historical inevitability also played a big part in creating the massive tragedy of the Civil War. For those whites in the Confederacy who held no slaves—though they were a majority, they lacked political and economic clout—there was the fear of being overwhelmed. The fear itself was fanned by politicians and editorial pages, warning that the "Black Republicans" of Lincoln and the "abolitionist Yankees" who owned the banks and set the prices for their crops would make them economic slaves. And once the war began, that fear was bolstered by a more basic human instinct: the powerful need to defend one's home and property.

Who Elected Jefferson Davis President?

Abraham Lincoln was on a train bound for his inauguration when word came that Jefferson Davis had been chosen as provisional president of the Confederate States of America on February 18, 1861, in Montgomery, Alabama, the Confederacy's provisional capital. Alabama "Fire-eater" William Lowndes Yancey (1814–1863), one of the most outspoken advocates of secession, introduced Davis to a cheering crowd: "The man and the hour have met. Prosperity, honor and victory await his administration." His inauguration was marked by boisterous applause, booming cannons, and incessant renditions of "Dixie" by the Montgomery Theater band. An actress named Maggie Smith danced on a flag of the United States.

Lincoln's opposite had a much better résumé and seemed much better prepared to lead his "country" into war. A former war secretary, Davis has been called one of the most effective to hold that position, even though he is best mockingly remembered for a proposal to create camel-mounted cavalry patrols in the southwestern desert. In fact, he was largely responsible for professionalizing and modernizing the U.S. Army. An experienced soldier, Davis was a West Pointer, a decorated veteran and hero of the Mexican War. Lincoln had none of these qualifications. His military experience was limited to leading a few volunteers in the Black Hawk War; Lincoln said he'd fought mosquitoes and led a charge on an onion patch.

Yet the two men, who never met, shared some interesting traits. Like the Union president, Jefferson Davis was not a popular leader, although time and Confederate mythology have made him seem more so. His handling of military affairs, like Lincoln's, was extremely controversial. Both men would have terrible times with their generals. Politically, Davis faced tremendous hostility from a large, well-organized faction, just as Lincoln had to contend with cranky fellow Republicans and virulent opposition Democrats. When Davis tried to manage the war by placing power in the hands of a central government, he was called a tyrant and dictator, just as Lincoln was. And on a personal level Davis, like Lincoln, had known tragedy and great personal loss. The war would bring even more. Each man lost a son as well as many friends during the war years. Each would grimly read the casualty reports, a heavy burden for both.

Although much has been made of Lincoln's 1860 victory with less than 40 percent of the vote, Davis's claim to the office was on even more shaky democratic ground. For a "country" that came into existence based on the protection of individual rights against a powerful central government, the Confederacy had allowed few people any say in electing a president. None of the six initial Confederate states had even chosen secession in popular voting. According to William C. Davis in *Jefferson Davis: The Man and His Hour*, their withdrawal from the Union had been decided in state conventions by 854 men, all of them selected by their legislatures. Of those, 157 had voted against secession. In other words, fewer than 700 mostly wealthy and upper-middle-class men had decided the destiny of 9 million people without benefit of an election. And Tennessee's secession was orchestrated by its governor after the popular defeat of a secession proposal.

Born in Kentucky on June 3, 1808, the president of the Confederacy graduated in the lower third of his 1828 West Point class. Posted to the infantry, Davis fell in love with Sarah Taylor, the daughter of his commander and a future president of the United States, Zachary Taylor. A brief first army career ended in 1835, distinguished by run-ins with superiors and fellow officers that showed Davis's streak of certainty. This was a man who didn't believe that he was always right; he *knew* he was always right. He returned to Mississippi as a plantation owner with his eldest brother, Joseph. By standards of the day, the Davises treated their slaves progressively; for instance, they allowed them to take what food they wanted rather than being rationed. The slaves were also permitted to choose their own names, and their housing was better than standard slave quarters. Nonetheless, they were still slaves, inferior beings fit only for servitude, and, in Davis's view, private property protected by the Constitution.

In 1835 Davis married Sarah Taylor without the full blessing of her parents, but they did not elope, as later myth suggested. Five weeks later both of them fell sick, and Sarah died at age twenty-one, the first of Davis's personal tragedies. Malaria or yellow fever contracted in the mosquito-infested swampland of Davis's plantation, Brierfield, felled the newlyweds, and Davis would be plagued by ill health brought on by malarial fevers for the rest of his life.

After ten years out of public life, Davis was elected to Congress as a Democrat in 1845. That same year he married Varina Howell, a nineteen-year-old beauty half his age. Serving one year in the House, he resigned to lead a group of Mississippi volunteers to fight in Mexico. Fate brought him to serve again under Taylor, who treated him warmly, their grief a shared sorrow. Wounded at the Battle of Buena Vista, Davis returned a war hero and parlayed his new fame into a Senate seat, which he held until he was named war secretary by President Franklin Pierce in 1853.

Davis reclaimed his Senate seat in 1857 and sat on the committee that investigated John Brown's raid on Harpers Ferry. The senators concluded that "it was simply the act of lawless ruffians, under the sanction of no public or political authority."

A staunch states' rights Democrat, Davis tested the waters for a run at the White House in 1860. At the Baltimore convention, his name was placed in nomination, ironically, by Benjamin Butler of Massachusetts. (During the Civil War, Butler became one of the most notorious Union generals, known throughout the Confederacy as "Beast Butler.") Davis didn't press hard because he knew he would never garner sufficient northern support to win the election.

When the Democrats split between Stephen Douglas and John Breckinridge, their defeat was guaranteed. Counseling caution and still seeking compromise, Davis backed Mississippi's secession. On

January 21, 1861, after making an emotional farewell speech on the Senate floor, he withdrew with four other senators. Named commander of Mississippi's militia, he became a compromise candidate for the provisional presidency of the six seceded states of Confederacy. At age fifty-three and in poor health, he would have preferred a military appointment. But Davis was elected on February 9 and inaugurated nine days later in Montgomery. (He was formally elected by popular vote to a six-year term and was inaugurated for a second time on February 22, 1862.)

Soon after his election, Davis told a northern visitor that slavery was not the cause of secession, explaining: "My own convictions, as to negro slavery, are strong. It has its evils and abuses. . . . We recognize the negro as God and God's Book and God's Laws, in nature, tell us to recognize him—our inferior, fitted expressly for servitude. . . . You cannot transform the negro into anything one-tenth as useful or as good as what slavery enables them to be."

In the flurry of organizing a government and an army, one of Davis's first acts was to dispatch three commissioners to Washington in an attempt to negotiate a settlement with the Union. Leading them was the Confederate vice-president, Alexander Hamilton Stephens of Georgia. Nicknamed "Little Ellick," he weighed about ninety pounds and was described by Lincoln as "a little, slim, pale-faced consumptive man."

Lincoln admired a speech that Stephens once gave in the House, and during their years as fellow Whigs in the Thirtieth Congress, the lanky Lincoln and the small Georgian became friends, although they later parted ways over slavery. A slaveowner, Stephens had a reputation as a "humanitarian master" because he didn't whip his slaves and never separated families. A moderate Unionist, he voted against Georgia's secession but felt honor-bound to remain loyal when the state seceded. When the Confederate Congress met in Montgomery, "Little Ellick" sought the presidency; losing out to Davis, he was chosen as vice-president. Throughout the war he and Davis would disagree, often vehemently, and Stephens became the center of much of the sentiment against Davis in the Confederate government.

At Christmas, before going to Washington, Lincoln had sent Stephens a letter marked "For Your Eyes Only," promising that his administration would not interfere with slavery. Now Stephens arrived in Washington, hoping to negotiate an end to the crisis. With the situation moving toward a showdown, Stephens and the other delegates met in secret with Lincoln's secretary of state, William Henry Seward. Speaking without presidential authority, Seward promised that Fort Sumter would be evacuated. But Lincoln refused to meet with Stephens, unwilling to legitimize a Confederate government he now viewed as a collection of rebels.

Civil War Voices

London *Times* correspondent William Russell on Jefferson Davis (1861).

> Mr. Davis is a man of slight, sinewy figure, rather over the middle height, and of erect, soldierlike bearing. He is about fifty-five years of age; his features are regular and well-defined, but the face is thin and marked on cheek and brow with many wrinkles, and is rather careworn and haggard. One eye is apparently blind, the other is dark, piercing, intelligent.

William Howard Russell, the world's first great war correspondent, had gained an international reputation for his battlefield reporting on the Crimean War (1853–1856). In 1861 Russell came to America to cover the Civil War for the London *Times*, but his dispatches were also published in both Union and Confederate papers. Although he was initially welcomed by both sides, his objectivity, his general scorn of things American, and his unwillingness to take sides eventually earned him few friends on either side of the conflict.

Songs of the Civil War

"Dixie" (1859)

> I wish I was in de land ob cotton,
> Ole times dar am not forgotten;
> Look away, look away, look away, Dixie Land.
> In Dixie Land whar I was born in,
> Early on one frosty mornin',
> Look away, look away, look away, Dixie Land.

> *Chorus*
> Den I wish I was in Dixie! Hooray! Hooray!
> In Dixie Land, I'll take my stand,
> To lib and die in Dixie.
> Away, away, away down south in Dixie
> Away, away, away down south in Dixie.

Composed in New York by Daniel Decatur Emmett (1815–1904), one of the originators of the blackface minstrel show, this song was widely popularized in the South. Played at Jefferson Davis's inauguration, it was a favorite on both sides of the conflict and was played at Lincoln's inauguration as well.

Henry Hotze, a member of the Mobile (Alabama) Cadets, recalled hearing "Dixie" when he first went to Virginia in the early days of the war: "It is marvelous with what rapid-fire rapidity this tune of 'Dixie' has spread over the whole South. Considered as an intolerable nuisance when first the streets re-echoed it from the repertoire of wandering minstrels, it now bids fair to become the musical symbol of the new nationality, and we shall be fortunate if it does not impose its very name on our country."

The source of the name "Dixie" is somewhat obscure. It has often been considered an abbreviation of the Mason and Dixon Line, which honors British astronomers Charles Mason and Jeremiah Dixon who in 1763 had surveyed the disputed boundary of the colonies of Maryland and Pennsylvania. During the Missouri Compromise debate in 1820, the Mason and Dixon Line came to mean the border dividing the free states from the slave states.

But according to Stuart Berg Flexner in *I Hear America Talking*, the name may actually derive from the French *dix* (ten). In bilingual Louisiana, "Dix" was printed on the back of ten-dollar notes issued by the Citizen's Bank in New Orleans. In time New Orleans and much of the South were called "the Land of Dixie," inspiring Emmett's song (whose actual title was "Dixie's Land"). A third suggestion is that slaves popularized the word with a song about the good life at "Dixie's," supposedly a kind slaveowner.

As Henry Hotze feared, "Dixie" indeed did "impose its very name on our country." Sung by Union soldiers as they advanced toward the war's first major battle, it was also played by military bands in Washington and Richmond after Lee's surrender to Grant in 1865.

Why Did Lincoln Sneak into Washington for His Inauguration?

President-elect Lincoln and his family broke their train journey from Illinois to Washington with a stop in Philadelphia. Late that February evening, Lincoln met in secret with private detective Allan Pinkerton.

The son of a Glasgow, Scotland, police sergeant, Pinkerton (1819–1894) had come to America and settled near Chicago. While cutting wood, he discovered and captured a gang of counterfeiters. The incident led to a career in police work, with Pinkerton ultimately becoming Chicago's first detective. In 1850 he opened Pinkerton's National Detective Agency, specializing in railroad security. An abolitionist, he also helped the Underground Railroad. During the 1850s his agency worked for the Illinois Central Railroad, the line for which Lincoln had provided legal services and where in 1857 Pinkerton encountered another Illinois Central executive, former army engineer George Brinton McClellan.

After the 1860 election, another of Pinkerton's clients, the Philadelphia, Wilmington & Baltimore Railroad, feared that secessionists in Maryland would cut its tracks in order to isolate the capital. Sent to Baltimore to protect the railroad, Pinkerton's men infiltrated se-

cessionist groups, finding the city seething with Confederate sentiment.

That night in Philadelphia, Pinkerton told Lincoln that his agents
had learned of an assassination plot: Lincoln was to be killed when
he changed trains in Baltimore. Though Pinkerton advised him to
depart for Washington a day early, Lincoln refused, having already
promised to attend Washington's Birthday festivities in Harrisburg
the next day. Later that night, however, he received confirmation of
the plot from General Winfield Scott.

After the following day's events, Lincoln secretly returned to Philadelphia. Wearing an overcoat and disguised by a brown hat, he
slipped into the last sleeping car of a train bound for Baltimore.
Squeezing his lanky frame into a small sleeping berth, he posed as
the invalid brother of one of Pinkerton's female detectives. Traveling
with Lincoln were Pinkerton and a trusted friend from Springfield,
Ward Hill Lamon. Acting as bodyguard for the president-elect, Lamon was a walking arsenal, with brass knuckles, a pair of revolvers,
derringers, and two large knives. At three-fifteen in the morning the
train reached Baltimore without incident, and Lincoln's coach was
hooked up with a train going to Washington. (Years later Lamon
claimed that Pinkerton had invented the entire assassination plot to
gain access and credibility with Lincoln. But there was certainly
enough evidence to support the detective's suspicions, including the
corroboration by General Scott. And four years later Lincoln's assassination would actually be plotted by Maryland natives in Baltimore.)

Lincoln reached the capital at dawn on February 23. His wife, who
had stayed on their original train with their sons, arrived later that
day, shaken by the journey and by the frenzied Baltimore crowds
who shouted Lincoln's name in anger. When the opposition newspapers heard of Lincoln's surreptitious arrival, the president-elect was
widely mocked as a country bumpkin and depicted by cartoonists as
slinking into the town. London correspondent William Howard Russell wrote, "The cold shoulder is given to Mr. Lincoln. People take
particular pleasure in telling how he came towards the seat of his
government disguised."

Civil War Voices

Abraham Lincoln, from his first Inaugural Address (March 4, 1861). (See Appendix IV for the complete text.)

> In your hands, my dissatisfied fellow countrymen, and not in mine, is the momentous issue of civil war. The government will not assail you. You can have no conflict without being yourselves the aggressors. You have no oath registered in heaven to destroy the government, while I shall have the most solemn one to "preserve, protect and defend it."
>
> I am loath to close. We are not enemies, but friends. We must not be enemies. Though passion may have strained, it must not break our bonds of affection. The mystic chords of memory, stretching from every battlefield and patriot grave to every living heart and hearthstone all over this broad land, will yet swell the chorus of the Union when again touched, as surely they will be, by the better angels of our nature.

Civil War Voices

Texas Governor Sam Houston: "I am for the Union without any 'if.' "

Heroic stature counted for little when it came to where a man stood on the question of secession. That reality was dramatically underscored in Texas. As the greatest hero of Texas's battle for independence from Mexico and the Republic's first president, Sam Houston had fought hard to bring Texas into the United States. Once Texas gained statehood, he was elected one of the state's first two senators and, in 1859, was elected governor of Texas.

But Houston was a Union man. Even if Texans chose to secede, Houston urged them to remain independent and not join the Confederacy. When the Texas convention voted to leave the Union, Governor Houston issued the formal statement of secession in February 1861. But he refused to take a required oath of loyalty to the Confederacy and was deposed as governor.

When Lincoln offered to send Union troops to maintain his governorship, Houston declined and spent his last two years in Huntsville, Texas, where he died in July 1863.

Did Lincoln Deliberately Provoke the Shots Fired at Fort Sumter?

Lincoln did not have to wait long to find out that "the better angels" would not prevail. "The mystic chords" would soon be drowned out by the sound of cannonballs exploding around a federal fort. Nor would it take long for America to realize that this was going to be a war of strange and horrifying coincidences, pitting friend against friend, former comrades against one another.

Just as people have wondered whether President Franklin Roosevelt provoked the attack on Pearl Harbor in 1941 to draw America into the war with Germany and Japan, some historians have accused Lincoln of pushing the Confederacy to fire the shots that started a civil war. While his inaugural address promised that the federal government would not start a war and that no attempt would be made to retake federal facilities already held by Confederate forces, Lincoln had also declared that he would "hold, occupy, and possess" installations still under federal control in the Confederacy. This seemed well within a president's call. But beyond that, Lincoln made no threats.

Everything changed on the day after his inauguration, when Lincoln received an alarming message from Major Robert Anderson, who commanded the U.S. troops holding Fort Sumter in Charleston Harbor. He reported that there was less than a six-week supply of food left in the fort.

Near the center of the entrance to Charleston Harbor, Fort Sumter was an unfinished, five-sided brick structure that rose directly from the water, its foundations built on a shoal. Named for Thomas Sumter, a South Carolina general in the Revolution, it was one of a series of forts begun when the War of 1812 had shown the inadequacy of America's coastal defenses. Designed to serve 650 men, Fort Sumter was now occupied by about 125 men, some 40 of whom were completing the fortifications. Surrounding it were smaller forts already held by the South Carolina militia under the command of the elegantly named Pierre Gustave Toutant Beauregard (1818–1893).

One of President Davis's first military appointments, Beauregard had been made a brigadier general in the Provisional Army of the Confederate States on March 1. A Louisianian nicknamed the "Little Creole" and "Little Napoleon," he was second in his West Point class (1838), where he studied artillery under the same Robert Anderson who now commanded Fort Sumter. In Mexico, Beauregard had served with Winfield Scott and was twice wounded. He probably set an unusual record when he served as the superintendent of West Point for only five days, January 23–28, 1861, before being removed for his Confederate sympathies.

When South Carolina seceded on December 20, 1860, Fort Sumter and the other two United States military installations in Charleston Harbor—Fort Moultrie and Castle Pinckney—became a symbol of "foreign" authority. Moultrie and Castle Pinckney were abandoned by Major Anderson following South Carolina's secession. President Buchanan believed that both the secession and "coercion" of states were unconstitutional and wanted only to finish his term in peace. When South Carolina seized Fort Moultrie and Castle Pinckney, along with the customhouse, post office, treasury, and an arsenal, Buchanan did nothing. When South Carolina fired on *The Star of the West*, an unarmed steamer coming into the harbor with supplies for the fort, he again stood pat. He had also pledged not to reinforce Fort Pickens, on the coast near Pensacola, Florida, in exchange for an agreement not to attack it. Uncharacteristically, however, he refused demands to remove the federal troops from Fort Sumter. "If I withdraw Anderson from Sumter, I can travel home to Wheatland by the light of my own burning effigies," said Buchanan, referring to his home in Pennsylvania.

Anderson's letter about his dwindling supply of food gave Lincoln his first crisis. Few presidents have ever faced a problem of similar proportion on the day after taking office. General-in-chief Winfield Scott advised Lincoln that relief of the fort was impossible; Secretary of State William H. Seward thought evacuating the fort was a good way to cool things off. Seward, the ambitious former governor of New York and U.S. senator who had helped found the Republican party, resented Lincoln's election and saw himself as a "prime minister," a puppet master who would eventually pull the strings for the inexperienced new president. It was Seward's certainty that Unionists in the Confederacy would come to their senses and revoke secession that led him to meet with the Confederate emissaries against Lincoln's specific instructions and secretly promise that Lincoln would abandon the fort.

Searching for a middle ground that would uphold federal authority but not provoke a war, Lincoln quickly demonstrated his independence from the calculating Seward. On March 29 he decided to resupply but not reinforce the fort, and to let South Carolina Governor Francis W. Pickens know the supplies were coming. Seward opposed the idea and again recommended Sumter's evacuation. But Lincoln held his ground, establishing his authority over Seward and the rest of the Cabinet. On April 8 the president notified Governor Pickens of his decision.

Civil War Voices

Editorial, the *Charleston Mercury:*

> The gage is thrown down and we accept the challenge. We will meet the invader, and God and Battle must decide the issue between the hirelings of Abolition hate and Northern tyranny, and the people of South Carolina defending their freedom and their homes.

After Lincoln refused to meet with the delegates of the Confederate states, the die was cast. Jefferson Davis ordered Beauregard to demand Sumter's surrender, "and if this is refused, proceed . . . to reduce it."

As the federal defenders scurried to complete Sumter's defenses, Beauregard delivered the Confederate ultimatum to Major Anderson. Anderson's reply: "Gentlemen, I will await the first shot and if you do not batter the fort to pieces about us, we shall be starved out in a few days."

Told to state the time of his evacuation, Anderson replied that he would leave by noon on April 15 barring other instructions or additional supplies from his government. Learning that his answer was unacceptable, Anderson shook the hand of Colonel James Chesnut, Jr., Beauregard's emissary (a former U.S. senator now a member of the Confederate Congress), and said, "If we never meet in this world again, I hope that we may meet in the next."

Anderson himself was a proslavery Kentuckian but a Unionist. Both he and Beauregard had been wounded in the Mexican War. Now Beauregard had his former teacher surrounded by cannons, perhaps eager to show that he had been taught well.

At 4:30 A.M. on April 12, 1861, a single mortar was discharged. It was the signal for forty-three Confederate guns around Fort Sumter which proceeded to fire some four thousand shells. The bloodiest war in American history had begun. Although Edmund Ruffin, a wild-eyed secessionist with a long mane of silvery-white hair, was said to have fired the first shot, that distinction actually belonged to Captain George S. James. Ruffin fired the first shot from another battery.

All of Charleston crowded the harbor shores to watch. Mary Boykin Chesnut, the wife of Beauregard's aide, wrote in her diary:

> There was a sound of stir all over the house, pattering of feet in the corridors. All seemed hurrying one way, I put on my double gown and a shawl and went too. It was to the housetop. The shells were bursting. . . . The regular roar of the cannon—there it was. . . . The women were wild there on the housetop. Prayers came from the women and imprecations from the men. And then a shell would light up the

> scene. . . . We watched up there, and everybody wondered
> that Fort Sumter did not fire a shot.

Inside the fort, Captain Abner Doubleday, Anderson's second-in-command, fired the first federal shot in reply. It bounced harmlessly off the iron wall of a nearby Confederate fortress. Although federal relief vessels were beginning to arrive, the expedition was not prepared for the crisis, and the ships rode at anchor. On the second day of the bombardment, the fort began to burn. Lacking supplies and troops sufficient to mount a true defense, Anderson ordered the fort's flag lowered and a bedsheet raised in its place. Fort Sumter was surrendered without the loss of any men on April 13.

During the next day's evacuation, Anderson ordered a cannon salute to the flag. One gun exploded, killing Private Daniel Hough. Another private, Edward Galloway, was wounded and died several days later. The war's first deaths were accidental.

On April 15, following the attack on Fort Sumter, Lincoln proclaimed a state of insurrection rather than a state of war. He issued a call for seventy-five thousand volunteers for three months to crush the rebellion.

For many years some historians—generally those sympathetic to the Confederate cause—argued that Lincoln had provoked the Confederates into firing the first shots in order to get world opinion on the side of the Union. But this theory ignores a great many facts. First, South Carolina had already fired the first shots against a federal ship and, by taking the other federal forts and arsenals, had begun the war. Jefferson Davis was ready to fire on Sumter before hearing of the relief expedition, as soon as Lincoln refused to meet with Vice-President Stephens and the other Confederate delegates. Davis had also written to General Braxton Bragg at Pensacola, ordering him to take Fort Pickens. Bragg was not ready, otherwise the war might have begun off the coast of Florida.

Lincoln, hoping to keep the wavering slave states from joining the Confederacy, was reluctant to take offensive action. He expected that these southern states—Virginia, in particular—would remain loyal if the Union were attacked. But instead of bringing back the undecided states, Fort Sumter and Lincoln's proclamation calling for volunteers prompted secession proceedings in four more states. Virginia, Arkansas, North Carolina, and Tennessee soon joined the Confederacy, now eleven states strong. (The others were South Carolina, Georgia, Alabama, Mississippi, Florida, Louisiana, and Texas.)

Civil War Voices

From the diary of Theodore Upson, a sixteen-year-old Indiana farmboy, who enlisted and later fought under Grant and Sherman (April 1861).

Father and I were husking out some corn. We could not finish before it wintered up. When William Cory came across the field he was excited and said, "Jonathan they have fired upon and taken Fort Sumter." Father got white and couldn't say a word.

William said, "The President will soon fix them. He has called for 75,000 men and is going to blocade (sic) thier (sic) ports, and just as soon as those fellows find out that the North means business will get down off thier (sic) high horse."

Father said little. We did not finish the corn and drove to the barn. Father left me to unload and put out the team and went to the house. After I had finished I went in to dinner. Mother said, "What is the matter with Father?" He had gone right upstairs. I told her what we had heard. She went to him. After a while they came down. Father looked ten years older. We sat down at the table. Grandma wanted to know what was the trouble. Father told her and she began to cry. "Oh my poor children in the South. Now they will suffer! God knows how they will suffer! I knew it would come. Jonathan I told you it would come!"

"They can come here and stay," said Father,

"No they will not do that. There is their home. There they will stay. Oh to think that I should have lived to see the day when Brother should rise against Brother."

She and Mother were crying. I lit out for the barn. Oh I do hate to see women cry.

We had another meeting at the schoolhouse last night; we are raising money to take care of the families of those who enlist. A good many gave money, others subscribed. The Hulper boys have enlisted and Steve Lampman and some others. I said I would go but they laughed at me and said they wanted men not boys for this job; that it would all be over soon; that those fellows down South are big bluffers and would rather talk than fight. I am not so sure about that.

What Was the Anaconda Plan?

Troubled by gout and too fat to ride a horse, seventy-five-year-old Winfield Scott (1786–1866) was head of the army. Born in Virginia ten years after American independence was declared, he was nicknamed "Old Fuss and Feathers" because of his reputation as a stickler for strict conformity to regulations. Scott had been in every American military action since the War of 1812, in which he had been captured once but emerged with a hero's reputation second only to Andrew Jackson's. He had fought in the Indian wars and then in the Mexican War, where he had led the army to Vera Cruz. Again he emerged a hero in that most political of American wars, becoming the first lieutenant general in the American army since George Washington. But unlike Washington and Jackson, Scott couldn't translate his military success into political victory. Becoming the Whig candidate after fifty-three ballots in 1852, he lost the presidential election to Democratic Senator Franklin Pierce of New Hampshire.

When he counseled the evacuation of Fort Sumter, some Republicans saw reason to question the Virginian's loyalty. A devoted Unionist, however, Scott responded to the "insurrection" with a strategic plan. In May he wrote to General George McClellan,

> We rely greatly on the sure operation of a complete block-ade of the Atlantic and Gulf ports soon to commence. In connection with such a blockade, we propose a powerful movement down the Mississippi to the ocean, with a cordon of posts at proper points ... the object being to clear out and keep open this great line of communication in connec-tion with the strict blockade of the seaboard, so as to en-velop the insurgent States and bring them to terms with less bloodshed than by any other plan.

As Scott pointed out, his strategy had two main elements: a block-ade of the South's Atlantic and Gulf ports and an expedition of eighty thousand men supported by navy gunboats down the Mississippi to New Orleans. The Confederacy would be split in two, and the em-bargo would lead to its economic strangulation.

Leaked to the press, Scott's approach was mocked as overly cau-tious and was soon derisively christened the Anaconda Plan, after the large, boa-like snake that kills its victims by constriction. But this was the post-Napoleonic era of visions of battlefield glory. The ro-mantic images of the recent Crimean War and its famed Charge of the Light Brigade, immortalized in Tennyson's 1854 poem, were what civilians and soldiers had in mind when they thought of war. Most people, including young General McClellan, wanted a grander, Napoleonic offensive, and even Lincoln thought Scott's plan was too mild. Still, Lincoln saw the wisdom of this strategy and ordered the blockade. Despite the considerable derision mounted against Scott's Anaconda Plan, the blockade of southern ports and control of the Mississippi provided the ultimate basis for the economic and military defeat of the Confederacy.

When Lincoln announced the blockade on April 19, 1861, it was bark without much bite. There was actually very little that the Union could do either to go to war or to prevent southern trade. The Union, faced by an adversary whose "country" was larger than all of western Europe, was woefully unprepared. Even though it comprised twenty-three states and possessed most of the nation's industry and agriculture as well as the vast majority of its banking and financial wealth along with canals, roads, and a rapidly growing railroad and telegraph network, the Union didn't have much of an army. Before the war, the U.S. Army was small: about 16,000 officers and men. At the outbreak of the war, there were 1,108 officers. Nearly 400 of them, including many with significant wartime experience in Mexico, remained loyal to their home states and resigned to join the Confederacy or were dismissed for suspected Confederate sympathies. The same was true of the navy. Of 1,554 officers, 373 were either dismissed or left to join the Confederacy.

While the challenge on land was daunting for both armies, Lincoln's call for a naval blockade seemed preposterous. With thirty-five hundred miles of Confederate coastline and 180 possible ports of entry to patrol, the blockade would be the largest such effort ever attempted. The U.S. Navy had only forty-two warships in operation, most of them patrolling distant oceans. Almost all of them were sailing vessels, which a new generation of steam-powered ships had rendered obsolete. The Union actually had only three warships suitable for blockade duty.

While the military situation faced by Lincoln and his administration was difficult, it was far worse for Jefferson Davis. Although the Confederate Congress authorized an army of a hundred thousand men, they could offer little to pay, clothe, feed, or arm them. The Confederate states had a small fraction of the Union's manufacturing capabilities and hardly more of the share of the nation's available cash. Dependent on its cotton trade for hard currency, the Confederacy was extremely vulnerable to a blockade. Davis responded to Lincoln's call by asking southern shipowners to help "resist aggression" by operating as privateers against the North's seagoing commerce. It was officially sanctioned piracy.

The Union blockade was ridiculed by people in the Confederacy and by sympathizers in Great Britain. But Lincoln had the last laugh and a secret weapon in Navy Secretary Gideon Welles who soon earned the nickname "Old Father Neptune." By calling back the navy ships, buying vessels from the large merchant fleet, and beginning a massive shipbuilding program, Welles put together a fleet of 136 new ships and had 52 more under construction by the end of the year.

Privateers and blockade running—remember, that was Rhett Butler's contribution to the Confederate effort in *Gone With the Wind*—enabled the Confederacy to survive as long as it did, bringing in 60 percent of the weapons used by its armies, along with shoes, food, blankets, and medicine. But as the war lengthened, the Union navy's increasing ability to reduce shipments into southern ports took a terrible toll on the Confederate economy. *Inflation* is one of those boring economic words, but it crippled the Confederacy. Salt—which before refrigeration was the only way to preserve meat—went from $2 a bag before the war to as high as $60 a bag in 1862. Despite the Confederacy's grandiose talk about brilliant generals and glorious causes, you can't feed an army—or the families they left behind—on idealism or troop movements. The blockade and the harsh realities it created meant suffering equal to that brought by any Union general.

Songs of the Civil War

"Maryland! My Maryland," a poem by southern schoolteacher James Ryder Randall (1839–1908), was later set to the tune of the familiar Christmas hymn "O Tannenbaum" (April 1861).

> The despot's heel is on thy shore,
> Maryland!
> His torch is at thy temple door,
> Maryland!
> Avenge the patriotic gore
> That flecked the streets of Baltimore.
> And be the battle queen of yore,
> Maryland! My Maryland!

I hear the distant thunder-hum,
  Maryland!
The Old Line's bugle, fife, and drum,
  Maryland!
She is not dead, nor deaf, nor dumb—
Huzza! she spurns the Northern scum!
She breathes! she burns! she'll come! she'll come!
  Maryland! My Maryland!

Who Were the "Plug Uglies"?

As Lincoln had learned from his miserable experience in Baltimore, Maryland's chief city was part of the Union in name only. The city was seething with secessionist sentiment, and the fate of Maryland as a member of the Union was still up in the air. With Virginia's secession and the Confederacy moving its capital to Richmond, the Confederacy was now literally in Lincoln's view. Confederate flags flew from the rooftops in Alexandria, directly across the Potomac from the Capitol. Washington was near panic as rumors of a Confederate attack raced through the city, which was defended only by a disorganized, ragtag Union army lacking effective leadership.

Rushing to the capital was the 6th Massachusetts militia under the impetuous Benjamin Butler (1818–1893), a paunchy, ambitious Massachusetts Democrat who had supported Jefferson Davis for the presidency in 1860. After secession, Butler became an enthusiastic War Democrat (later a Republican), commanding the first northern troops to reach Washington. As Butler's men passed through Baltimore on April 19, 1861, they were taunted by pro-secession civilians known as "plug uglies." (According to *New York Times* columnist and word maven William Safire in his Civil War novel *Freedom*, "their name came from the plug hats as well as from the spikes studded in the front of their boots, worn by the hoodlums to do greater injury with a kick.") When the mob began to throw bricks, paving stones, and rocks, the troops fired. No one ever took responsibility for ordering or firing the first shot, but federal guns were turned on civilians armed only with clubs and stones. The ensuing melee left at least twelve civilians dead and dozens more injured. Four soldiers were also killed. Weeks passed before a real clash between armies led to a comparable number of casualties.

The governor of Maryland and the mayor of Baltimore begged Lincoln to forbid any more federal troops to pass through the city, but Lincoln refused. That night, Marylanders burned railroad bridges to prevent any more troops from moving through the state. Bands of "plug uglies" roamed over Maryland, assaulting travelers. The area was close to anarchy. And in Washington rumors of a secessionist mob intent on attacking and burning the city filled the air. For the moment, Maryland appeared to be on the verge of joining the Confederacy, which would completely surround the capital with rebellious states. When he met some of the wounded Massachusetts men, Lincoln was beyond worried, saying, "I don't believe there is any North. The [New York] Seventh Regiment is a myth. Rhode Island is not known in our geography any longer. *You* are the only Northern realities."

Civil War Voices

Warren Lee Goss of Massachusetts, who enlisted after the Baltimore riots, describes the transition from civilian to soldier (1861):

My first uniform was a bad fit; My trousers were too long by three or four inches; the flannel shirt was coarse and unpleasant, too large at the neck and too short elsewhere. The forage cap was an ungainly bag with pasteboard top and leather visor; the blouse was the only part which seemed decent; while the overcoat made me feel like a little nubbin of corn in a large preponderance of husk. Nothing except "Virginia mud" ever took down my ideas of military pomp quite so low.

... The first day I went out to drill, getting tired of doing the same thing over and over, I said to the drill sergeant: "Let's stop fooling and go over to the grocery." His only reply was addressed to a corporal: "Corporal, take this man out and drill him like hell"; and the corporal did! I found that suggestions were not so well appreciated in the army as in private life, and that no wisdom was equal to a drill-master's "Right face," "Left wheel," and "Right, oblique, march." It takes a raw recruit some time to learn that he is not to think or suggest, but obey. Some never do learn. I acquired it at last, in humility and mud, but it was tough. Yet I doubt if my patriotism, during my first three weeks' drill, was quite knee high. Drilling looks easy to a spectator, but it isn't. After a time I had cut down my uniform so that I could see out of it, and had conquered the drill sufficiently to see through it. Then the word came: On to Washington!

Why Did Robert E. Lee Resign from the U.S. Army?

Barely able to climb a set of stairs, aging General Winfield Scott was physically incapable of commanding in the field. But he still had his wits. Besides proposing the Anaconda Plan, he also advised the new president to appoint a new army commander and suggested his fellow Virginian Robert E. Lee. On April 18, Lee met with Frank Blair, Sr., a powerful backroom politician (and the father of Postmaster General Montgomery Blair) who unofficially offered Lee command of the Union army.

Born on January 19, 1807, in Stratford Hall, a plantation on the banks of the Potomac River in Virginia, Robert E. Lee descended from the line of Virginia Lees that had been among the country's most influential families. It was clear that he had the credentials and bloodlines to lead the nation's military in a crisis. One of his ancestors, Richard Henry Lee, issued the motion calling for independence at the Continental Congress in 1776. Another, Francis Lightfoot Lee, had also signed the Declaration of Independence. A third, Arthur Lee, was a chief minister to France during the Revolution and served in the Continental Congress.

Robert E. Lee's father, Major General Henry "Light-Horse Harry" Lee, had been one of George Washington's most skilled cavalry officers and trusted aides. The man who eulogized Washington as "first in war, first in peace, and first in the hearts of his countrymen," Henry Lee also served as Virginia's governor and as a U.S. congressman. A great soldier, he was a terrible businessman. Twice jailed for failing to pay his debts, Lee also became involved in Aaron Burr's notorious misadventure to set up an empire in the Louisiana territory. During a political controversy in Virginia just before the War of 1812, Lee was helping a friend defend his printing press against an angry mob when he was stabbed and left for dead. Broke, disfigured, and crippled, Henry Lee was sent to Barbados by President Monroe. Robert's two brothers didn't restore much of the luster to the family name. His half brother, Henry, married a morphine-addicted woman and then fathered a child with his wife's teenage sister. Known as "Black Horse Harry," he then embezzled his in-laws' money. Robert's other brother, Carter, used up what little cash their father had left behind.

Forced from the family home, Robert E. Lee lived with his mother's family until he went to West Point, emerging second in his class (1829). He later married Mary Anna Randolph Custis, a granddaughter of Martha Custis Washington, the wife of the first president, whose son by her first husband had been adopted by Washington. The connection to Washington was a powerful one. After the Mexican War, Lee was superintendent of the the U.S. Military Academy at West Point from 1852 to 1855. Following a two-year leave, he commanded the troops sent to put down John Brown's raid at Harpers Ferry. He once said, "Duty is the sublimest word in our language. Do your duty in all things. You cannot do more. You should never wish to do less."

After Virginia seceded from the Union on April 17, Lee was torn between duty and home. Declining Lincoln's offer of command, on the twentieth of April he resigned his commission and received command of Virginia's state forces on April 22. Lee said he personally opposed slavery as "a moral and political evil," even though he was part of the plantation aristocracy, and later in the war did nothing to prevent captured blacks from being returned to slavery. His views of the Union were also contradictory. Lee supported the preservation of the Union that his father and uncles had helped create. But his deepest loyalty was to his native Virginia, a common feeling in a time when there was not yet an American identity. To Lee and others, state meant more than country, and Lee said he could not "raise my hand against my relatives, my children, my home." Lincoln later observed that he could never quite understand Lee and all the other southern officers who professed loyalty to America but still broke their oaths of allegiance to join the Confederacy.

Two days after Lee resigned, Arlington House, the Custis family residence overlooking the Potomac, was occupied by Union forces, and General Irvin McDowell took the mansion as his headquarters. A wartime law required that property owners in areas occupied by federal troops appear in person to pay their taxes. The Lees were obviously unable to comply, and the estate was confiscated. In 1864 two hundred acres of the grounds around Arlington House were set aside as a military cemetery for the Union dead. In 1882 Lee's son George Washington Custis Lee sued for return of the land. By then the hills were covered with graves, and he accepted the government's offer of $150,000 for the property that is now Arlington National Cemetery.

Civil War Voices

Robert E. Lee's letter to General Winfield Scott, explaining his decision to resign his command in the U.S. Army (April 20, 1861).

Since my interview with you on the 18th inst. I have felt that I ought no longer to retain my commission in the Army.

I therefore tender my resignation, which I request you will recommend for acceptance. It would have been presented at once but for the struggle it has cost me to separate myself from a service to which I have devoted the best years of my life, and all the ability I possessed.

During the whole of that time—more than a quarter of a century—I have experienced nothing but kindness from my superiors and a most cordial friendship from my comrades. To no one, General, have I been as much indebted as to yourself for uniform kindness and consideration, and it has always been my ardent desire to merit your approbation. I shall carry to the grave the most grateful recollections of your kind consideration, and your name and fame shall always be dear to me.

Save in defense of my native State, I never desire again to draw my sword.

Be pleased to accept my most earnest wishes for the continuance of your happiness and prosperity, and believe me most truly yours,

R. E. Lee

While Lee would one day become a legendary general of Napoleonic stature, one of his first encounters in the Civil War was nearly disastrous. Cheat Mountain in western Virginia was a critical position, controlling the traffic on a major turnpike and several mountain passes. On September 11, 1861, in a steady, drenching rain, Lee attacked two thousand Union troops on Cheat Mountain. Some captured Union soldiers convinced Lee that he was outnumbered when in fact he had a large advantage. Union reinforcements arrived, and the skirmish continued for two days. Having lost the element of surprise, Lee withdrew. His failure was severely criticized, and he was dubbed "Granny Lee" and "Evacuating Lee" by the Richmond newspapers.

Songs of the Civil War

"The Battle Cry of Freedom," by George Root. Written in response to Lincoln's call for volunteers, this was the most popular of the more than two hundred songs he composed (others include "Tramp, Tramp, Tramp, the Boys Are Marching" and "Just Before the Battle, Mother").

> Yes, we'll rally round the flag, boys, we'll rally once again,
>> Shouting the battle cry of Freedom!
> We will rally from the hillside, we'll gather from the plain,
>> Shouting the battle cry of Freedom.

> *Chorus*
> The Union, forever, hurrah! boys, hurrah!
> Down with the traitor, up with the star;
> While we rally round the flag, boys, rally once again.
> Shouting the battle cry of Freedom.

> We will welcome to our numbers the loyal, true and brave.
>> Shouting the battle cry of Freedom;
> And although they may be poor, not a man shall be a slave,
>> Shouting the battle cry of Freedom.

MILESTONES IN THE CIVIL WAR: 1861

January Following Lincoln's election and the secession of South Carolina, *Florida, Alabama, Georgia, Mississippi,* and *Louisiana* secede from the Union.

January 9 The merchant vessel *Star of the West,* en route to Fort Sumter with supplies, is fired on by South Carolina forces. The ship returns to New York.

January 29 Following years of bitter fighting over the status of slavery in the territory, *Kansas* is admitted as the thirty-fourth state with a constitution prohibiting slavery.

Throughout the month, various state militias take over a series of federal forts and arsenals throughout the South. These include the valuable arsenal at *Baton Rouge, Louisiana; Forts Morgan* and *Gaines,* guarding the harbor at *Mobile, Alabama;* and the U.S. arsenal at *Mount Vernon, Alabama.*

February 4 In *Montgomery, Alabama,* the six seceding states form a provisional government of the Confederate States of America.

On the same day, a Peace Convention called by Virginia's leaders and with former President John Tyler presiding convenes in *Washington, D.C.* With 131 delegates from twenty-one states—though none from the seceded states—the convention fails to reach a compromise.

February 9 The Confederate Provisional Congress elects Jefferson Davis president. The vice-president is Alexander Stephens, a relatively moderate senator from Georgia. In Davis's Cabinet are Secretary of State Robert A. Toombs (who served a year, then became a brigadier general), Treasury Secretary Christopher G. Memminger, War Secretary Leroy Walker, Navy Secretary Stephen Mallory, and Attorney General Judah P. Benjamin, who would later serve as secretary of war and of state.

Tennessee voters reject a secession convention by nearly ten thousand votes.

February 23 *Texas* secedes from the Union, the seventh Confederate state.

March 4 Abraham Lincoln is inaugurated as the sixteenth president. The major figures in his Cabinet are Secretary of State William H. Seward, Secretary of War Simon Cameron (to be replaced by Edwin M. Stanton), Secretary of the Treasury Salmon P. Chase, Postmaster General Montgomery Blair, Navy Secretary Gideon Welles, and Attorney General Edward Bates.

March 16 The Confederate Congress, meeting in *Montgomery, Alabama,* adopts a constitution modeled on the U.S. Constitution but which prohibits the passage of any law impairing slavery.

April 12 After months of threats, negotiations, and attempts at supply and reinforcement, *Fort Sumter,* in the harbor of *Charleston, South Carolina,* is bombarded by South Carolina troops. Union Major Robert Anderson surrenders the fort on *April 13.*

April 15 Following the attack on *Fort Sumter,* Lincoln proclaims "a state of insurrection." He issues a call for seventy-five thousand three-month volunteers. Black volunteers are rejected.

April 17 Balking at Lincoln's request for troops, *Virginia* secedes from the Union, the eighth and most important state to do so. One month later, its capital, *Richmond*, is designated the capital of the Confederacy.

April 19 In *Baltimore*, Union troops marching to Washington from Massachusetts are stoned by Confederate sympathizers. During the ensuing riot, twelve civilians and four soldiers die.

Lincoln orders the naval blockade of all Confederate ports; Confederate blockade runners and privateers will operate successfully during the war, but this blockade cripples the southern economy.

May 6 *Arkansas* and *Tennessee* secede from the Union, the ninth and tenth Confederate states. *Tennessee* joins without a vote and remains divided throughout the war, with the eastern areas remaining loyal and contributing troops to the Union.

May 13 Queen Victoria announces Great Britain's neutrality. The Confederacy is not officially recognized but granted "belligerent status."

May 20 *North Carolina*, which will suffer the heaviest death toll of any Confederate state, secedes, the eleventh Confederate state.

May 24 When Union troops move into *Alexandria, Virginia*, Elmer Ellsworth, a close friend of Lincoln's, becomes the first combat fatality of the war, shot while removing a Confederate flag from a hotel roof. James T. Jackson, the hotelkeeper who shot Ellsworth, is killed by Union troops. Ellsworth's body lies in state at the White House, and both men become instant martyrs to their respective sides.

July 2 Lincoln authorizes the suspension of the constitutional right of habeas corpus in exceptional cases in a limited area between *Washington* and *New York*.

July 21 *Battle of 1st Bull Run (1st Manassas)*. Union forces are routed by Confederates in a battle watched by Washington residents, who had come by carriage, expecting to see the war end in a day.

July 22 A congressional resolution states that the war is being fought to "preserve the Union," not to abolish slavery.

July 27 Lincoln replaces General Irvin McDowell as head of the Army of the Potomac with thirty-four-year-old General George B. McClellan.

August 5 The first income tax is passed to finance the war.

August 30 In *Missouri*, Union General John C. Frémont institutes martial law and frees the slaves of any secessionists. In *September*, Lincoln overrules the order and transfers Frémont.

August The Confiscation Act authorizes the appropriation of any property, including slaves, of rebel slaveholders.

September U.S. Navy Secretary Gideon Welles authorizes the enlistment of slaves.

October 21 *Battle of Ball's Bluff.* Another one-sided Union defeat near Leesburg, Virginia. Lincoln's good friend Senator Edward Baker is killed in the fighting.

November 1 After a battle of wills, General-in-Chief Winfield Scott is retired by Lincoln and replaced by General George B. McClellan.

November 8 *The* Trent *Incident.* Confederate agents Mason and Slidell are taken from the British ship *Trent*, setting off a diplomatic crisis between the United States and Great Britain and bringing the two countries to the brink of war.

What Is Habeas Corpus and What Did Lincoln Do with It?

From the White House, Lincoln could see Confederate flags flying in Alexandria and Confederate campfires burning at night. Baltimore was ready to explode. And vastly fewer Union militiamen had arrived in the nation's capital than Lincoln and his staff expected. In this atmosphere of fear and uncertainty, the president called a Cabinet meeting. One of his first decisions was a limited suspension of the writ of habeas corpus along the rail lines between Philadelphia and Washington, essentially targeting Maryland.

Habeas corpus literally means "[that] you have the body"; a writ of habeas corpus is issued by a court and orders the authority to release a person being held in custody. Derived from English law, it is an essential part of the Constitution, intended to protect individuals from arbitrary imprisonment. Basically, it protects everyone from being arrested and held without reasonable charges. It was and is one of the individual protections that separates America from monarchies and other governments in which soldiers or sheriffs can literally knock on your door and haul you away without explanation.

With the suspension of habeas corpus, Lincoln authorized General Scott to make arrests without specific charges to prevent secessionist Marylanders from interfering with communications between Washington and the rest of the Union. In the next few months, Baltimore's Mayor William Brown, the police chief, and nine members of the Maryland legislature were arrested to prevent them from voting to secede from the Union.

When an otherwise obscure secessionist named John Merryman was arrested, the ancient Chief Justice, Roger Taney, went into action. A Marylander himself and the author of the Dred Scott decision, Taney issued a writ of habeas corpus for Merryman, demanding that the authorities give a reason for his detention. The military refused, and Taney, who thought that Lincoln might have even him arrested, issued an argument stating that only Congress could suspend the privilege of the writ and that Lincoln had broken the laws. Protection from arbitrary arrest became the first serious constitutional crisis of the war.

But in an address to Congress in July, Lincoln responded to Taney by asking "whether all the laws, but one, [were] to go unexecuted, and the government itself to go to pieces, lest that one be violated?" In other words, if Taney was so worried about the Constitution, why hadn't he done anything to prevent secession? Lincoln further argued that the Constitution states that "the Privileges of the Writ of Habeas Corpus shall not be suspended, unless when, in Cases of Rebellion or Invasion the public Safety may require it." The Constitution never specified which of the three branches of government could suspend the writ, so Lincoln argued that any of the three equal branches could do so.

Twice more during the war Lincoln, suspended habeas corpus, including the suspension "throughout the United States" on September 24, 1862. Although the records are somewhat unclear, more than thirteen thousand Americans, most of them opposition Democrats, were arrested during the war years, giving rise rise to the charge that Lincoln was a tyrant and dictator.

Civil War Voices

Commanding a unit of Illinois volunteers, Ulysses S. Grant, having resigned his army commission, applied for reinstatement into the regular army (May 24, 1861).

> Having served for fifteen years in the regular army, including four years at West Point, and feeling it the duty of every one who has been educated at the Government expense to offer their services for the support of the Government, I have the honor, very respectfully, to tender my services until the close of the war, in such capacity as may be offered. I would say, in view of my present age and length of service, I feel myself competent to command a regiment, if the President, in his judgment should see fit to entrust one to me.
>
> Since the first call of the President, I have been serving on the staff of the Governor of this State, rendering such aid as I could in organization of our State militia, and am still engaged in that capacity.
>
> U. S. Grant

A few weeks after Robert E. Lee made his fateful decision to cast his lot with his home state of Virginia and the Confederacy by resigning from the army, another soldier was trying desperately to get himself back into the army.

The two men, who had briefly crossed paths in Mexico, couldn't have been more different. Lee was a courtly, patrician, southern gentleman; Ulysses S. Grant, the son of a flinty, tough Ohio tanner and self-made businessman. Cast against the elegant, brilliant Lee, Grant was the "Rodney Dangerfield" of the Civil War: he got little respect, both early in the war and from later historians. A blue-collar sort of soldier, Grant had been an abject failure at almost everything he tried, including his early army career, until it came to making modern war.

Born in Ohio on April 27, 1822, the future general and president was the first child of Jesse Root Grant and Hannah Simpson Grant. The Grants named their son Hiram Ulysses and moved soon after his birth to Georgetown, Ohio, where the boy spent the first sixteen years of his life.

Jesse Grant, whose own father had been unable to support his children, had been apprenticed to a tradesman at the age of eleven, and a hard pioneer childhood had made him mean-spirited as well as ambitious. Always disappointed with Ulysses, Jesse Grant never failed to let his son know that he showed so little potential. On the puny side, Ulysses detested working in his father's tannery and had no head for business. To escape, he worked on a farm owned by his father.

After a year at a Kentucky boarding school, the seventeen-year-old Grant was sent to West Point, an appointment secured with the help of one of his father's politically connected acquaintances. When he arrived at "the Point," Grant stood at five-foot-one and weighed 120 pounds. In a momentous twist of fate, his name was changed forever. Attempting to enroll as Ulysses Hiram Grant, he discovered his appointment had already been made in the name of Ulysses S. Grant, because the congressman who appointed him thought his middle name was his mother's maiden name. Unable to correct the error, Grant took the new name as his own.

If it hadn't been for the Civil War, Grant might have been relegated to history's dustbin. Though he had served in Mexico, his postwar army career in the depressing northwest frontier had been clouded by his resignation under a charge of drunkenness. After that, his every business venture, every investment as a civilian, even a small farm, turned sour. Grant was back working as a clerk in his father's tannery in Galena, Illinois—a humiliating personal defeat—

when the war broke out and "rescued" him. He immediately saw a return to service as the only road for his future and wrote letters seeking a commission.

Grant also tried the personal approach, going to the Cincinnati headquarters of George B. McClellan, who had been named a general of the Ohio volunteers. But Grant's infamous reputation preceded him: McClellan had a recollection of his being on a drinking spree when they had crossed paths at Fort Vancouver in 1853. The general avoided an interview with Grant, who had to settle for the command of a group of Illinois volunteers.

As in Lee's first Civil War battle, Grant's first encounter was also less than glorious. Early in the morning of November 7, 1861, some three thousand Union troops under Grant were transported by boat from their camp at Cairo, Illinois, and met the Confederate forces under the inept General Gideon Pillow, one of Jefferson Davis's worst political appointees to a military position. Though the Confederates fought stubbornly, they were pushed back to their camp at Belmont, Missouri, on the banks of the Mississippi.

Grant's troops were celebrating and looting the Confederate camp when they suddenly came under heavy fire from cannon on the high bluff across the river. These troops were commanded by General Leonidas Polk (1806–1864), a friend of Davis's. A West Pointer, Polk had traded his sword for the robes of an Episcopal bishop but then returned to the Confederate army. Now he ferried twenty-seven hundred Confederate troops across the river and attacked Grant. When an aide cried that they were surrounded, Grant is said to have replied calmly, "Well, we must cut our way out as we cut our way in."

Forced to leave behind his wounded and the captured Confederate materials, Grant was fortunate to escape with his command intact. He was also fortunate to be alive. "Bishop" Polk could see Grant and had invited nearby soldiers to "try your marksmanship on him if you wish." Perhaps Grant was already out of range because none took up the challenge. Grant called the action a "raid" and said he'd won. Polk called it a "battle" and said he'd won. It was an inconclusive waste of lives, but it gave Grant his first taste of command and seeing the men under him die.

Documents of the Civil War

Pierre G. T. Beauregard, who had ordered and led the attack on Fort Sumter, took command of a Confederate army guarding an important train junction at Manassas, Virginia, about thirty miles south of Washington, D.C. He issued this proclamation aimed at rousing the citizens to defend their state (June 1, 1861).

> A reckless and unprincipled tyrant has invaded your soil. Abraham Lincoln, regardless of all moral, legal, and constitutional restraints, has thrown his Abolitionist hosts among you, who are murdering and imprisoning your citizens, confiscating and destroying your property, and committing other acts of violence and outrage, too shocking and revolting to humanity to be enumerated.
>
> All rules of civilized warfare are abandoned, and they proclaim by their acts, if not on their banners, that their war-cry is "BEAUTY AND BOOTY." All that is dear to man—your honor and that of your wives and daughters— your fortunes and your lives, are involved in this momentous contest.

What Happened at Manassas?

When the Disney Company lost "the third battle of Manassas" in 1994, a lot of attention was focused on the area where a little stream called Bull Run meanders through the Virginia countryside. That modern battle was fought over whether a theme park devoted to Disney's version of American history, along with extensive real estate and commercial development, belonged near the site of two of the most important battles of the Civil War, including the very first significant meeting of the two armies.

The first three months of war saw only minor fighting. On June 10, 1861, the Confederates beat back a Union attack at Big Bethel on the Virginia coast. Other clashes took place in western Virginia (soon to become the state of West Virginia), where, according to newspaper reports, the victor at the Battle of Rich Mountain was a young Union general named George Brinton McClellan—even though he had little to do with the victory. But McClellan understood that the Civil War would be fought in the newspapers as well as the trenches, and he made certain that telegraph lines followed him wherever he went.

Events were now moving toward the war's first great battle. At Alexandria, Virginia, General McDowell's federal army was being pressured to heed the Union newspapers and politicians demanding "On to Richmond!"

A West Pointer, Irvin McDowell (1818–1885) had assumed command somewhat by default. He was one of the few regular army officers available after the mass defections to the Confederacy. A man with a legendary appetite—it was said he could eat an entire watermelon for dessert—McDowell was modest, honest, and didn't drink alcohol, which set him apart from many of the Union generals who would follow him. With strong political pressure behind him, he was forced to act before his army was ready. When given his assignment, he did not even have maps of Virginia.

Responsible for blocking the federal approach to the Confederate capital and holding the railroad junction at Manassas was General Pierre G. T. Beauregard, the hero of Fort Sumter and McDowell's classmate at West Point. Beauregard had deployed his army along Bull Run, the stream near Manassas about thirty miles southwest of Washington.

Some fifty miles away, at the northern end of Virginia's rich Shenandoah Valley, two more armies faced each other. The Union troops were under General Robert Patterson, an aging army veteran; the Confederates commanded by General Joseph E. Johnston (1807–1891). Patterson's orders from Washington were to block Johnston so he could not slip south to Manassas to reinforce Beauregard.

Marching from Alexandria on July 16, the Union army began to move into Confederate territory. Many of the men were three-month volunteers, assembled immediately after Fort Sumter, and their discharge date was approaching rapidly. Singing "Dixie" as they marched and with regimental bands playing, these green soldiers had no foreboding of what lay ahead.

Adding to the pomp and almost festive mood were crowds of civilians and politicians accompanying the army. Among them was Lincoln's private secretary, John G. Nicolay, who later wrote:

> The business of the war was such a novelty that McDowell's army accumulated an extraordinary number of camp-followers and noncombatants. The vigilant newspapers of the chief cities sent a cloud of correspondents to chronicle the incidents of the march and conflict. The volunteer regiments carried with them . . . companionship unknown to regular armies . . . Senators and representatives in several instances joined in what many rashly assumed would be a mere triumphal parade.

A Confederate observer said the procession included "gay women and strumpets" and that they carried picnic baskets, opera glasses, champagne, and tickets that had been printed for "a grand ball in Richmond." As John Nicolay reported, there were at least fifty correspondents from Union newspapers along with another twenty-six journalists from the Confederate press, making it perhaps the best-reported battle of any war before CNN came along.

McDowell ignored the reporters and spectators. He was far more concerned about his undisciplined troops. While his plan of attack was sound, he was relying on untested troops who lacked experience with the rigors of a forced march, and had little or no acquaintance with combat. The march had the air of a country outing as the soldiers broke ranks to pick berries or fill canteens. Their overconfidence was only bolstered when the first Confederate sentries retreated before their celebratory advance. But then came a brief exchange of fire between the Union troops and some Confederates under James K. Longstreet; like his fellow Confederate generals, Beauregard and Johnston, Longstreet was still wearing his Union army uniform. At this sign of a fight, some of the Union soldiers began to have second thoughts. Volunteers nearing the end of their enlistment period quickly decided that this was a good time to request an early discharge.

When McDowell's army delayed for two days at Centreville, Virginia, Johnston used the time to move about two thirds of his Confederate troops from the Shenandoah Valley to Bull Run by train, giving the armies almost equal strength. In so doing, he made military history: it was the first time that troops used the railroad for strategic mobility—one of many historic firsts in the Civil War.

On Sunday, July 21, the battle began in full fury. Initially, the Union forces seemed justified in their confidence as the Confederates retreated. But among those troops fresh from the Shenandoah area were a brigade of Virginians commanded by Thomas J. Jackson (1824–1863). Born in western Virginia, Jackson was the son of a debt-ridden lawyer who died of typhoid when the boy was two years old. When his mother died five years later, Thomas was separated from his brother and sister and raised by a bachelor uncle. With the equivalent of only a fourth-grade education, he was admitted to West Point in 1842; rising steadily in the class rankings, he graduated in 1846, seventeenth in a class of fifty-nine.

After West Point, Jackson was commissioned a second lieutenant and joined an artillery unit in Mexico. For bravery during the siege of Vera Cruz, he was promoted to first lieutenant and earned wide praise. He stayed in the army until 1851, when he left to join the faculty of the Virginia Military Institute in Lexington. Even then, Jackson demonstrated some of the eccentricities for which he was later famed. Daring, calm, and tactically brilliant on the battlefield, he was also a hypochondriac, very concerned about his digestive system and his diet. Among his idiosyncrasies was a refusal to eat pepper because he thought it made his left leg hurt. A tall, rigid man, he never let his back touch a chair, always sitting bolt upright to keep his internal organs in "alignment." Aides said they often saw him raise his right arm and hold it aloft for many minutes. He never explained whether he did so to engage in silent prayer or to cause blood to flow downward and "establish equilibrium," which he considered essential to good health. Stern and silent, with little sense of humor, Jackson didn't drink, smoke, dance, curse, play cards, or attend the theater. Instead, he strolled around camp handing out Sunday school leaflets. He refused to write a letter that would be in transit on Sunday, and he habitually sucked lemons, spoke in a voice "so shrill it seemed feminine," and napped before battle. He also believed that Yankees were devils.

This day at Manassas, General Barnard Bee told Jackson he was being beaten back, but Jackson said he would stop the Union advance with bayonets if necessary. What happened next belongs to Civil War mythology. Bee supposedly called out, "Oh men, there are Jackson and his Virginians, standing behind you like a stone wall! Let us determine to die here, and we will conquer. Follow me."

Thus Bee supposedly gave Jackson his immortal nickname. But there are many reported versions of this speech. Although the name seemed complimentary, it was not clear if it had been meant as an insult. Was Bee praising Jackson's immovable resolve or complaining that Jackson was not coming to his support? Bee never said; he was wounded in the fighting moments later and died the next day. Some have labeled the entire episode a fabrication. Almost everything about Bee's words—if he said them at all—is subject to dispute. In fact, Bee's men may have been behind Jackson's. But all that matters is that the Richmond papers reported the heroic version of Jackson's stand, and a legend was created. Thomas Jackson became "Stonewall" and his 1st Virginia Brigade the "Stonewall Brigade."

At a moment when both armies were exhausted from the day's fighting and the fate of the battle hung in the balance, fresh forces reinforced the thinning Confederate lines. Their arrival had an extraordinary effect. What first seemed like a sure Union victory quickly turned into a massive rout of the inexperienced Union volunteers, who wilted under the Confederate surge. Jackson then issued the order "Charge, men and yell like the furies." This was the first Union experience of the blood-curdling "Rebel Yell," a shrieking, high-pitched scream that has entered Civil War folklore. Self-assured and confident three days earlier, the Union army turned back toward Washington in a riotous dash of soldiers, horses, and all those civilians who had come to watch.

As one Union newsman reported, "All sense of manhood seemed to be forgotten. . . . Even the sentiment of shame had gone. . . . Every impediment to flight was cast aside. Rifles, bayonets, pistols, haver-sacks, cartridge-boxes, canteens, blankets, and overcoats lined the road."

Civil War Voices

New York World correspondent Edmund Clarence Stedman saw two congressmen attempting to stem the rout at Manassas.

> Both these Congressmen bravely stood their ground till the last moment. . . . But what a scene! . . . For three miles, hosts of Federal troops . . . were fleeing along the road, but mostly through the lots on either side. Army wagons, sut-ler's teams [merchants who followed the army with wagons full of goods], and private carriages choked the passage, tumbling against each other amid the clouds of dust and sickening sights and sounds. Hacks containing unlucky spectators of the late fray were smashed like glass, and the occupants were lost sight of in the debris.
>
> Wounded men lying along the banks . . . appealed with raised hands to those who rode horses . . . but few regarded such petitions.
>
> Then the artillery, such as was saved, came thundering along, smashing and overpowering everything. . . .
>
> Who ever saw such a flight? . . . It did not slack in the least until Centreville was reached.

In Washington, where the Associated Press had already issued a premature report of victory, the news changed around 5:00 P.M. As word came in of the devastated Union army in full retreat, the city was sent reeling into a panicky shock.

When a jubilant President Davis arrived at the Manassas battlefield from Richmond, Jackson asked him for ten thousand men to follow the fleeing Union troops right into Washington and end the whole thing. But his request was ignored, for the Confederates lacked sufficient men and supplies. This was the first chance for the Confederate press to turn on the Davis administration, which also suffered its first casualty. Criticized because the troops were unable to pursue the defeated Union army, Leroy Walker resigned as the Confederate secretary of war. His place was taken by Judah Benjamin, who was emerging as Davis's most reliable and efficient administrator. In spite of the emerging criticism, this first decisive Confederate victory was hailed wildly throughout the Confederacy and taken as a sign that the second "War of Independence" would end quickly. The powers of Europe and the Lincoln administration would surely recognize the legitimacy of the Confederate cause and negotiate a peaceful settlement.

In the wake of the battle, Stonewall Jackson sent off an envelope to his pastor. Expecting a battle report, the preacher discovered a contribution for his church's "colored Sunday school," which Jackson had forgotten to send the day of the battle.

Wilmer McLean, a farmer who had granted the use of his home as Beauregard's Manassas headquarters, had seen enough of the war. After his house had been shelled, McLean decided to move his family farther south, where "the sound of battle would never reach them." They settled in Appomattox Court House, Virginia, and in the McLean parlor, on an April day nearly four years later, Lee would surrender his army to Grant.

Civil War Voices

Generals Joseph E. Johnston and Beauregard spoke to the soldiers of the Confederate states (July 1861).

> One week ago, a countless host of men, organized into an army, with all the appointments which modern art and practiced skill could devise, invaded the soil of Virginia. Their people sounded their approach with triumphant displays of anticipated victory. Their generals came in almost royal state; their great Ministers, Senators and women came to witness the immolation of our army and subjugation of our people, and to celebrate the result with wild revelry.
>
> It is with the profoundest emotions of gratitude to an overruling God, whose hand is manifest in protecting our homes and liberties, that we, your Generals commanding, are enabled, in the name of our whole country, to thank you for that patriotic courage, that heroic gallantry, that devoted daring exhibited by you in the actions of the 18th and 21st, by which the hosts of the enemy were scattered, and a signal and glorious victory obtained.

Rebuffed in his desire to carry the war to Washington, Beauregard took his feud with his civilian superiors to the press. It was the beginning of an unhappy relationship that was the equal of any of Lincoln's more widely discussed problems with his generals. Beauregard was transferred west, but the hero of Manassas was already being touted as an alternative to President Davis.

Who Was "Little Mac"?

The Confederacy had its Little Napoleon in Beauregard. But the Union could boast of one as well. On July 27, 1861, six days after the Union defeat at Bull Run, Lincoln replaced General Irvin McDowell, the defeated Union leader, with General George B. Mc-

Clellan. Not yet thirty-five, McClellan had gained national prominence by taking the credit for two of the few Union victories in the
first year of the war, the relatively minor battles of Rich Mountain
and Carrick's Ford in western Virginia.

If history is indeed often determined by character—the so-called
Great Man theory—McClellan may provide clear proof of it. In the
negative. He seemed to have everything going for him and blew it.
Brilliant, vain, superior, egotistic, paranoid: These are just a few of
the words used to describe the man who—after Lincoln and Grant—
influenced the course of the war more than anyone else in the Union.
Unfortunately, in McClellan's case it was for the worse.

In his biography of McClellan, *The Young Napoleon*, Stephen W.
Sears wrote:

> When making war, General George Brinton McClellan was
> a man possessed by demons and delusions. He believed
> beyond any doubt that his Confederate enemies faced him
> with forces substantially greater than his own. He believed
> with equal conviction that enemies at the head of his own
> government conspired to see him and his army defeated so
> as to carry out their traitorous purposes. He believed him-
> self to be God's chosen instrument for saving the Union.
> When he lost the courage to fight, as he did in every battle,
> he believed he was preserving his army to fight the next
> time on another and better day. . . . While he basked in the
> appellation given him by his admirers—the Young Napo-
> leon—he was called by a legion of derisive opponents
> Young McNapoleon.

McClellan was admitted to West Point before he turned sixteen,
having already begun his university education. The son of a Philadelphia doctor, a haughty young aristocrat who looked down his nose
at his social inferiors, he served with distinction in Mexico. Trained
as an engineer, he went into the booming world of railroading and
became an executive for the Illinois Central, where he encountered

two people who would change his life. Allan Pinkerton, the private detective under contract to the railroad, and Abraham Lincoln, then serving the railroad as an attorney. A conservative Democrat, Mc-Clellan differed sharply with the Republican Lincoln in politics and held a low opinion of the future president, whom he considered his social and intellectual inferior. He later wrote that Lincoln "was not a man of very strong character, & as he was as destitute of refinement—certainly in no sense a gentleman—he was easily wrought upon by the coarse associates whose style of conversation agreed so well with his own."

Now Lincoln had summoned him to take command. And a few days later McClellan noted with characteristic egotism to his wife of one year, "Who would have thought, when we were married, that I should so soon be called upon to save my country?"

Civil War Voices

General George B. McClellan's recollections of his assuming command of the troops in the Washington, D.C., area following the defeat at Bull Run.

All was chaos, and the streets, hotels, and bar-rooms were filled with drunken officers and men absent from their regiments without leave. . . .

The first and most pressing demand upon me was the immediate safety of the capital and the government. This was provided for by at once exacting the most rigid discipline and order. . . .

I lost no time in acquiring an accurate knowledge of the ground in all directions, and by frequent visits to the troops made them personally acquainted with me, while I learned about their condition, and their needs, and thus soon succeeded in inspiring full confidence and good morale in place of the lamentable state of affairs which existed on my arrival.

Thus I passed long days in the saddle and my nights in the office—a very fatiguing life, but one which made my power felt everywhere and by everyone.

At this moment of crisis, McClellan, the professional soldier, was exactly what the demoralized Union army desperately needed. Initially he proved a superb organizer and had an extraordinary rapport with his troops, who dubbed him "Little Mac." By the end of August, McClellan's army had grown to seventy-five thousand men.

Civil War Voices

London banker Baron Rothschild, when asked who would win the war.

"The North."
"Why?"
"Because it has the largest purse."

What Was the Trent Affair?

One of the basic oversights of a typical American education has been the tradition of teaching American history as if it happened in a vacuum and that nothing else of interest or importance was happening elsewhere in the world at the same time. For instance, during the Civil War years, the Taiping Rebellion (1850–1864) in China was responsible for the loss of as many as *twenty million lives*. In 1848 Karl Marx wrote *The Communist Manifesto*. This cultural myopia is due partly to a powerful streak of American egotism: the only important things are what happened in America. That's one reason our schoolbooks tended to underplay, for instance, the crucial role of the French in winning America's independence. And it is certainly true of teaching the Civil War. In 1860 the world was getting smaller, for steamships, railroads, and telegraphs had created a nineteenth-century revolution in communications and transportation. More than ever before, events in Europe had an impact on events in America.

Throughout the early months and years of the war, one of the most critical questions facing both sides was the reaction of Europe. Would Queen Victoria's Great Britain, in particular, recognize the Confederate cause? Such recognition would give the Confederacy political legitimacy and allies capable of openly furnishing weapons and other supplies.

This question was complicated by the messy relations among Europe's three great powers—England, France, and Russia—at the time. England and France, the victors over Russia in the Crimean War to control Central and Eastern Europe, salivated at the prospect of the downfall of the United States government. Having lost the war, Russia needed the United States to balance its European enemies. Soon after the surrender of Fort Sumter, Queen Victoria officially announced Britain's neutrality but acknowledged the Confederates as belligerents. This status meant that the Confederacy could buy arms from neutral nations and seize ships on the high seas. Though short of recognizing the Confederacy, this gave it hope and was viewed as an unfriendly act by Lincoln's administration.

Throughout the Union, the assumption was that England, which had abolished slavery and the slave trade, would side with the Union. (Russia had also freed its serfs in 1861.) But this question, like most every other great question between nations, ultimately fell to dollars and pounds and international trade, not morality. The British foreign minister, Lord Palmerston, was sympathetic to the Confederacy. England's manufacturers needed cotton, and the merchants knew that the Union blockade would close lucrative ports to trade. Despite the government's neutrality, British shipbuilders were soon producing blockade runners designed to slip past the Union ships to deliver English goods to the South and carry cotton back to England.

European recognition was critical to the Confederacy, and the issue nearly blew up for the Union in the fall of 1861 over an incident known as the *Trent* Affair. During a storm on the night of October

11, a Confederate blockade runner slipped out of Charleston, South Carolina, carrying James Mason and John Slidell to London and Paris to seek official recognition of the Confederacy.

A Virginia lawyer who had served in both houses of Congress, James Murray Mason (1798–1871) was a tobacco-chewing, staunch states' rights Democrat who had drafted the controversial Fugitive Slave Act in 1850. Jefferson Davis chose his old friend Mason, a ten-year veteran of the Senate Foreign Relations Committee, to serve as the Confederacy's envoy to England. Born in New York and educated at Columbia, John Slidell (1793–1871) moved to New Orleans and set up a law practice. He joined the Senate in 1853 but resigned along with his law partner and fellow senator Judah Benjamin when Louisiana seceded in February 1861.

Reaching Havana, Mason and Slidell transferred to the *Trent*, a British mail steamer bound for England. The next day the *Trent* was met by the Union sloop *San Jacinto*. In a flagrant violation of international law, Captain Charles Wilkes (1798–1877), one of the first Americans to sail around the world, ordered two shots fired across the *Trent*'s bow. He then boarded the *Trent* and demanded the surrender of Mason and Slidell. Outraged, the British captain had no choice; the *San Jacinto* sailed to Boston, where Mason and Slidell were jailed while the *Trent* continued to England.

While the Union rejoiced over the capture of the Confederate diplomats and Congress thanked Wilkes for his "brave, adroit, and patriotic conduct in the arrest of the traitors," the British had an entirely different reaction. Queen Victoria herself was outraged, and London's cries for war were heard across the Atlantic. Lord Palmerston issued an ultimatum: release the diplomats or face war. Prince Albert, the queen's influential husband, counseled moderation and toned down the outraged diplomatic correspondence from London. But an army of eight thousand British soldiers set sail for Canada, and sympathy for the Confederate cause grew in the British business community. On both sides of the Atlantic, the rhetoric grew hot. The *Times* of London wrote, "By Capt. Wilkes let the Yankee breed be judged. Swagger and ferocity, built on a foundation of vulgarity and cowardice, these are the characteristics, and these are the prominent marks by which his countrymen, generally speaking, are known all over the world."

Although Secretary of State Seward thought that war with England might bring the seceded states back to the Union, the president realized that America could not fight London and the Confederacy at the same time. Lincoln said, "One war at a time," and sought a face-saving reconciliation. He had another compelling reason to seek peace: The Union needed the 2,300 tons of saltpeter, a key ingredient in gunpowder, that had been secretly purchased from the British. By late December passions had cooled between the two countries, and Lincoln agreed to release the Confederate emissaries, recognizing that they had been taken illegally. On January 1, Mason and Slidell were turned over to the British.

Mason's job was to convince the British government to support an independent Confederate nation. This effort was part of "cotton diplomacy." Although never announced as an official policy of the Confederate government, planters had begun to withhold cotton for export. The idea went back to an 1858 speech by South Carolina's James Hammond, who suggested that if the South's cotton planters held back their crop, "England would topple headlong and carry the whole civilized world with her."

It was an ambitious scheme, but it never really worked. In 1860, before the war, a large harvest had pushed prices down and allowed British mill operators to stockpile a two-year inventory of cotton. In addition, new sources of cotton were emerging in Egypt and India, England's expanding empire on the other side of the world. Finally, the loss of southern cotton was far less troublesome to Britain than the loss of the lucrative northern wartime market.

What Was a "Radical Republican"?

It sounds like a contradiction in terms, doesn't it? Radical and Republican simply don't fit. In modern politics, even the most moderate Republicans would profess to be conservatives, and few would suffer to be called liberals. But the picture was different in 1861. The "Radical Republicans" were those men who viewed the war as a crusade against slavery, pure and simple. Although a minority in the Republican party, they held great sway, and, from the war's outset, they often made life difficult for Lincoln.

If Lincoln was going to do away with constitutional guarantees, such as habeas corpus, Congress—and a group of Radical Republicans, in particular—decided to wreak equal havoc on due process and individual rights.

As the war's first year drew to a close, Lincoln faced political opposition not just from Democrats in Congress and the press opposed to the war but also from members of his own Republican party who were unhappy with the course of the war. A few months after the debacle at Bull Run, another Union military disaster only increased the pressure on Lincoln. Once more, it was the result of the inexperience of the Union commanders. And, sadly for Lincoln, it would mean a personal tragedy as well as another humiliating Confederate victory.

Born in London, Edward D. Baker came to America with his family in 1816 and moved to the Utopian community of New Harmony, Indiana. After studying law and serving in the brief Black Hawk War, he moved to Springfield, Illinois, in 1835 to open a law office. He was elected to the state legislature, where he met Lincoln. Though they squabbled over local politics, the men became close friends, and Lincoln named his second son, Eddie (who died in 1850, at age four), after Baker.

As a congressman, Baker broke ranks with his party over the war with Mexico and volunteered to fight. A handsome man with a fondness for champagne, cards, and poetry, Baker moved to Oregon, becoming its senator in 1860, the first Republican elected to high office on the West Coast. He introduced Lincoln at his first inauguration and following Fort Sumter's fall, he issued a call for "sudden, bold, forward determined war; and I do not think anybody can conduct war of that kind as well as a dictator." When Lincoln offered him an appointment as a brigadier general, Baker chose instead a commission as colonel in order to retain his Senate seat.

Despite his early military experience, Baker was no military genius, as was true of a good many of the "political" generals and officers produced by both Union and Confederate sides in the war's early days. On October 21, 1861, Baker's 1,700-man brigade crossed the Potomac River from Maryland to Virginia on a reconnaissance mission.

With only three boats available, the crossing was ill planned and slow. On the Virginia side of the river, the Union troops faced a hundred-foot-high bank called Ball's Bluff, which could be reached only by walking up a narrow cowpath. When the troops reached the top, they found themselves on open ground and confronted by four concealed Confederate regiments. Baker was killed instantly by a bullet through the brain. In a frenzied panic, the Union men raced backward, tumbling over the cliff onto the bayonets of their comrades below. Dozens tried to scramble into the boats, which capsized, drowning many more soldiers.

Civil War Voices

Randolph A. Shotwell, a seventeen-year-old private in the Virginia army, was among the Confederates on Ball's Bluff.

A kind of shiver ran through the huddled mass upon the brow of the cliff; it gave way; rushed a few steps; then, in one wild, panic stricken herd, rolled, leaped, tumbled over the precipice! The descent is nearly perpendicular, with ragged, jutting crags, and a water-laved base. Screams of pain and terror filled the air. Men seemed suddenly bereft of reason; they leaped over the bluff with muskets still in their clutch, threw themselves into the river without divesting themselves of their heavy accouterments, hence went to the bottom like lead. Others sprang down upon the heads and bayonets of those below. A gray-haired private of the First California was found with his head mashed between two rocks by the heavy boots of a ponderous "Tammany" man, who had broken his own neck by the fall! The side of the bluff was worn smooth by the numbers sliding down.

. . . As it happened, the two larger bateaux were just starting with an overload when the torrent of terror-stricken fugitives rolled down the bluffs—upon them. Both boats were instantly submerged. . . . The whole surface of the river seemed filled with heads, struggling, screaming, fighting, dying! Man clutched at man, and the strong, who might have escaped, were dragged down by the weaker. . . . Captain Otter, of the First California . . . was found a few days later with two men of his company clutching his neckband. Had he attempted to save them, or had they seized him and dragged him down? One officer was found with $126 in gold in his pocket; it had cost his life.

The Union losses included more than two hundred killed and wounded, with more than seven hundred captured. Told of Baker's death, Lincoln was deeply shaken. His close friends Elmer Ellsworth, who died in Alexandria, Virginia, the war's first Union combat fatality, and now Baker were dead. One who saw Lincoln at this moment reported, "His hands were clasped upon his heart; he walked with a shuffling, tottering gait, reeling as if beneath a staggering blow. He did not fall, but passed down the street, carrying not only the burden of the nation but a load of private grief which, with the swiftness of the lightning's flash, had been hurled upon him."

Ball's Bluff was immediately denounced by newspapers and politicians as another fiasco. They demanded accountability for the disaster and for the loss of a patriot like Baker. A few days later, three Radical Republicans met with the president. Fellow party members, they were critical of Lincoln's conciliatory attitude toward the South, the appointment of McClellan, a Democrat, and his slow pace, the sharp defeats the Union had suffered at Bull Run and Ball's Bluff, and Lincoln's assumption of almost dictatorial powers. They told Lincoln that they planned to set up a congressional Joint Committee on the Conduct of the War, ostensibly to oversee the president and specifically charged with rooting out corruption and inefficiency in the Union army. The committee was made up of these abolitionist politicians known as Radical Republicans; it was led by Senators Benjamin Franklin Wade (Ohio), who served as chairman, and Zachariah Chandler (Michigan) and from the House Congressman George Washington Julian (Indiana). Another member was the pro-Union Democrat from Tennessee, Andrew Johnson.

Working in secret session, the committee's main target was McClellan, whom they considered a coward. Some even suggested that he was a traitor. But they avoided a frontal attack on the young general, going after his subordinate generals. The first victim was General Charles P. Stone. Although the overall commander of the forces defeated in the Ball's Bluff disaster, Stone had had little to do with the battle and had been nowhere near the scene. But McClellan knew the wolves needed a lamb, and he happily sacrificed Stone. Brought before a secret session of the committee, Stone faced an inquisition. He was without benefit of counsel, was not told the charges against him, and did not know his accusers or their testimony. On the committee's orders, Stone was arrested at midnight on February 8, 1862, and whisked away to Fort Lafayette in New York Harbor, where he was imprisoned for 189 days. Never charged and never cleared, his military career was ruined, although he eventually prospered as a civilian engineer. (Ironically, in 1887 he was the engineer who constructed the pedestal for the Statue of Liberty in the same harbor where he had been held.)

Later victims of the committee included Major General Fitz-John Porter, one of the Union's most successful generals on the battlefield—and unfortunately for him, an avid supporter of McClellan's. A cousin of the naval hero David Dixon Porter, he became the scapegoat for the Union defeat in the Second Battle of Bull Run, in 1862. Charged with disloyalty, disobedience, and misconduct, Porter was court-martialed and cashiered from the army in disgrace. (An 1878 inquiry completely exonerated him and he was recommissioned. He later served as New York City's police, fire, and public works commissioner.)

Although the committee continued its political vendettas throughout the war, it had little influence on Lincoln. And the generals it tended to support usually had little success with Lincoln or on the battlefield. Myth has it that Mrs. Lincoln was scrutinized by the committee for suspected secessionist sympathies. There was even a report that Lincoln was forced to appear before the committee to defend her, but the story was an unfounded rumor.

Was It Really "the Brothers' War"?

The Civil War has been given many names, but perhaps the most fitting and poignant is "the Brothers' War." The war is filled with the stories of families divided and former comrades facing one another over their different loyalties. Sadly, it was no exaggeration.

Divided loyalties went all the way to the White House. Mary Todd Lincoln was from an aristocratic family in the border state of Kentucky, and the Washington gossips spoke of her as being proslavery and in favor of southern secession. Four of Lincoln's brothers-in-law served the Confederate cause. One of them, Ben Hardin Helm, was a West Pointer who had turned down Lincoln's personal offer of a commission in the Union army; becoming a Confederate general, he was killed in the Battle of Chickamauga.

The border states provided the most frequent source of divided loyalties. Kentucky's Henry Clay, "the Great Compromiser," who tried for years to overcome the sectional strife that led to the war, had grandsons serving on both sides. John J. Crittenden, the former governor and U.S. senator from Kentucky who had proposed the Crittenden Plan in December 1860 as a way of avoiding the war, had sons serving on opposite sides: Major General Thomas L. Crittenden for the Union and Major General George B. Crittenden for the Confederacy.

And the saga of the Civil War is peppered with stories of friends, family members, and former comrades who encountered one another on the battlefield. Major A. M. Lea was part of the Confederate boarding party that captured the U.S.S. *Harriet Lane* during a naval battle off Galveston, Texas. On deck he found a dying Union lieutenant—his son.

James McQueen McIntosh, a brigadier general in the Confederate army who was killed in action in 1862, had a younger brother, John B. McIntosh, who reached the same rank in the Union army.

During the famous battle of the first ironclads, the *Virginia* (previously known as the *Merrimac*) was commanded by Franklin Buchanan. On board the *Congress*, a Union vessel sunk during that battle, was his brother Paymaster McKean Buchanan. And in one of the most extreme and notorious instances of families divided, Union Brigadier General Philip St. George Cooke was assigned to hunt down the daring Confederate Cavalry General Jeb Stuart. Stuart was Cooke's son-in-law and supposedly said of his father-in-law's decision to remain in the Union army, "He will regret it but once, and that will be continuously." Cooke's own son, John Rogers Cooke, was also a Confederate brigadier general. And his nephew John Esten Cooke, a well-known novelist before the war, also rode with Jeb Stuart and was responsible for making Stuart one of the war's great romantic heroes.

chapter five

1862
"Let Us Die to Make Men Free"

*I have put you in motion to offer battle to the invaders of your country. . . .
You can but march to a decisive victory over . . . mercenaries sent to
subjugate and despoil you of your liberties, property, and honor.*

—CONFEDERATE GENERAL ALBERT SIDNEY JOHNSTON

APRIL 3, 1862

*If I could save the Union without freeing any slave, I would do it—if I
could save it by freeing all the slaves, I would do it—and if I could do it by
freeing some and leaving others alone, I would also do that.*

—PRESIDENT ABRAHAM LINCOLN

AUGUST 22, 1862

*The truth is, when bullets are whacking against tree trunks and solid shot
are cracking skulls like eggshells, the consuming passion in the breast
of the average man is to get out of the way.*

—UNION SOLDIER DAVID L. THOMPSON

AFTER THE BATTLE OF ANTIETAM

✳

* What Happened at Fort Donelson?

* How Did "a Tin Can on a Shingle" Make History?

* What Was the Peninsular Campaign?

* What Happened at Shiloh?

* Who Fought for the Confederacy?

* Why Did the Defeated Citizens of New Orleans Call Benjamin Butler a "Beast"?

* Why Did Lincoln Fire General McClellan?

* What Was "The Prayer of Twenty Millions"?

* What Happened at Antietam?

* Why Did Lincoln Fire General McClellan Again?

"All quiet along the Potomac" became the catchphrase for journalists in Washington as the new year opened with an almost eerie stillness. Following the Union disasters of the war's first months, volunteers poured into the capital, reaching nearly 200,000 men by midwinter and turning Washington into an armed camp. But as one soldier noted, there were three types of people, the first group being soldiers; "the other two classes are politicians and prostitutes, both very numerous, and about equal in . . . honesty and morality."

The sight of McClellan's doing little more than overseeing parades was not a big hit with most of the politicians. Even President Lincoln, who desperately wanted McClellan to succeed, began to doubt his abilities and resolve. But Lincoln had few options. Most of the Union generals seemed more concerned with politicking for power, promotion, and glory. Lincoln's headaches were compounded by a Cabinet crisis. The charges of corruption in the War Department had become so overwhelming that Secretary of War Simon Cameron, a powerful Pennsylvanian who was also pressing for the emancipation of slaves, against Lincoln's wishes, was pushed out and made ambassador to Moscow. His place was taken by Edwin Stanton, a no-nonsense lawyer Lincoln had known since they butted heads in Springfield. A Democrat, Stanton did not think highly of Lincoln, calling him an "imbecile."

Lincoln probably took some pleasure from a proposal he received from the king of Siam, who offered to send the president a herd of fighting elephants to aid the Union war effort. Lincoln politely declined, stating that the weather "does not . . . favor the multiplication of the elephant."

What Happened at Fort Donelson?

While McClellan paraded, Ulysses S. Grant remained a rather obscure brigadier general with a reputation for hard drinking. But he was far less interested in parades and drills than he was in taking action. Operating in coordination with Flag Officer Andrew H. Foote and his small flotilla of gunboats, Grant attacked two Confederate forts on the Tennessee River. First, the gunboats forced the fall of Fort Henry, one of a string of small forts—little more than piles of dirt—thrown up along the Tennessee to protect the Confederacy's borders.

But the Confederate forces in nearby Fort Donelson proved more resistant. The Union gunboats failed to dislodge them, and in sharp fighting the Union forces were pushed back when Grant took charge. Leading a counterattack, he sent the Confederate defenders reeling. Commanding Fort Donelson was John B. Floyd, President Buchanan's war secretary, who had been accused of misusing funds meant for Indians and sending federal arms south during the Buchanan years. Next in command was Gideon Pillow, a man Grant knew and disdained since serving with him in Mexico. A law partner of former President Polk's, Pillow was a political appointee who had performed miserably during the Mexican War. Inept as a soldier, Pillow had been Polk's "spy" inside Winfield Scott's army in Mexico. He had compounded his battlefield shortcomings by attempting to take credit in the press for Scott's military success, and he was tried for insubordination. At Fort Donelson, both Floyd and Pillow offered sharp rebuttals to the notion of Confederate command superiority. During the war, each side would have its share of incompetent, corrupt, ill-equipped, cowardly, and drunken officers, often political appointees. For the Confederacy, Floyd and Pillow were such men.

Facing defeat, they were afraid to face also the ignominy of being the first Confederate officers to surrender. So they fled, leaving Simon B. Buckner in command of the hopeless situation. When Buckner later told Grant that Pillow had made a hasty departure, Grant said, "If I had got him, I'd let him go again. He will do us more good commanding you fellows."

Civil War Voices

General Grant, replying to a request for terms of surrender from General Simon B. Buckner (February 16, 1862): "Sir: Yours of this date, proposing armistice and appointment of Commissioners to settle terms of capitulation, is just received. No terms except an *unconditional and immediate surrender* can be accepted. I propose to move immediately upon your works."

A year behind Grant at West Point and a friend during the Mexican War, Simon B. Buckner had once loaned Grant money to pay a hotel bill when he resigned from the army. Expecting his old comrade to communicate in a more gentlemanly fashion, Buckner called Grant's terms "ungenerous and unchivalrous." But to Grant, honor among friends was one thing. War was another. His terms were accepted.

The fall of Fort Donelson meant Grant's name had been made. The newspapers quickly decided that his initials stood for "Unconditional Surrender." Other generals were not so thrilled. In particular, his immediate superior, Henry Halleck, fearful that Grant's rising star might damage his own career, demoted him and sought to smear his name by resurrecting the drinking charges. But with the help of Elihu Washburne, an influential Illinois congressman, the case went directly to Lincoln, who saw in Grant what he was looking for— aggressive command. He immediately nominated Grant for promotion to major general.

Pillow and Floyd were suspended for their actions, and Floyd died in 1863. Pillow held command once more but was accused of hiding behind a tree during a battle and never led troops again. Simon Buckner was held briefly as a prisoner, although Grant did return his former comrade's earlier favor by putting his own purse at the disposal of the captive.

Another Confederate officer at Donelson accepted neither surrender nor slinking off in the night. Nathan Bedford Forrest (1821– 1877), a hot-tempered Tennessean who had amassed a small fortune as a slave trader and plantation owner, enlisted in the Confederate army as a private and rose rapidly. With his own money, he mounted a cavalry battalion. When Fort Donelson was about to fall, he organized a breakout, successfully removing five hundred men along with horses and supplies to Nashville. Forrest eventually emerged as one of the most daring Confederate leaders, but he was also among the most controversial. Later in the war, he was accused of ordering the massacre of black Union troops.

For Lincoln, this brief glimmer of hope brought by the fall of Fort Donelson, the first major blow to the Confederacy, was tempered by personal tragedy. As Grant was assaulting the Tennessee forts, Lincoln's two youngest sons, Willie and Tad, were ill with what doctors called a bilious fever, probably typhoid. It was contracted from water contaminated by the overcrowding and poor sanitary conditions in Washington, where the troops' sewage ran straight into the Potomac, the source of the White House drinking water. Lapsing into a coma, eleven-year-old Willie died on February 20. First Eddie in 1850 and now for a second time, the grief-stricken president had to bury a son. Lincoln told an aide, "He was too good for this earth. It is hard to have him die."

Mary Todd Lincoln took the news even harder, collapsing with a nervous breakdown, and did not leave her room for three months. One day, according to Elizabeth Keckley, a seamstress who had become her confidante, the president led his wife to a window and pointed to a mental hospital in the distance, saying, "Try and control your grief or it will drive you mad and we may have to send you there." (Mary Todd's biographer Jean Baker refutes this story in its details but not its characterization of Mrs. Lincoln's depression.)

Although she recovered, Mary Lincoln never went into Willie's rooms again and began seeing spiritualists and attending séances in an attempt to contact her dead child. Spiritualism and mediums had experienced a major revival in midcentury America, and attempts to contact the dead were commonplace among Lincoln's Cabinet members and other politicians. Long after Willie's death, Mary Lincoln would tell her half sister Emili, the wife of Confederate General Ben Hardin Helm, that both her dead boys came into her rooms at the White House.

Lincoln himself also had to fight through the sorrow brought on by the loss. Having struggled with fits of depression—he called it the "hypo," short for hypochondria—all his life, he drew on tremendous reserves of personal strength during the war, and the melancholy that had sometimes disabled him in earlier years apparently never affected his wartime behavior.

Songs of the Civil War

Julia Ward Howe's poem, which would become "The Battle Hymn of the Republic," first appeared in *The Atlantic Monthly* (February 1862).

Mine eyes have seen the glory of the coming of the Lord;
He is trampling out the vintage where the grapes of wrath
are stored;
He has loosed the fateful lightning of his terrible swift
sword;
His truth is marching on.

I have seen Him in the watch fires of a hundred circling
camps;
They have builded Him an altar in the evening dews and
damps;
I have read his righteous sentence in the dim and flaring
lamps;
His day is marching on.

I have read a fiery gospel, writ in burnished rows of steel;
"As ye deal with my contemners, so with you my grace shall
deal";
Let the hero, born of woman, crush the serpent with his
heel,
Since God is marching on.

He has sounded forth the trumpet that shall never call re-
treat;
He is sifting out the hearts of men before his judgment seat;
Oh, be swift, my soul, to answer Him! Be jubilant, my feet!
Our God is marching on.

In the beauty of the lilies Christ was born across the sea.
With a glory in His bosom that transfigures you and me;
As he died to make men holy, let us die to make men free,
    While God is marching on.

Born into an affluent New York family, Julia Ward Howe (1819–1910) had written several books of poetry and, with her husband, Samuel Gridley Howe, one of abolitionist John Brown's six financial backers, published the *Abolitionist*, a Boston antislavery journal. In the capital to meet with Lincoln, Howe and a group of abolitionist friends sang snatches from popular army songs during a long carriage ride. One common tune, in homage to John Brown, started:

John Brown's body lies a-moldering in the grave,
His soul goes marching on.

After a companion suggested that Howe could write better words to the tune, that night in her hotel room she turned to the book of Isaiah and wrote all five verses. *The Atlantic Monthly*, the nation's leading literary magazine, paid her $5 for the poem.

But it was Army Chaplain Charles Caldwell McCabe who is credited with popularizing the poem. He first taught it to fellow prisoners of war in Richmond's Libby Prison. Later released, McCabe began to sing the song at fund-raisers for the chaplains corps. Following the battle at Gettysburg in 1863, he sang the song while the president was in the audience. With tears in his eyes, Lincoln stood and shouted, "Sing it again." The song soon became the unofficial anthem of the Union.

How Did "a Tin Can on a Shingle" Make History?

Throughout human history, violence and technology have gone hand in hand, and weapons grew more deadly as men became more "civilized." Stone replaced bone. Steel replaced stone. Gunpowder overcame steel. In the Civil War, the rapid advances of the Industrial Age changed warfare more dramatically than in any previous war. War would not change again so radically until the atomic bomb was dropped on Japan eighty-five years later.

The revolutionary changes in warfare in the Civil War became clear in the early weeks of March 1862. The railroad and telegraph, both relatively recent inventions, had already proven their value, but it was naval warfare that was about to change most drastically.

Hampton Roads is a channel through which three of Virginia's rivers—the James, Nansemond, and Elizabeth—empty into Chesapeake Bay. Union forces in Fort Monroe and Newport News held the northern shore of the channel; on the shores to the south were Norfolk and the Gosport Navy Yard, occupied by Confederate forces since the Union navy abandoned the port at the beginning of the war. The channel itself was controlled by a Union fleet, which blocked the water route to Richmond.

Secretly, the Confederate navy had begun work on a warship whose sides would be covered with metal armor. They had raised the frigate *Merrimac*, scuttled by the Union navy, and were covering its wooden sides with iron prepared at Richmond's Tredegar Iron Works. (At this arsenal of the Confederacy, slaves were employed to cast the cannons that were meant to keep them in chains.) But word of this secret soon reached Washington, which also launched a mission to build an ironclad. The plan was for the Union's ironclad to sail into the Norfolk Navy Yard and destroy the dock and the *Merrimac* before its conversion was completed.

The contract to design and build the Union's *Monitor* went to New Yorker John Ericsson, a Swedish immigrant and internationally renowned engineer. His design was not simply a wooden warship covered in steel. He drew plans for a flat, raftlike ship with a revolving turret equipped with two eleven-inch guns. Although the Union navy wanted the ship ready in a hundred days, that deadline came and went.

Late in February, a free black woman from Norfolk passed through enemy lines and went to the Navy Department. Inside her dress was hidden a letter from a Union sympathizer who worked in the Confederate navy yard reporting that the *Merrimac* was nearly finished. The race to complete the *Monitor* became more urgent.

Finished at a Brooklyn shipyard for $275,000, the *Monitor* was a long ocean voyage away from Hampton Roads. The strange-looking vessel moved into New York Harbor on March 6; barely seaworthy, it nearly missed its appointed date with history. It leaked badly, and in heavy seas, waves broke over the top of the vessel, threatening to put out its engine boilers. When the rough seas subsided, the *Monitor* was towed through relatively calm coastal waters.

In the meantime, the renamed *Virginia* was ready to steam out of port under the command of Franklin Buchanan (1800–1874). A Baltimorean who was the first superintendent of the U.S. Naval Academy and who had accompanied Commodore Matthew Perry on the expeditions to the Orient that opened Japan to the Western world, Buchanan had resigned from the U.S. Navy to enter Confederate service. On Saturday, March 8, he led a small Confederate fleet out to do battle with the Union's blockading fleet and shore batteries.

One Union observer recalled, "We saw what to all appearances looked like the roof of a very big barn belching forth smoke as from a chimney on fire. We were all divided in opinion as to what was coming. The boatswain's mate was the first to make out the Confederate flag. And then we all guessed it was the *Merrimac* (*Virginia*) come at last."

On the Union's *Cumberland*, pilot A. B. Smith said, "As she came ploughing through the water . . . she looked like a huge half-submerged crocodile. Her sides seemed of solid iron, except where the guns pointed from the narrow ports. . . . At her prow I could see the iron ram projecting straight forward, somewhat above the water's edge."

The Union shore batteries and ships fired as fast as they could. Aboard the *Cumberland*, Smith recalled,

> Still she came on, the balls bouncing upon her mailed sides
> like India-rubber, apparently not making the least impres-

sion, except to cut off her flag-staff and thus bring down the Confederate colors.... We had probably fired six or eight broadsides when a shot was received from one of her guns which killed five of our marines. It was impossible for our vessel to get out of her way, and the *Merrimac* soon crushed her iron horn, or ram, into the *Cumberland*... knocking a hole in the side ... and driving the vessel back upon her anchors with great force. The water came rushing into the hold.

The renamed *Virginia*'s attack on the wooden ships was like that of a tank assaulting a phalanx of Roman archers or a modern jet fighter encountering a World War I biplane. It was war between two different eras. As the *Cumberland* went down fighting, the tradition of wooden warships, dating from the ancient empires of Egypt, Greece, and Rome, sank with it.

The *Virginia* then turned its attention to the *Congress*, a Union vessel grounded in the channel's shallow waters. On board was Mc-Kean Buchanan, the brother of the *Virginia*'s commander, Franklin Buchanan. Set afire, the *Congress* surrendered. Franklin Buchanan ordered another Confederate vessel to board the *Congress* and remove the wounded. But the Union batteries onshore continued to fire at the Confederates, even though they were clearly rescuing Union sailors, among them McKean Buchanan. Enraged, Commodore Franklin Buchanan took up a rifle and returned fire, in the process suffering a severe leg wound that took him out of the action. He was replaced by Lieutenant Jones.

By day's end, the Union's situation looked grim. The *Virginia* was hardly damaged, even though it had been the target of more than a hundred Union guns. Some of its guns were damaged, and its ram had been sheared off and left in the side of the *Cumberland*, but the Confederate ironclad had lost little of its fighting ability. Both sides knew it could simply sail out the next day and obliterate the Union fleet. In Washington there was panic as rumors flew that the *Virginia* would move up the Potomac and level the capital. A dispirited emergency Cabinet gathered for prayers.

Late in the day they were answered. The scenario and naval history were altered with the arrival at Hampton Roads of the Union's *Monitor*. Lieutenant John L. Worden (1818–1897) and his crew of fifty-seven sailors had weathered the sea voyage and the very likely threat of sinking. The *Monitor* took up position alongside the *Minnesota*, another crippled Union vessel, to await the light of day. In another coincidence, the *Monitor*'s second-in-command, Dana S. Greene, knew that he would be facing his naval academy roommate, Walter R. Butt, aboard the *Virginia*.

When morning came, observers on both sides got their first look at the Union's entry in the ironclad arms race. Sitting only 18 inches above the water, at 172 feet long and 41½ feet wide, it was a small and rather unimpressive sight. One Confederate onlooker called it "a tin can on a shingle!"

Curious spectators crowded the shores on both sides. They did not have long to wait. The two ironclads locked onto each other, guns blasting. Watching with astonishment aboard the helpless *Minnesota*, Commander G. J. Van Brunt later recalled, "Gun after gun was fired by the *Monitor*, which was returned with whole broadsides from the rebels, with no more effect, apparently, than so many pebble-stones thrown by a child. . . . the shot glanced off . . . clearly establishing the fact that wooden vessels cannot contend successfully with iron-clad ones, for never before was anything like it dreamed of by the greatest enthusiast in maritime warfare."

Like prizefighters exchanging blows, the two vessels battled for hours. Aboard the *Monitor*, Commander Worden was blinded when a shot hit the pilothouse, and his vessel was left momentarily out of control. Lieutenant Jones, now commanding the *Virginia*, thought the Union ship was withdrawing. His boat now leaking, his crew exhausted by two days of continuous action, and short on powder and shot, Jones ordered the *Virginia* to return to Norfolk. However, the *Monitor* was undamaged; seeing the *Virginia* depart, it took up position once again by the grounded *Minnesota*, whose crew had been prepared for the worst. Now a hero, the "savior" of Hampton Roads, the *Monitor*'s temporarily blinded Commander Worden was taken to Washington to meet with a joyous but tearful Lincoln.

Inconclusive in the sense that neither ironclad emerged a clear victor, the long-term advantage went to the Union. The *Monitor*'s arrival prevented the Confederates from breaking the Union's effective blockade of the Confederate capital as well as another potential Union military disaster, which Lincoln could ill afford. The Confederate navy would soon abandon Norfolk, and the Union would be far more capable of stepping up its production of ironclads.

The two days of fighting at Hampton Roads attracted worldwide attention and dramatically demonstrated the superiority of ironclads. Naval warfare would never be the same. In England, then the world's greatest naval power, it was clear that the "tin can on a shingle" could wreak havoc on its mostly wooden navy. (England had two experimental ironclads in development.)

Ironically, neither vessel ever figured prominently in the war again. Forced out of Norfolk, the *Virginia* was run aground by her crew on May 11 and set afire to prevent her capture. The *Monitor* lasted only a few more months. While being towed off Cape Hatteras, North Carolina, she foundered in heavy seas and went down with sixteen crew members on December 31, 1862. (Rediscovered in 1976, the *Monitor* is a popular diving site. Her anchor was recovered in 1983, but the hull is too badly corroded to raise.)

Civil War Voices

Abraham Lincoln, in a note written but never sent regarding General George B. McClellan (April 9, 1862): "It is called the Army of the Potomac but it is only McClellan's bodyguard. . . . If McClellan is not using the army, I should like to borrow it for a while."

The relationship between Lincoln and McClellan grew testy. Though Lincoln offered his qualified support, McClellan barely concealed his disdain for the president, disparaging him as "the original gorilla" and "a well-meaning baboon," characterizations that were actually first used by Edwin Stanton and later given racial overtones in the anti-Lincoln press. In a notorious incident, Lincoln called one evening at McClellan's residence to find the general away. Lincoln waited patiently for him, but when McClellan returned, he ignored the president and went to bed. It was not an isolated incident. However, Lincoln was willing to overlook McClellan's contempt as long as the general gave him victories.

What Was the Peninsular Campaign?

Lincoln's patience finally ran out. Having issued presidential War Orders calling for an offensive that McClellan simply ignored, the president was exasperated and on the brink of replacing the young commander before he had actually led troops into battle. After nearly eight months of organizing and training the Army of the Potomac, McClellan finally yielded to the calls for action by Lincoln, Congress, and the public.

Lincoln wanted a drive directly south from Washington, straight to Richmond. But McClellan, fearing the Confederate armies camped at Manassas, had concocted a much more ambitious plan. Moving his large army by water, he hoped to bypass the Confederate lines and start his drive only sixty miles from Richmond. Reluctantly, Lincoln approved but ordered McClellan to leave behind a sufficient force to defend the Union capital. This diversion of troops would lead to trouble, for McClellan repeatedly used Lincoln's order as an excuse for his failures.

Before moving, McClellan and the Union received another embarrassment. When the Confederate army pulled out of its trenches at Manassas and moved south to the Rappahannock River, closer to Richmond, the abandoned artillery at Centreville proved to be nothing more than logs painted black, which the press dubbed "Quaker guns." It was also clear that the Confederate army at Manassas was about half the size McClellan had claimed.

Undeterred, the young general simply altered his plan by taking the army farther south to Fort Monroe, near the site of the *Monitor-Merrimac* battle. From there he would march up the narrow peninsula between the York and James rivers and capture Richmond. This approach gave the plan its name, the Peninsular Campaign.

An armada of 400 boats was pulled together near Washington and Alexandria to move the Army of the Potomac. Along with the troops, this fleet had to haul more than 1,200 wagons and ambulances, 300 pieces of artillery, tons of ammunition, 15,000 horses, and tons of rations for the troops and food for the animals. This logistical nightmare was accomplished by Quartermaster General Montgomery Meigs, one of the obscure and unsung heroes of the Union effort. Brought into the War Department after Cameron's dismissal, Meigs was a genius at supply and distribution, one of the major factors behind the Union's ultimate success.

On April 4, McClellan's 112,000 men landed and began slogging through the rain-soaked, flat, muddy Virginia countryside. Despite months of training, many of the recruits were unwilling to carry all their equipment, and the Army of the Potomac's route was soon littered with castoffs. One soldier recalled, "Castaway overcoats, blankets, parade-coats, and shoes were scattered along our route in reckless profusion. . . . The colored people along our route occupied themselves in picking up this scattered property. They had on their faces a distrustful look, as if uncertain of the tenure of their harvest."

Civil War Voices

P. Regis de Trobriand, a French-born New York officer, on the Union's march up the peninsula:

> The small number of houses . . . which were on the line of our march were all abandoned. Their occupants had left on our approach. . . . Near a deserted hut we met four children crouched at the side of the road. . . . Their mother was dead, and their father had abandoned them. They wept while

asking for something to eat. The soldiers immediately gave them enough provisions to last them several days. . . . But what became of these children? This is the horrible side of war.

The army continued to Yorktown, the site of the historic surrender of the British army to George Washington's ragged rebels eighty years earlier, sealing America's victory in the War of Independence. The powerful symbolism of this nearly sacred Virginia site was not lost on the opposing Confederate troops. They saw themselves as heirs of the patriot army of Washington, the greatest Virginian of all, pitted in a struggle for liberty against a powerful government in a second war of independence.

Under General Joseph E. Johnston, one of the victorious commanders at Bull Run, the Confederates had only 15,000 men entrenched at Yorktown. Lincoln urged McClellan to attack. But the general was stalled by his natural caution, reinforced by the faulty intelligence provided by his secret service head, Allan Pinkerton.

Following Lincoln's notorious incognito arrival in Washington, Pinkerton had remained in Washington, hoping for a formal appointment from the president whose life he had apparently saved. But when no job was offered, he joined McClellan as chief of the secret service in the Department of the Ohio. In this role Pinkerton proved highly successful in his own spying expeditions into the seceded states, but he also kept McClellan informed of behind-the-scenes politicking in Washington, specifically, inside Lincoln's Cabinet.

When McClellan was promoted, Pinkerton followed him to Washington and then to the peninsula, where he was in charge of collecting information about Confederate troop strength. With much of his information coming from the escaping slaves who trickled into the Union lines, Pinkerton proved totally inept at consolidating the duplicated or erroneous reports he was receiving, and he consistently overestimated the Confederate strength, often by as much as two or three times the actual number. Known by his code name of Major E. J. Allen, Pinkerton catered to McClellan's need to feel that he was outnumbered, enabling the general continually to call for additional men and supplies.

At Yorktown, McClellan was convinced that he was outnumbered by the Confederates, when his numbers were vastly superior. Instead of attacking, he ordered the army to dig in and began a siege. His caution and hesitation gave the Confederate armies time to reinforce both Yorktown and Richmond.

In Richmond, President Davis was now receiving military advice from Robert E. Lee. At the war's outset, Lee had failed in what may have been the impossible task of securing pro-Union western Virginia for the Confederacy.

As McClellan settled in, Lee wrote to Jefferson Davis: "I am preparing a line that I can hold with part of our forces out front, while with the rest I will endeavor to make a diversion to bring McClellan out." When Lee ordered his soldiers to build earthwork fortifications to protect Richmond, he was called "the King of Spades." All along the sixteen-mile line, Lee's troops complained that they had not joined up to dig holes.

In early May, McClellan finally moved to attack. But the Confederate army had anticipated his move. Having forced the Union army to delay for a month, ample time to fortify the Richmond area with troops, it now withdrew from Yorktown back toward the capital, leaving McClellan holding an empty bag. He nonetheless claimed a great victory. A small action against the Confederate army's rear guard at the historic colonial capital of Williamsburg, Virginia, was inconclusive. But McClellan trumpeted what seemed like consecutive triumphs, and the Union press enthusiastically agreed. The real picture was far less rosy.

Civil War Voices

Confederate General Albert Sidney Johnston to his troops before attacking General Grant's Army at Pittsburg Landing, Tennessee (April 3, 1862).

> Soldiers of the Army of the Mississippi: I have put you in motion to offer battle to the invaders of your country.... You can but march to a decisive victory over ... mercenaries sent to subjugate and despoil you of your liberties, property, and honor. Remember the precious stake involved; remember the dependence of your mothers, your wives, your sisters, and your children on the result; remember the fair, broad, abounding lands, the happy homes that will be desolated by your defeat. The eyes and hopes of eight million rest upon you.

What Happened at Shiloh?

While "Young Napoleon" inched toward Richmond, the Union armies were moving more rapidly in the West. Following the capture of Forts Henry and Donelson, Ulysses Grant was taking the war to the Confederacy while the Union navy attempted to control the Mississippi River.

With 42,000 men, Grant moved to Pittsburg Landing, on the west bank of the Tennessee River only a few miles from Tennessee's border with Mississippi. As Grant marched, his waterborne ally Foote attacked Island Number Ten, a Confederate fort on the Mississippi near New Madrid, Missouri. At the same time, a Union army led by Major General Don Carlos Buell was moving from the east to link up with Grant.

Facing Grant, General Albert Sidney Johnston (no relation to General Joseph E. Johnston, the Confederate commander in Virginia) was building a Confederate army in Corinth, Mississippi. He was joined by P.G.T. Beauregard, the hero of Fort Sumter and Bull Run, whose pronouncements to the press about Jefferson Davis's administration had won him few fans in Richmond. He had been sent west to keep him quiet.

At Pittsburg Landing, Grant made one of his most costly mistakes. Confident that the Confederates would remain at Corinth, he failed to order adequate defenses. His troops did not dig defensive entrenchments; pickets—or sentry lines—were not placed far enough from the camp. Cavalry patrols were not sent out to serve as an alarm in case of attack. The Union soldiers—many of them recent recruits with little training—were casually camping in tents near Shiloh Church, a simple log meetinghouse a few miles from the river. Named for the biblical village whose name means, ironically, "place of peace," it was this primitive Methodist church that would give the coming battle its awful name in history.

When the Confederates attacked unexpectedly on the morning of April 6, chaos reigned once more in the Union ranks. Suddenly another disaster loomed as the Union soldiers ran in panicked retreat, many of them hiding near the river. Under William Tecumseh Sherman, one of Grant's commanders and a veteran of the Bull Run debacle, some of the Union troops began to regroup. Of Sherman, Grant later said, "He inspired a confidence in officers and men that enabled them to render services on that bloody battlefield worthy of the best of veterans."

Civil War Voices

New York Tribune war correspondent Junius Henry Browne.

> Hotter and hotter grew the contest.... The light of the sun was obscured by the clouds of sulphurous smoke, and the ground became moist and slippery with human gore.... Men glared at each other as at wild beasts; and when a shell burst with fatal effect among a crowd of the advancing foe, and arms, legs, and heads were torn off, a grim smile of pleasure lighted up the smoke-begrimed faces of the transformed beings who witnessed the catastrophe....
>
> ... There was no pause in the battle. The roar of the strife was ever heard. The artillery bellowed and thundered, and the dreadful echoes went sweeping down the river, and the paths were filled with the dying and the dead. The

sound was deafening, the tumult indescribable. No life was worth a farthing. . . . Yonder a fresh regiment rushed bravely forward, and ere they had gone twenty yards a charge of grape sent the foremost men bleeding to the earth. . . . Death was in the air, and bloomed like a poison-plant on every foot of soil.

During the fierce fighting around a peach orchard near the Shiloh meetinghouse, General Albert Sidney Johnston was wounded. Having led a charge that captured a critical position, he had apparently emerged without a scratch even though there were many bullet holes in his clothes. He had been hit in the leg but thought it a minor wound and sent his surgeon to tend to the Union wounded, ignoring the blood that was trickling into his boot. When Johnston turned pale and fainted, his aides finally realized that he was wounded and assisted him off his horse. There was no chance to treat what was actually a severed artery, and Johnston died a few minutes later.

Born in Kentucky, Johnston had graduated eighth in his class from West Point (1826) and served in the Black Hawk War. Following his wife's death in 1836, he went to the Republic of Texas, enlisting as a private. His skills as a soldier were obvious, and a year later he was named a brigadier general and then Texas's secretary of war. At the beginning of the Civil War Johnston, who had returned to the U.S. Army as a brigadier general in California, was a highly esteemed officer, and Grant "expected him to prove the most formidable man that the Confederacy could produce." Both sides had offered him a high command. But as a Texan and a close friend of Jefferson Davis's, Johnston followed Texas out of the Union.

Despite the loss of their commander, the Confederates still seemed to have the upper hand, and a large number of Union soldiers under General Benjamin M. Prentiss was captured. Although Prentiss was later scapegoated for the Union losses at Shiloh, the heroic stand of his heavily outnumbered men at a spot so frenzied with fighting it was called the Hornet's Nest had actually gained the Union valuable time. The additional time needed to collect his men as prisoners further delayed the Confederate attack.

Driven back to the river, the Union forces were now supported by artillery and the heavy weapons aboard two gunboats. They were able to withstand the final Confederate push when the first of General Buell's Union reinforcements arrived in the late afternoon, ferried from the opposite side of the river.

With nightfall came a heavy rain, and Beauregard, now in command, made his headquarters in the same tent that the Union's Sherman had occupied that morning. He wired Richmond a confirmation of Johnston's death along with a premature report of victory. But on the following day the Union troops regained the advantage, strengthened by Buell's reinforcements and the arrival of another force under General Lew Wallace, which had been lost in the nearby woods. Many of the men who had run from the battle and hidden by the river during the first day's combat now returned to fight. Few were prepared for the scenes of carnage they would witness.

Civil War Voices

Sixteen-year-old John A. Cockerill, a Union regimental musician.

> I passed . . . the corpse of a beautiful boy in gray who lay with his blond curls scattered about his face and his hands folded peacefully across his breast. He was clad in a bright and neat uniform, well garnished with gold, which seemed to tell the story of a loving mother and sisters who had sent their household pet to the field of war. His neat little hat lying beside him bore the number of a Georgia regiment. . . . He was about my age. . . . At the sight of the poor boy's corpse, I burst into a regular boo-hoo, and started on.
>
> Here beside a great oak tree I counted the corpses of fifteen men. . . . The blue and the gray were mingled together. . . . It was no uncommon thing to see the bodies of

Federal and Confederate lying side by side as though they
had bled to death while trying to aid each other.

In General Grant's own recollections of Shiloh: "I saw an open
field . . . over which the Confederates had made repeated charges the
day before, so covered with dead that it would have been possible
to walk across . . . in any direction, stepping on dead bodies without
a foot touching the ground."

On this second day of intense fighting, the Union army finally
broke the Confederate resistance, and Beauregard decided to withdraw back to Corinth. The impact of the fighting at Shiloh was staggering, with the casualty figures outstripping anything that had yet
been witnessed. More than 13,000 Union men had been killed,
wounded, or captured; the Confederate losses were nearly 11,000.
More Americans were killed in these two days of fighting than in all
three previous American wars combined. Besides General Johnston,
the Confederacy lost Mary Lincoln's half brother Samuel B. Todd to
a sharpshooter's bullet in his head as he led a charge. As bad as the
situation looked to Lincoln in Washington, it was worse for Jefferson
Davis in Richmond. In Johnston, he had lost the man he considered
his best general and a dear friend. He wept at the news and said,
"The cause could have spared a whole State better than that great
soldier." And he still had to contend with the increasingly difficult
Beauregard, who was already fixing blame on the loss at Shiloh on
the dead Johnston.

If anyone still retained romantic notions of a short war won by
grand charges and heroic actions, the illusions were shattered at Shiloh. Along with the rest of the country, Grant realized that the war
would not come to a quick, bloodless end. As he later wrote,

Up to the battle of Shiloh I, as well as thousands of other
citizens, believed that the rebellion against the Government
would collapse suddenly and soon if a decisive victory could
be gained over any of its armies. Donelson and Henry were
such victories. . . . But when Confederate armies were col-

lected which not only attempted to hold a line farther south
. . . but assumed the offensive and made such a gallant effort to regain what had been lost, then, indeed, I gave up
all idea of saving the Union except by complete conquest.

Back in Washington, news of the twin victories at Shiloh and Island
Number Ten, which had fallen to naval bombardment on April 7,
was tempered by word of the horrendous losses. The casualty lists
were accompanied by calls for Grant's removal. Once again, rumors
flew that the general had been drunk during the battle, backbiting
most likely perpetrated by his superior Henry Halleck, who arrived
at the battlefield and assumed direct command, in essence demoting
Grant. Disgusted by this treatment and the savaging he received in
the press, Grant considered resigning. Charles Dana, a newsman
working for Secretary of War Stanton, met with Grant and found no
evidence of a drinking problem. Lincoln said, "I can't spare this man.
He fights." Grant was restored to his position.

Civil War Voices

Belle Reynolds, a housewife from Peoria, Illinois, who had followed
her husband into the field at Shiloh, became a battlefield nurse and
an eyewitness to the carnage.

We climbed the steep hill opposite the Landing, picked our
way through the corrals of horses, past the long line of
trenches which were to receive the dead, and came to an
old cabin, where the wounded were being brought. Outside
lay the bodies of more than a hundred, brought in for recognition and burial—a sight so ghastly it haunts me now.
. . . And that operating table! These scenes come up before me now with all the vividness of reality . . . one by one,
they would take from different parts of the hospital a poor
fellow, lay him out on those bloody boards, and administer
chloroform; but before insensibility, the operation would

begin, and in the midst of shrieks, curses, and wild laughs, the surgeon would wield over his wretched victim the glittering knife and saw; and soon the severed and ghastly limb, white as snow and spattered with blood, would fall upon the floor—one more added to the terrible pile.

Until three o'clock I had no idle moments; then, having done all in my power to minister to so much wretchedness, I found my long-taxed nerves could endure no more. One of the surgeons brought me a spoonful of brandy, which revived me. Feeling that my labors were at an end, I prepared to leave, and had just turned to go in the direction of the boat, when a hand was laid upon my shoulder. The shock was so sudden I nearly fainted. There stood my husband! I hardly knew him—blackened with powder, begrimed with dust, his clothes in disorder, and his face pale. We thought it must have been years since we parted. It was no time for many words; he told me I must go. There was a silent pressure of hands. I passed on to the boat....

At night I lived over the horrors of the field hospital and the amputating table. If I but closed my eyes, I saw such horrible sights that I would spring from my bed; and not until fairly awakened could I be convinced of my remoteness from the sickening scene. Those groans were in my ears! I saw again the quivering limbs, the spouting arteries, and the pinched and ghastly faces of the sufferers.

The conditions described by Belle Reynolds were not unusual but commonplace during the war. Medicine had barely emerged from its dark ages. (Harvard Medical School did not own a microscope or a stethoscope until 1868.) Anyone who saw the film *Dances with Wolves*, in which Kevin Costner pulls his boot back on over a bloody foot rather than face a surgeon, has seen an example of Civil War battlefield medicine and a soldier's reaction to it. As Belle Reynolds witnessed, surgery basically meant amputation. "Sawbones" as a nickname for *doctor* was grimly accurate, for at that time amputation was the only medical treatment for a fracture or severe laceration; the procedure accounted for 75 percent of all the operations performed by Civil War doctors.

Although ether, chloroform, and nitrous oxide had come into use as anesthetics, few doctors were trained to use them. Deprived of supplies by the Union blockade, Confederate doctors couldn't obtain any anesthetics. Then, a good surgeon meant a fast surgeon; the best could remove a limb in a few minutes. When the demand for ether and chloroform during the hellish battles sometimes outstripped the supplies, soldiers were given a drink of whiskey—the most widely administered medicine—and a gun cartridge and were grimly and literally told to "bite the bullet."

Adding to the ghastly statistics was the fact that surgery was not the worst horror the wounded had to face. To survive the operating table only meant the likelihood of getting gangrene or other little-understood infections, thought to be caused by bad air. The doctors had no understanding of antiseptic conditions in hospitals, let alone on a battlefield. Combat surgery was performed in the open on tables made of doors or rough wooden boards laid upon boxes, with tubs underneath to catch the blood. The surgeons used the same knives and saws all day, wiping them on a bloody apron. Unfortunately for many on the battlefield, a revolution in medicine lay just ahead. Building on the work of French bacteriologist Louis Pasteur (1822–1895) and his germ theory of disease, English surgeon Joseph Lister (1827–1912) had begun his work in antiseptics around 1865, too late to have any impact on the Civil War. (By 1869 Lister's use of carbolic acid on surgical instruments, wounds, and dressings had reduced surgical mortality from 50 to 15 percent.)

Yet as primitive as Civil War medical conditions were, the majority of amputees were probably saved by the saw. According to fairly well-kept Union records, of some 29,000 amputations performed, a little more than 7,000 resulted in death. Operations performed within forty-eight hours of a wound were twice as likely to be successful as those performed after that length of time. Union medical records—the Civil

War was the first bureaucratic war, and very good records exist, at least on the Union side—show that amputation was far from a death sentence, depending on what was amputated. For instance, of nearly 8,000 finger amputations performed, there were fewer than 200 deaths. One of the most deadly procedures was amputation of the thigh. Of some 6,300 Union cases there were 3,411 fatalities (a mortality rate of 53.6 percent). Of some 29,000 surgical amputations recorded by the Union, there were 7,283 fatalities. (Confederate records were not kept as scrupulously, and most were later destroyed. It is safe to assume that Confederate medical conditions were worse than those of the Union, given the severe shortages created by the blockade.)

The horrors of amputation and its bleak aftermath were compounded by diseases that killed approximately twice the number of soldiers who fell in battle. Americans who watched the modern civil war in the African nation of Rwanda during the summer of 1994 saw the deadly effects of dysentery and other intestinal illnesses: Massive numbers of people died in very short periods of time. The situation was not so different during the Civil War.

Simple hygiene, as we understand it, could have saved thousands of lives. Crowded into filthy camps where open-air latrines often ran directly into sources of drinking water, the soldiers were subject to a medical textbook gallery of diseases. Even simple ailments for modern Americans—measles, mumps, chicken pox—were often deadly to the recruits, especially those from rural, unpopulated areas where they were less likely to have been exposed to common diseases.

Farmboy soldiers were even more likely to succumb to sexually transmitted diseases, known at the time as "the ailments of Venus." Venereal disease ran rampant, particularly among the troops that spent long periods near cities like Washington, Richmond, New Orleans, and Nashville, where bordellos and red-light districts flourished. In *The Story the Soldiers Wouldn't Tell: Sex in the Civil War*, Thomas P. Lowry told of more than 180,000 cases of venereal disease among the white Union troops. (The Confederate rate was probably lower because the troops spent less time near cities.) Untreated venereal diseases—or, more accurately, venereal diseases treated only with medieval combinations of herbs, poultices, and whiskey—sent many veterans home to infect their wives with a scourge that lasted well after the last shots of the war were fired.

Early in 1862, before most of the war's major engagements, the Union had lost 2 percent of its force to disease. By far the worst of the Civil War ailments was dysentery, as well as severe diarrhea. To the soldier, these ailments went by several names—"Tennessee Trots," "Virginia quick steps," or simply "the bowel complaint." The massive outbreaks of dysentery actually influenced military events on several occasions. Its debilitating effects were one of the reasons McClellan was ultimately forced to withdraw from the peninsula. After the battle at Shiloh, General Beauregard had to abandon Corinth, Mississippi, because of an epidemic of dysentery that put a third of his army on the sick lists. Bluntly stated, many more soldiers died from diarrhea than were killed in battle.

Civil War Voices

From Herman Melville's poem "Shiloh," written after the bloody two-day battle (April 1862).

> Skimming lightly, wheeling still,
>   The swallows fly low
> Over the field in clouded days,
>   The forest field of Shiloh—
> Over the field where April rain
> Solaced the parched one stretched in pain
> Through the pause of night
> That followed the Sunday fight
>   Around the church of Shiloh—
> The church so lone, the log-built one,
> That echoed to many a parting groan

And natural prayer
  Of dying foemen mingled there—
Foemen at morn, but friends at eve—
  Fame or country least their care:
(What like a bullet can undeceive!)
  But now they lie low,
While over them the swallows skim,
  And all is hushed at Shiloh.

Who Fought for the Confederacy?

Draft-dodging is something most people associate with cowardice—
or high principle, depending on your point of view—and they think
it came into vogue during the Vietnam War. But it is actually a long
tradition in America, one that dates from the Civil War. In every
previous American war, from the Revolution to the War of 1812 and
the conflict with Mexico, volunteers had supplied sufficient numbers
of troops. But with McClellan's Army of the Potomac practically
banging on the doors of Richmond, the Confederate government was
forced to turn to a military draft. It was unpopular then and has been
a difficult issue for Americans ever since.

The spring of 1862 looked pretty disastrous for the Confederacy
after the defeats at Forts Henry and Donelson, Shiloh, and the fall
of Island Number Ten. With a large Union army nearing Richmond
and the one-year enlistments about to run out for many of its men,
the vastly outmanned Confederacy passed the Conscription Act on
April 16. The measure seemed to be completely at odds with the
Confederacy's fundamental existence. Even more than in the Union
(which would institute a draft a year later), the Confederate draft was
hated. To states' rights hard-liners, conscription amounted to the
trampling of individual rights by a central government, exactly why
so many had left the Union. Coming on the heels of controversy over
its few victories, as well as several more recent defeats, the draft
created another crisis for Jefferson Davis's government.

Under the Conscription Act, all healthy white men between the ages of eighteen and thirty-five were liable for a three-year term of service, and the enlistment for all one-year soldiers was extended to three years. An amendment in September 1862 raised the age limit to forty-five. Late in the war, in February 1864, when the Confederate armies had suffered huge losses, the age limits were extended to between seventeen and fifty. At the time, Jefferson Davis remarked, "We're about to grind the seed corn of the nation."

But not everyone had to go, and there was the rub, as it would continue to be. The Confederate draft law exempted men in certain occupations considered to be valuable for the home front, such as railroad and river workers, civil officials, telegraph operators, miners, druggists, and teachers. In many Confederate states, there was suddenly a booming interest in the teaching profession. Far more controversial was an amendment passed on October 11, the so-called "twenty-nigger" law, which exempted the overseers of twenty or more slaves. Fearing the possible danger of hundreds of thousands of unsupervised slaves, this was one of the most hated parts of the law, as wealthy young planters used the exemption for themselves.

Equally onerous was the policy of allowing substitutes. For a price, a wealthy young Southerner could get a farmboy to take his place as a conscript. In the Confederacy, a new call arose that would later be repeated in the northern states: "A rich man's war, a poor man's fight." The clerks who administered the Conscription Act in Richmond would be called the Bureau of Exemptions.

Many in the Confederacy, including Vice-President Stephens and Governors Joseph Brown of Georgia and Zebulon Vance of North Carolina, vehemently opposed the draft as unconstitutional and openly worked to thwart it in their states. Brown believed that Georgia's troops should fight only in Georgia. Though North Carolina contributed the most volunteers to the Confederate armies, its Governor Vance fraudulently exempted men by adding them to civil servant rolls or by having them enlist in the state militias.

Civil War Voices

With the Confederate army in Corinth, Mississippi, Sam R. Watkins recalls the aftermath of Shiloh and the passage of the draft law in his memoir, *Co. Aytch.*

Well, here we were, again "reorganizing," and after our lax discipline on the road to and from Virginia, and after a big battle, which always disorganizes an army, what wonder is it that some men had to be shot, merely for discipline's sake? And what wonder that General Bragg's name became a terror to deserters and evil doers? Men were shot by scores, and no wonder the army had to be reorganized. Soldiers had enlisted for twelve months only . . . and they naturally looked upon it that they had a right to go home. . . . War had become a reality; they were tired of it. A law had been passed by the Confederate States called the Conscript Act. . . . From this time till the end of the war, a soldier was simply a machine, a conscript. It was mighty rough on rebels. We cursed the war, we cursed Bragg, we cursed the Southern Confederacy. All our pride and valor had gone, and we were sick of war and the Southern Confederacy.

A law was made by the Confederate States Congress about this time allowing every person who owned twenty negroes to go home. It gave us the blues; we wanted twenty negroes. . . . The glory of the war, the glory of the South, the glory and pride of our volunteers had no charms for the conscript.

Why Did the Defeated Citizens of New Orleans Call Benjamin Butler a "Beast"?

"Success has a hundred fathers and failure is an orphan" goes an old saying. And the capture—or fall—of New Orleans in April 1862 may prove the maxim. The largest city in the Confederacy, its wealthiest and most active port, the home of many Europeans sympathetic to the Confederacy who felt comfortable in the city's Continental atmosphere, New Orleans was crucial to the Confederate cause. For the Union, taking the port meant control of the lower Mississippi. Just as numerous Union Navy and War Department officials would later take credit for hatching the plan to attack New Orleans, many Confederate officers and politicians would point accusing fingers long after the city was lost.

But the credit for carrying out the Union attack belonged to one man, sixty-year-old Flag Officer David Glasgow Farragut (1801–1870), one of the Civil War's indomitable spirits. Born in Tennessee, young Farragut had been taken in as a foster child by Commodore David Porter (1780–1843). At the age of nine he became a midshipman, serving under his adoptive father. Barely a teenager in the War of 1812, Farragut was captured while fighting the British. He later served in the Mediterranean and in the Mexican War. Having left Norfolk, Virginia, when secession was announced, Farragut initially balked at taking a command that might force him to attack Norfolk, but he accepted command of the fleet attacking New Orleans.

One of the greatest naval battles in North American history commenced early on the morning of April 24, 1862, when Farragut led his seventeen ships past a waiting Confederate fleet. The Union navy also faced numerous obstructions, including burning boats and an immense chain placed across the river as well as the two fortresses overlooking the mouth of the Mississippi. For three days these forts had been pounded by mortars on flatboats, the brainchild of David Dixon Porter (1813–1891), Farragut's adoptive brother. A major contributor to the Union's naval success, Porter was one of those men who would aggressively try to take credit for everything that worked for the Union at New Orleans. While Navy Secretary Gideon Welles—nicknamed "Old Father Neptune"—respected his ideas and energy, he was suspicious of his overwhelming need for glory. Of Porter, Welles once recorded, "I did not always consider David to be depended upon if he had an end to attain, and he had no hesitation in trampling down a brother officer if it would benefit himself."

As Farragut's armada advanced, the crash of hundreds of cannon and mortars could be heard sixty miles away. When it was over, Farragut had run the gantlet of the forts and Confederate fleet with the loss of only a single ship, 37 men killed, and another 147 wounded— a small fraction of the numbers who died in the great land battles of the war. On the morning of April 25, Farragut's warships steamed into New Orleans and aimed their guns at the defenseless city. Twenty-nine thousand bales of cotton, along with rice, corn, tobacco, and sugar, had been torched to prevent their capture. A substantial number of supplies had already been taken by looters as New Orleans, in a panic, fell into near chaos. In the nearby shipyards, miles of cordwood stacked for steamboat fuel were also set afire. As he sailed into New Orleans, Farragut saw "desolation, ships, steamers, cotton, coal, etc., were all in one common blaze. . . . The *Mississippi*, which was to be the terror of the seas, and no doubt would have been to a great extent . . . soon came floating by us, all in flames, and passed down the river." The uncompleted *Mississippi*, the Confederacy's great hope of destroying Union blockade vessels, was also scuttled to prevent its capture.

Historians have debated whether the loss of New Orleans might have been averted. Most requests for men and ships to defend the city had been rejected by Jefferson Davis and his Cabinet. In one of his worst military miscalculations, Davis believed the two forts guarding the mouth of the river could stop any Union fleet. Instead of confronting Farragut, available ships were ordered upriver to stop the Union gunboats coming down the Mississippi from the north. Confederate troops and guns had been stripped from New Orleans's defenses to be sent to other armies. And two Confederate ironclad battleships under construction, more powerful than anything then on the seas, might have created havoc for the Union fleet, but they were delayed by shortages and incompetence. Everything conspired against a Confederate success. But the great injury of the loss of New Orleans would soon be compounded by insult.

Documents of the Civil War

Major General Benjamin Franklin Butler's General Orders No. 28
(May 15, 1862).

> As the Officers and Soldiers of the United States have been
> subject to repeated insults from the women calling them-
> selves ladies of New Orleans, in return for the most scru-
> pulous non-interference and courtesy on our part, it is
> ordered that hereafter when any Female shall, by word, ges-
> ture, or movement, insult or show contempt for any officer
> of the United States, she shall be regarded and held liable
> to be treated as a woman of the town plying her avocation.

When the contents of a chamber pot was dumped on the head of
Flag Officer Farragut, Benjamin Butler, the commander of the Union
troops occupying New Orleans, was not amused. For Butler, it was the
final indignity that his troops would have to take from the defiant ci-
vilians. Having led the first troops into Baltimore after Fort Sumter,
Butler had made no friends when he clamped martial law on that city.
He became even more reviled when, faced with the disdain of New
Orleans, he issued his notorious order that the women of the city be
treated like prostitutes if they showed any sign of contempt toward the
Union officers. The Confederate press christened him "Beast" Butler.

Another Confederate response came from Louisiana's own General
P.G.T. Beauregard, who had Butler's order read to his troops, then
said, "Men of the South! shall our mothers, our wives, our daughters
and our sisters, be thus outraged by the ruffianly soldiers of the
North, to whom is given the right to treat, at their pleasure, the ladies
of the South as common harlots? Arouse friends, and drive back from
our soil, those infamous invaders of our homes and disturbers of our
family ties."

Butler's portrait was later found adorning the inside of chamber
pots used throughout the South long after the war. Butler was given
a second nickname when the rumors spread that he and his troops

were looting the grand homes of New Orleans, stealing the silver. Soon political cartoonists called him "Spoons" Butler and depicted him carrying off Confederate silverware.

Far more significant than blackening the name of "Beast" Butler, however, the fall of New Orleans was a devastating military and economic loss to the Confederacy.

Civil War Voices

From the diary of Mary Chesnut: "New Orleans gone—and with it the Confederacy. Are we not cut in two? The Mississippi ruins us if it is lost.... I have nothing to chronicle but disasters.... The reality is hideous."

The famed lady diarist was succinct and accurate. The loss of New Orleans devastated Confederate hopes, for it meant Union control of the Mississippi and the loss of one of the Confederacy's chief ports. The Confederacy was now cut off from the Texas beef and Louisiana salt that its armies desperately needed. At a relatively small cost in terms of casualties—particularly when set against the huge losses at Shiloh and in other battles that lay ahead—the capture of New Orleans was also a severe blow to any hope for recognition by either Great Britain or France, for both had come to see the Confederacy as a losing cause. When news of New Orleans reached London, Charles Francis Adams, the American ambassador to Great Britain, literally danced across the floor.

MILESTONES IN THE CIVIL WAR: 1862

January 11 Edwin Stanton replaces Simon Cameron as war secretary. Under Cameron, a Pennsylvania politician with an unsavory reputation, the War Department has been riddled by corruption and mismanagement.

January 27 In General War Order No. 1, Lincoln calls for a Union offensive, but it is ignored by McClellan.

February 13 The West Virginia Constitutional Convention decides that "no slave or free person of color should come into the state for permanent residence."

February 16 *Fort Donelson* (*Tennessee*) surrenders to the Union forces under Grant.

February 25 Union troops take *Nashville, Tennessee,* without a struggle, and the city remains in Union hands for the rest of the war. The loss is a blow to Confederate morale and means the loss of tons of Confederate supplies stockpiled there.

March 9 *Battle of the* Monitor *and the* Merrimac, *Hampton Roads, Virginia.* The first battle between two ironclad ships ends inconclusively. The *Merrimac,* a captured Union ship which has been refitted by the Confederates and rechristened the *Virginia,* is later blown up to prevent its capture.

March 11 Because of McClellan's failure to act, Lincoln demotes him from general-in-chief to command only the Army of the Potomac, allowing him to concentrate on the Richmond campaign. All the Union commanders now report directly to the secretary of war.

April 4 *The Peninsular Campaign.* In its first major offensive, the Army of the Potomac advances toward *Yorktown, Virginia,* on the peninsula between the James and York rivers. It will take *Yorktown* on *May 4* and *Williamsburg* on *May 5.*

April 6–7 *Battle of Shiloh.* At *Pittsburg Landing, Tennessee,* two days of furious fighting and tremendous casualties end when the Union forces under Grant force the Confederate army under Beauregard to withdraw.

April 7 Island Number Ten, a Confederate fortress on the Mississippi, falls to combined Union land and naval forces. The Mississippi is in Union hands all the way to *Memphis.*

April 10 Lincoln signs a congressional resolution calling for gradual, compensated emancipation. Slavery is also abolished in *Washington, D.C.*

April 16 President Davis signs the Conscription Act, the first draft in American history.

April 25 *New Orleans* surrenders to Union Flag Officer David Farragut.

May 14 Despite his successes and overwhelming superiority, McClellan halts the Union army's advance six miles from *Richmond*.

May 9 General David Hunter, who had organized a black regiment without official approval and was nicknamed "Black David," issues a proclamation freeing the slaves of rebels in *Georgia, Florida,* and *South Carolina* and arming able-bodied blacks in those states. Lincoln revokes this order.

May 12 *Natchez, Mississippi,* surrenders to Farragut, who is pushing up the Mississippi.

May 20 Lincoln signs the Homestead Act, guaranteeing 160 acres of land to anyone who will settle and improve it for five years.

June 1 Robert E. Lee replaces the wounded Joseph Johnston.

June 6 *Memphis* falls to Union forces.

June 25–July 2 *Seven Days' Battle.* Under Lee's command, the Confederates force McClellan's Army of the Potomac to retreat, ending the Union threat to *Richmond* and the possibility of an early end to the war. But the casualties are high again. Union losses are about 16,000 dead and wounded; the Confederates lose some 20,000 dead and wounded, a much higher percentage of their forces.

July Congress authorizes the acceptance of blacks into military service and passes a second Confiscation Act, freeing the slaves of all rebels.

July 11 Looking for an effective commander, Lincoln names Henry W. "Old Brains" Halleck general-in-chief.

August 28–30 *2nd Bull Run* (*2nd Manassas*). The Union forces under General Pope are defeated by the Confederates for a second time at this spot, a few miles south of *Washington*. McClellan is brought back to replace Pope, who is sent west to quell an Indian uprising in *Minnesota*.

September 17 *Antietam.* The Union forces under McClellan meet Lee's army in *Maryland* in the bloodiest single day of the war and American history. There are combined casualties of 23,110 dead, wounded, or missing. Although technically a draw, the battle forces Lee to abandon a general invasion of the North.

September 23 The preliminary text of the Emancipation Proclamation is published. It will take effect on *January 1, 1863*.

September 27 Composed of free blacks from *New Orleans*, the First Regiment Louisiana Native Guards is formed, the first officially recognized black regiment.

October 8 *Battle of Perryville (Kentucky)*. A second Confederate invasion of the North under Braxton Bragg is checked by Union General Don Carlos Buell.

November 5 After McClellan's failure to pursue Lee after *Antietam*, Lincoln dismisses McClellan for a second time, replacing him with Ambrose E. Burnside. McClellan returns to New Jersey and does not command again, but he will run against Lincoln in 1864.

December 13 *Battle of Fredericksburg, Virginia*. Under Burnside, the Union army is routed with large casualties.

December 31 *Battle of Murfreesboro or Stones River (Tennessee)* begins. It will be fought until *January 2, 1863*. Inconclusive, it produces combined casualties of nearly 25,000.

Documents of the Civil War

General David Hunter's order freeing the slaves in his military department (May 9, 1862).

General Orders No. 11. The three states of Georgia, Florida, and South Carolina, comprising the Military Department of the South, having deliberately declared themselves no longer under the protection of the United States of America, and having taken up arms against the said United States, it becomes a military necessity to declare them under martial law. This was accordingly done on the 25th day of April, 1862. Slavery and martial law in a free country are altogether incompatible; the persons in these three states—Georgia, Florida, and South Carolina—heretofore held as slaves are therefore declared forever free.

Lincoln still wanted this to be a war for the Union, not emancipation. Any action that changed the Union's goals might cause the secession of the border states and the loss of politically shaky congressmen who had already declared that the war was being fought to preserve the Union. Ten days after Hunter's order was issued, Lincoln overruled Hunter on the grounds that he lacked the authority to issue such an order. The exercise of such power, declared Lincoln, "I reserve to myself."

Why Did Lincoln Fire General McClellan?

It took God seven days to create the earth. It took Robert E. Lee that long to force George McClellan out of Virginia. With his army, moreover, went the Union's hopes for an early victory.

The Union's successes in the West in the early months of 1862—the fall of Forts Henry and Donelson, the costly battle at Shiloh and the victory at Island No. Ten, the fall of New Orleans, Natchez, Mississippi, and Memphis, and Beauregard's withdrawal in June from Corinth, Mississippi—seemed to be astonishingly good news. War Secretary Stanton was so optimistic, he ordered recruitment offices closed. It seemed that the very backbone of the Confederacy might be broken with a victory in Virginia, where McClellan was only six miles from Richmond with a large, well-equipped army.

But McClellan had little of the strength of character and single-mindedness that men like Grant and Farragut brought to the war. The Confederate leaders seemed to have his number. At Yorktown, McClellan was tricked by General John Magruder, who simply had his men march around in circles, letting the Union outposts see this seemingly endless line of troops. Lincoln's private secretary, John Hay, summed up the feelings about McClellan when he wrote, "The little Napoleon sits trembling before the handful of men at Yorktown afraid either to fight or run. Stanton feels devilish about it. He would like to remove him if he thought it would do." Stanton, a War Democrat and previously a supporter of McClellan's, was also growing impatient. When McClellan asked for 40,000 additional men, he stormed, "If he had a million men he would swear the enemy has two millions, and then he would sit down in the mud and yell for three." Those words were echoed by Treasury Secretary Salmon Chase, also a disillusioned McClellan supporter, who wrote to Horace Greeley, "McClellan is a clear luxury—fifty days—fifty miles—fifty millions of dollars—easy arithmetic but not satisfactory. If one could have some faith in his competency in battle—should his army ever fight one— . . . it would be a comfort." When decisiveness and a willingness to commit to all-out war might have ended the Civil War in the summer of 1862, McClellan chose caution. It was a terribly expensive decision in more ways than Salmon Chase was calculating.

While McClellan played the Union's Hamlet, fitfully wondering whether "to fight or not to fight," three Confederate legends were about to be made through the audacious leadership McClellan so sorely lacked. In the spring and summer of 1862 the brilliance of Robert E. Lee, Thomas "Stonewall" Jackson, and Jeb Stuart would dazzle and embarrass the vastly larger Union forces in Virginia. Lee had served as an adviser to President Davis, but when Joseph Johnston was wounded on May 31 at the Battle of Fair Oaks (also called Seven Pines), Lee was put in charge of the army defending Richmond, which he renamed the Army of Northern Virginia.

Before Lee took that command, he had already made one important decision as Davis's adviser. Attempting to keep the Union troops from reinforcing McClellan in the assault on Richmond, Lee dispatched Stonewall Jackson into the Shenandoah Valley, where, over the course of a few months, he made military history.

The pious, Bible-thumping Jackson, who believed his troops were "an army of the living God," embarked on a two-month campaign that mystified the Union commanders. Vastly outnumbered, Jackson marched his 17,000 troops around the Shenandoah Valley in the spring of 1862 like a man possessed. Enduring forced marches of twenty-five miles, his men moved more quickly than any army should have been able to, earning them the nickname "Jackson's foot cavalry." In a month they marched more than four hundred miles, threw panic into the Union, and stole many tons of goods from

Union depots. Food and supplies were taken from Union General Nathaniel Banks in such quantity that the Confederates christened him "Commissary" Banks. The Confederate press soon lionized Jackson. One paper said he was "the idol of the people and is the object of greater enthusiasm than any other military chieftain of our day." Another Southerner wrote, "Hurrah for old Stonewall!!!! . . . What can save their army from annihilation. Isn't it delightful to think about?"

But this romantic vision of a somewhat eccentric Jackson and the glorious exploits of his troops masked another image, that of a relentless Confederate counterpart to John Brown, whose brave demeanor at his hanging Jackson had witnessed and admired. Like Grant and Sherman, the Union commanders who would be vilified for their devastating approach to the conflict, Jackson believed in total, aggressive war. In *The Destructive War*, Charles Royster summed up Jackson:

> He ordered hard marches; he denied applications for furloughs; he severely punished infractions of discipline; he arrested officers who departed from his instructions; he ordered soldiers absent without leave to be brought back into the army in irons; he had deserters shot—during three days in August 1862, thirteen of his men were executed; he tried to kill as many of the enemy as possible, and he did not shrink from getting his own men killed doing it. Jackson did not go through the Civil War's often-described transition from notions of chivalric gallantry to brutal attrition. For him the war was always earnest, massed, and lethal.

Lee's plan worked perfectly; the threat Jackson posed to Washington, D.C., meant that Lincoln desired more troops for the capital's defense. Instead of assisting McClellan at Richmond, General McDowell's 40,000 men were pulled back to defend Washington. Here again McClellan would find reason to claim that he was being sabotaged. He was still under the misapprehension caused by Pinkerton's reports of the Confederate troops' numbering as high as 150,000; Johnston, later Lee, actually had about 50,000 men defending Richmond. McClellan had more than 150,000.

During this time, the heroics of Jackson's "foot cavalry" would soon be complemented by the antics of twenty-nine-year-old Jeb Stuart (1833–1864). Flamboyant in his ostrich-plumed hat, red-lined cap, and gold spurs, Stuart commanded with a style that was equally flamboyant. He kept a group of black minstrels and a banjo player on his headquarters staff. He wrote dispatches for the London newspapers that made him an international celebrity. But it was his military daring that thoroughly embarrassed the Union commanders, especially his notorious ride around McClellan in June 1862. For four days, June 12–15, Stuart led 1,200 cavalry troopers—including another young officer named John S. Mosby and writer John Esten Cooke, who would provide much of the literary embellishment that made Stuart's ride so remarkable—on a reconnaissance that completely encircled the Union positions in Virginia. While providing Lee with valuable information about the federal troops and their positions, Stuart's ride was also a propaganda coup. Confederate newspaper headlines boasted of his "Magnificent Achievement" and "Unparalleled Maneuver," boosting the sagging Confederate morale and once again embarrassing McClellan. However, it also alerted McClellan to the weakness of his position.

A few days after Stuart's return, the so-called Seven Days' Battle began in earnest on June 26 with the Battle of Mechanicsville (also called Beaver Dam Creek or Ellerson's Mill). In a pattern that would repeat itself, the Union army technically won in Lee's first battle as commander of the Army of Northern Virginia. Nevertheless, McClellan ordered a retreat to a better defensive position, asked for more troops, and blamed Lincoln for his defeat. The week's worth of fighting around Richmond concluded at Malvern Hill, where the Confederates again suffered greater losses, but still succeeded in pushing McClellan back from the gates of Richmond. He was ultimately forced to withdraw from the peninsula, and in August the

Army of the Potomac was loaded onto steamers for the depressing ride back to Washington.

Lincoln decided he had seen enough of McClellan. General John Pope, an unbearably cocky soldier who had taken credit for the fall of Island Number Ten and brought an unduly optimistic spirit with him, was given command of all Union troops north and west of Richmond. (Henry Halleck, made general-in-chief of the U.S. Army, would stay in Washington.) Pope made few friends, either among his own troops or with his adversaries. First he disparaged the men of his new command when he arrived:

> I have come to you from the West, where we have always seen the backs of our enemies; from an army whose business it has been to seek the adversary and to beat him when he was found. . . . Dismiss from your minds certain phrases, which I am sorry to find much in vogue amongst you. I hear constantly of "taking strong positions and holding them," of "lines of retreat" and "bases of supplies." Let us discard such ideas. . . . Let us look before us and not behind. Success and glory are in the advance, disaster and shame lurk in the rear.

Then Pope raised Confederate hackles by enforcing the harsh treatment of Virginians. He ordered food taken from farms and threatened to hang captured Confederate officers and civilian sympathizers as traitors. Lee would take particular pleasure in dealing with Pope. And later that summer Pope would get a taste of McClellan's humiliation by Jeb Stuart when Stuart's cavalry rode into Pope's camp and stole his hat, uniform, and dress cloak along with $35,000 in cash and notes regarding Union troop positions—information that would come in handy for Robert E. Lee.

Civil War Voices

Katharine Wormley, a U.S. Sanitary Commission volunteer serving on a Union transport ship used to move wounded from the peninsula battles to hospitals (June 1862).

We went on board; and such a scene as we entered and lived for two days I trust never to see again. Men in every condition of horror, shattered and shrieking, were being brought in on stretchers borne by "contrabands," who dumped them anywhere, banged the stretchers against pillars and posts, and walked over the men without compassion. There was no one to direct what ward or what bed they were to go into. Men shattered in the thigh and even cases of amputation, were shoveled into top berths without thought or mercy. The men had mostly been without food for three days, but there was nothing on board for them; and if there had been, the cooks were only engaged to cook for the ship, and not for the hospital.

We began to do what we could. The first thing wanted by wounded men is something to drink (with the sick, stimulants are the first thing). Fortunately we had plenty of lemons, ice, and sherry on board, and these were available at once. Dr. Ware discovered a barrel of molasses, which, with vinegar, ice and water made a most refreshing drink. After that we gave them crackers and milk. . . . It was hopeless to try and get them into bed; indeed, there were no mattresses. All we could do at first was to try and calm the confusion, to stop some agony, to revive the fainting lives, to snatch, if possible, from immediate death with food and stimulants. Imagine a great river or sound steamer filled on every deck—every berth and every square inch of room covered with wounded men; even the stairs and gangways and guards filled with those who are less badly wounded; and then imagine fifty well men, on every kind of errand, rushing to and fro over them, every touch bringing agony to the poor fellows, while stretcher after stretcher came along, hoping to find an empty place; and then imagine what it was to keep calm ourselves, and make sure that every man on both boats was properly refreshed and fed. We got through about one A.M.

When Lincoln called for troops to put down the rebellion, there was an immediate outpouring of food, clothing, medical supplies, and money from individual citizens. In April 1861 the Reverend Henry Bellows of New York organized a number of women's aid organizations into the Women's Central Association of Relief. But seeing an urgent need for coordinating these volunteer groups, Bellows proposed a commission of civilians, medical men, and military officers to oversee the Union's aid activities. The administration named a Commission of Inquiry and Advice in Respect of the Sanitary Interests of the United States Forces, which later became known as the U.S. Sanitary Commission, the forerunner of the American Red Cross.

Some of the country's most accomplished citizens joined the Sanitary Commission, including Elisha Harris, the sanitarian in charge of the Staten Island quarantine station; abolitionist Samuel Howe, the husband of the author of "The Battle Hymn of the Republic"; and Frederick Law Olmsted, the chief architect and superintendent of New York's Central Park. Mary Livermore, the wife of a Chicago minister, left her family in a housekeeper's care and went to work for the commission, eventually organizing three thousand chapters in the Midwest. Mary Ann Bickerdyke, a commission agent, went farther, joining the Union troops on the battlefield and assisting them for four years. A trained nurse, "Mother Bickerdyke," as she came to be known, won the admiration of William Sherman, becoming the only woman he allowed in his camps.

Mother Bickerdyke was one of thousands of Union nurses. They were officially led by Dorothea Lynde Dix (1802–1887), a professional nurse who had fought to reform the treatment of prisoners and the mentally ill and who was appointed head of the nursing corps by

Simon Cameron. In temperament Bickerdyke's opposite, she was called "Dragon Dix." Fighting the Victorian notion that women did not belong with men in hospitals, Dix had to ensure the moral rectitude of her nurses. She accepted only women over thirty and stated unequivocally that "all nurses are required to be very plain-looking women. Their dresses must be brown or black, with no bows, no curls, no jewelry and no hoopskirts."

Another volunteer was Louisa May Alcott, not yet famous as the author of *Little Women*. In the fall of 1862 she applied to serve as a nurse, although her only experience had been nursing her dying sister and reading *Notes on Nursing*, Florence Nightingale's landmark 1860 book about her experiences in the Crimean War. Working twelve-hour shifts amid the pain and screams of an army hospital, Alcott bore up as many other women did. She later contracted typhoid herself and returned to Massachusetts, where she collected letters about her experience in *Hospital Sketches*, her first published book.

The entry of women into the nursing profession as well as the opening of certain federal clerk's jobs was a foot in the door for many women. While much has been made of the women who worked in the factories during World War II—the "Rosie the Riveters"—the first openings actually came in the Civil War, when thousands of women were forced to fend for themselves when the men went off to war. It was the beginning of a sea change in how American women saw themselves; many who had committed their energy to abolition would channel their protest into the suffrage movement after the war.

Among those women who clerked in the Patent Office was Clara Barton (1821–1912). Even though she had not been trained as a nurse, she left the Patent Office to become a volunteer nurse (not affiliated with Dix's nursing corps). While she feared that people would view her as a "camp follower," Barton was undeterred and took wagonloads of supplies to the battlefields. During the fighting at Antietam (see below), she took over from a surgeon who was killed as he drank from a cup she had handed him. Although she would cross swords in a rivalry with Dix, Barton emerged from the Civil

War as the "Angel of the Battlefield" and went on to found the American Red Cross.

What Was "The Prayer of Twenty Millions"?

President Lincoln was walking a tightrope. The crisis precipitated by General Hunter's order freeing the slaves demonstrated the precarious balancing act he was attempting to perform. Focused solely on winning the war, Lincoln had to contend with his powerful abolitionist supporters, who wanted him to make it clear that the war was being fought to end slavery. Knowing that emancipation was not why most men fought for the Union and fearful of losing border state support, Lincoln was reluctant to turn the war into a struggle for the freedom of the slaves. As long as he maintained preservation of the Union and an end to the rebellion as the cause, he could control opinion in Congress and the country.

But the voices calling for emancipation were growing louder. Before the war, the influential newspaper editor Horace Greeley (1811–1872), editor of the *New York Tribune,* had advocated allowing the South to secede. But once the war began, he was a staunch Unionist who joined the Radical Republicans impatient with Lincoln's delaying of emancipation.

One of the most influential journalists of his century, Greeley was born in New Hampshire and apprenticed to a Vermont printer at the age of fourteen. At twenty he moved to New York and became the editor of several papers, gaining a reputation as a political writer and allying himself with two powerful Whig leaders, William Seward (then governor of New York) and Thurlow Weed, another prominent New York politician and journalist. In 1841 Greeley founded the *New York Tribune* as an alternative to the sensational tabloids of the day. The paper was successful and influential—Greeley would remain its editor for thirty-one years—espousing the issues of social fairness and decrying the monopolies and land being given to speculators and the wealthy. (For several years, Greeley employed Karl Marx, who wrote *The Communist Manifesto,* as a European correspondent.) Greeley was also an enthusiastic supporter of migration to the West. But the quote most often associated with him—"Go west, young man," his advice to a young minister who had lost his voice—was coined by John Soule, an Indiana newsman who wrote the words in 1851. Greeley repeated the phrase in print and became so identified with it that he later reprinted Soule's article to acknowledge the proper credit.

While opposed to slavery, Greeley had never been an outright abolitionist, preferring a moderate course like the one adopted by Lincoln. (He was influential in Lincoln's nomination.) Once the slave states seceded, however, he became passionate about the return of the Confederate states and was one of the first to voice the cry "On to Richmond." Once the war began, there was no room for moderation. To Greeley, it was a war for freedom. And by the summer of 1862 he publicly chided Lincoln over the faltering war effort. With the claim that he spoke for "twenty millions," Greeley wrote an open letter to Lincoln on August 19, 1862.

> To Abraham Lincoln, President of the United States:
>
> Dear Sir: I do not intrude to tell you—for you must know already—that a great proportion of those who triumphed in your election, and of all who desire the unqualified suppression of the rebellion now desolating our country, are sorely disappointed and deeply pained by the policy you seem to be pursuing with regard to the slaves of rebels. I write only to set succinctly and unmistakably before you what we require, what we think we have a right to expect, and of what we complain.
>
> I. We require of you, as the first servant of the Republic, charged especially and preeminently with this duty, that YOU EXECUTE THE LAWS. . . .
>
> II. We think you are strangely and disastrously remiss in the discharge of your official and imperative duty with regard to the emancipating provisions of the new Confiscation Act. Those provisions were designed to fight Slavery with Liberty. They prescribe that men loyal to the Union, and

willing to shed their blood in her behalf, shall no longer be held, with the nation's consent, in bondage to persistent, malignant traitors, who for twenty years have been plotting and for sixteen months have been fighting to divide and destroy our country. Why these traitors should be treated with tenderness by you, to the prejudice of the dearest rights of loyal men, we cannot conceive.

III. We think you are unduly influenced by the councils, the representations, the menaces, of certain fossil politicians hailing from the Border Slave States. Knowing well that the heartily, unconditionally loyal portion of the white citizens of those States do not expect nor desire that Slavery shall be upheld to the prejudice of the Union ... we ask you to consider that Slavery is everywhere the inciting cause and sustaining base of treason. . . .

IV. We think timid counsels in such a crisis calculated to prove perilous, and probably disastrous. It is the duty of a Government so wantonly, wickedly assailed by rebellion as ours has been, to oppose force in a defiant, dauntless spirit. It cannot afford to temporize with traitors, nor with semi-traitors. . . .

V. We complain that the Union cause has suffered, and is now suffering immensely, from mistaken deference to rebel Slavery. Had you, sir, in your Inaugural Address, unmistakably given notice that, in case the rebellion already commenced, were persisted in, and your efforts to preserve the Union and enforce the laws should be resisted by armed force, you would recognize no loyal person as rightfully held in Slavery by a traitor, we believe the rebellion would therein have received a staggering if not fatal blow. . . .

VI. We complain that the Confiscation Act which you approved is habitually disregarded by your generals, and that no word of rebuke for them from you has yet reached the public ear. Frémont's Proclamation and Hunter's Order favoring Emancipation were promptly annulled by you;

while Halleck's Number Three, forbidding fugitives from slavery to rebels to come within his lines—an order as unmilitary as inhuman, and which received the hearty approbation of every traitor in America—with scores of like tendency, have never provoked even your remonstrance. ... And finally, we complain that you, Mr. President, elected as a Republican, knowing well what an abomination Slavery is, and how emphatically it is the core and essence of this atrocious rebellion, seem never to interfere with these atrocities, and never give a direction to your military subordinates, which does not appear to have been conceived in the interest of Slavery rather than of Freedom. ...

VIII. On the face of this wide earth, Mr. President, there is not one disinterested, determined, intelligent champion of the Union cause who does not feel that all attempts to put down the rebellion and at the same time uphold its inciting cause are preposterous and futile—that the rebellion, if crushed out to-morrow, would be renewed within a year if Slavery were left in full vigor. ...

IX. I close as I began with the statement that what an immense majority of the loyal millions of your countrymen require of you is a frank, declared, unqualified, ungrudging execution of the laws of the land, more especially of the Confiscation Act. ... We cannot conquer ten millions of people united in solid phalanx against us, powerfully aided by Northern sympathizers and European allies. We must have scouts, guides, spies, cooks, teamsters, diggers, and choppers from the blacks of the South, whether we allow them to fight for us or not, or we shall be baffled and repelled. As one of the millions who would gladly have avoided the struggle at any sacrifice but that of principle and honor, but who now feel that the triumph of the Union is indispensable not only to the existence of our country but to the well-being of mankind, I entreat you to render a hearty and unequivocal obedience to the law of the land.

Yours,

Horace Greeley.

Still looking to reassure the majority of those in the Union who didn't want to fight for abolition, Lincoln responded to Greeley on August 22.

The sooner the national authority can be restored, the nearer the Union will be the Union it was.

If there be those who would not save the Union unless they could at the same time save slavery, I do not agree with them.

If there be those who would not save the Union unless they could at the same time destroy slavery, I do not agree with them.

My paramount object is to save the Union and not either to save or destroy slavery.

If I could save the Union without freeing any slave, I would do it—If I could save it by freeing all the slaves, I would do it—and if I could do it by freeing some and leaving others alone, I would also do that.

What I do about slavery and the colored race I do because I believe it helps to save this Union; and what I forbear, I forbear because I do not believe it would help to save the Union. . . .

. . . I have stated my purpose according to my views of official duty; and I intend no modification of my oft-expressed personal wish that all men everywhere could be free.

Civil War Voices

Robert E. Lee, to the people of Maryland, encouraging them to secede (September 8, 1862).

The government of your chief city has been usurped by armed strangers—your Legislature has been dissolved and by the unlawful arrest of its members—freedom of the press and of speech has been suppressed—words have been declared offenses by an arbitrary decree of the Federal Executive—and citizens ordered to be tried by military commissions for what they may dare to speak.

Believing that the people of Maryland possess a spirit too lofty to submit to such a government, the people of the South have long wished to aid you in throwing off this foreign yoke, to enable you again to enjoy the inalienable rights of freemen, and restore the independence and sovereignty of your state.

In obedience to this wish our army has come among you, and is prepared to assist you with the power of its arms in regaining the rights of which you have been so unjustly despoiled.

This is our mission, so far as you are concerned. No restraint upon your free will is intended—no intimidation will be allowed within the limits of this army at least.

Marylanders shall once more enjoy their ancient freedom of thought and speech. We know no enemies among you, and will protect all of you in every opinion.

It is for you to decide your destiny, freely, and without constraint.

Documents of the Civil War

From Robert E. Lee's "Lost Order" (September 9, 1862).

The army will resume its march tomorrow, taking the Hagerstown Road. General Jackson's command will form the advance, and after passing Middletown, with such portions as he may select, take the route toward Sharpsburg, cross the Potomac at the most convenient point, and by Friday night take possession of the Baltimore and Ohio Railroad, capture such of the enemy as may be at Martinsburg, and intercept such as may attempt to escape from Harpers Ferry.

General Longstreet's command will pursue the same road as far as Boonsboro, where it will halt with the reserve, supply, and baggage trains of the army. . . .

The commands of General Jackson, McLaws, and Walker, after accomplishing the objects for which they have been detached, will join the main body of the army at Boonsboro or Hagerstown.

Each regiment on the march will habitually carry its axes in the regimental ordnance wagons, for use of the men at their encampments, to procure wood, etc.

By command of General R. E. Lee

What Happened at Antietam?

When John Pope arrived in Virginia, he warned his men about "disaster and shame." It was a pretty good description of what he had to live with after Lee, Jackson, and Longstreet finished with him at the Manassas battlefield in late August. Cut off from communications with Washington, unaware of his enemy's location, and expecting help from McClellan that would never come, Pope suffered a devastating defeat at the hands of Lee's army in the Second Battle of Bull Run. Only this time the casualties were five times as high as they had been the first time the two armies met. With Pope falling from grace, McClellan's star began to rise again, and in the aftermath of the defeat, McClellan was given command of the two armies, to merge them again into a single Army of the Potomac.

The vainglorious McClellan wrote to his wife: "Again I have been called upon to save the country. . . . I still hope for success, and I will leave nothing undone to gain it. . . . It makes my heart bleed to see the poor, shattered remnants of my noble Army of the Potomac . . . and to see how they love me even now. I hear them calling out to me as I ride among them, 'George don't leave us again.' "

By early September, Lee was on the move north to the Potomac, on Maryland's southwestern border. His army numbered about 55,000 men. Many were shoeless, short on ammunition, and exhausted from a rigorous march. McClellan's forces approached 90,000 men. Again, however, McClellan would be convinced by scouts that he was outnumbered. He thought Lee had more than 100,000 men.

A strange occurrence changed events. Special Orders No. 191, Lee's master plan for opening an invasion of the North, was sent to his generals. Stonewall Jackson copied the orders he received and sent them on to Harvey Hill, his brother-in-law. Hill, in the meantime, had received his own set, so the additional copy was treated casually. On September 13, when the Union troops took over the campground vacated by Hill, Private W. B. Mitchell found the copy of Lee's orders wrapped around some cigars. Presented with the dispatch, McClellan said, "Here is a paper with which, if I cannot whip Bobby Lee, I will be willing to go home."

But even then caution took hold. Instead of pressing forward, McClellan waited overnight. In pursuit of the Confederates, he encountered heavy resistance at South Mountain. Although the Confederates were driven back, the delay allowed Lee to set up a defensive line at Sharpsburg, in western Maryland, behind a creek called Antietam. McClellan followed Lee and spent two days preparing an assault on the Confederate lines. Once again his caution allowed Lee to summon reinforcements and strengthen his defenses.

The Battle of Antietam took place on September 17, with McClellan's artillery preceding an infantry assault on Lee's lines at two sites that would come to be known as "the cornfield" and "the Dunkard Church." One Confederate participant was Brigade Commander John B. Gordon, who recalled that

> the artillery of both armies thundered. McClellan's compact columns of infantry fell upon the left of Lee's lines with the crushing weight of a landslide. . . . Pressed back . . . the Southern troops, enthused by Lee's presence, re-formed their lines, and, with a shout as piercing as the blast of a

thousand bugles, rushed in countercharge upon the exulting Federals [and] hurled them back in confusion. . . .

Again and again . . . by charges and counter-charges, this portion of the field was lost and recovered, until the green corn that grew upon it looked as if it had been struck by a storm of bloody hail. . . . From sheer exhaustion, both sides, like battered and bleeding athletes, seemed willing to rest.

By ten-thirty in the morning the scene had shifted to a new Union attack that struck several Confederate brigades stationed at a sunken road. As the blue-clad troops of the Union formed a column four lines deep, John Gordon, who would be struck by five separate Union bullets in the following action, again described the scene: "The brave Union commander, superbly mounted, placed himself in front, while his band in rear cheered them with martial music. . . . As we stood looking upon that brilliant pageant, I thought . . . 'What a pity to spoil with bullets such a scene of martial beauty!' But . . . Mars is not an aesthetic god."

Mars that day was also an insatiable god. The fighting at "the sunken road" was devastating for both sides, but still it went on. When some of the Union troops finally moved around the right side of the Confederate lines, they opened up a withering fire at a spot that would all too aptly be called "the Bloody Lane."

Civil War Voices

Union Staff Officer D. H. Strother, watching from Union headquarters across Antietam Creek.

As the smoke and dust disappeared, I was astonished to observe our troops moving along the front and passing over what appeared to be a long, heavy column of the enemy without paying it any attention whatever. I borrowed a glass from an officer, and discovered this to be actually a column of the enemy's dead and wounded lying along a hollow road—afterward known as Bloody Lane. Among the prostrate mass I could easily distinguish the movements of those endeavoring to crawl away from the ground, hands waving as if calling for assistance, and others struggling as if in the agonies of death.

I was standing beside General McClellan during the progress and conclusion of this attack. The studied calmness of his manner scarcely concealed the underlying excitement, and when it was over he exclaimed, "By George, this is a magnificent field, and if we win this fight it will cover all our errors and misfortunes forever."

Civil War Voices

Enlisted New Yorker David L. Thompson, cut off after a Union retreat.

There was nothing to do but lie there and await developments. Nearly all the men in the hollow were wounded, one man . . . frightfully so, his arm being cut short off. He lived a few minutes only. All were calling for water, of course, but none was to be had.

. . . We heard all through the war that the army "was eager to be led against the enemy." It must be so, for truthful correspondents said so, and editors confirmed it. But when you came to hunt for this particular itch, it was always the next regiment that had it.

The truth is, when bullets are whacking against tree trunks and solid shot are cracking skulls like eggshells, the consuming passion in the breast of the average man is to get out of the way. Between the physical fear of going forward and the moral fear of turning back, there is a predicament of exceptional awkwardness.

The Battle of Antietam (Sharpsburg) would become known as "the bloodiest day of the war." The Confederates suffered about eleven thousand casualties, the Union attackers more than twelve thousand. But the Confederate army could not afford such losses.

After the grisly fighting McClellan, who continued to believe in the inflated estimate of Lee's army, decided once more to play it safe. An opportunity for what might have been a decisive and finishing blow to the Army of Virginia and the hopes of the Confederacy was lost. Despite possessing superior numbers as well as Lee's plans, McClellan acted tentatively, allowing Lee to escape across the Potomac and back into Virginia.

Lee would later attribute the failure of his invasion to the Lost Orders. Private Mitchell, wounded at Antietam, was discharged after eight months in the hospital. He died three years later, leaving his family in poverty.

Why Did Lincoln Fire General McClellan Again?

Lincoln was at turns angry and gloomy over Lee's escape. But the victory at Antietam gave him the opportunity he had been looking for. Still torn over the political wisdom and constitutionality of emancipation, Lincoln had decided that freeing the slaves in the Confederacy—while exempting those in the border states—would help shorten the war. He just needed a military victory to provide the right moment to make such a controversial announcement. On September 23 the Union newspapers published a preliminary version of the Emancipation Proclamation. It would not take effect until the following January.

With the announcement of the victory at Antietam and eventual emancipation, Confederate hopes for European recognition were further dashed. Both England and France had outlawed slavery years earlier, and support for the Confederacy was politically impossible, particularly in England, now that slavery had emerged as the underlying issue of the war. It was also more apparent in Europe that the Union would inevitably emerge the victor.

Yet McClellan still did little to follow up on his costly victory. Exasperated at his "slows," Lincoln finally ordered McClellan to "cross the Potomac and give battle to the enemy, or drive him south." One time, responding to his claim that his horses were weary, Lincoln wrote, "I have just read your dispatch about sore-tongued and fatigued horses. Will you pardon me for asking what the horses of your army have done since the Battle of Antietam that fatigues anything?"

Though McClellan finally started to move the army out of Washington late in October, his pace was glacial. Disappointed and angry, Lincoln finally had had enough. On November 5 he dismissed McClellan as the head of the Army of the Potomac and replaced him with General Ambrose E. Burnside.

Now best remembered as the man whose muttonchop whiskers on the side of his face became known as "sideburns," Burnside may have been one of the most hapless Union commanders; however, unlike others who preceded and followed him, Burnside was under no illusions about his talents. A native of Liberty, Indiana, Ambrose Everett Burnside (1824–1881) was the son of a South Carolina slave-owner who had moved to Indiana after selling his slaves. A West Pointer (1847), he served garrison duty in Mexico and was wounded by the Apaches. He tried to design a new type of breech-loading rifle but, failing to win a government contract, went bankrupt. Turning his patent over to creditors, he went to work for McClellan at the Illinois Central Railroad. With the outbreak of war he volunteered, serving at Bull Run, but left when his three-month enrollment was up. He then returned with a commission and served under McClellan. A reluctant commander, Burnside accepted the offer from Lincoln after twice refusing it because he felt inadequate to the job. He was right; that he was an unfortunate choice quickly became apparent.

Burnside took his Army of the Potomac to Warrenton, Virginia, just north of the Rappahannock River, and divided it into three divisions commanded by Generals Sumner, Hooker, and Franklin. Lee's army lay about thirty miles away, south of the Rappahannock, divided into two corps: one commanded by Stonewall Jackson; the other by General James Longstreet. In mid-November, Burnside began the next Union "drive to Richmond" by sending 30,000 men under Sumner toward Fredericksburg, a town on the south bank of the Rappahannock. The Union army stopped and made camp on some hills north of the river. Longstreet moved quickly and took up a position in some hills south of the town. Reinforcements from both sides began to stream in. Soon the Army of Northern Virginia was facing the Army of the Potomac, and it was clear that Fredericksburg would be the scene of a major battle.

The two armies were on opposing heights separated by the town and the river. Burnside had more than 100,000 men to Lee's 78,000, but the Confederates occupied the stronger position. Lee took ample time to prepare a strong defense and he wanted nothing more than for the Union army to attack the positions he had been fortifying for weeks. To attack, Burnside would have to cross the river on pontoon bridges (which had been delayed in arriving), move through the town, then cross a wide plain.

The Union advance on Fredericksburg began with an artillery barrage that would allow the placement of the pontoon bridges that would be used for the river crossing. The cannon fire was devastating, and houses and chimneys soon tumbled.

Civil War Voices

Confederate officer Robert Stiles of the Richmond Howitzers, a Yale man, described the scene in Fredericksburg during the Union bombardment.

> I saw walking quietly and unconcernedly along the same street I was on, and approaching General Barksdale's headquarters from the opposite direction, a lone woman. She apparently found the projectiles which were screaming and exploding in the air, and striking and crashing through the houses, and tearing up the streets, very interesting....

[H]aving reached the house I rode around back of it to put my horse where he would at least be safer than in front. As I returned on foot to the front, the lady had gone up on the porch and was knocking at the door.

One of the staff came to hearken, and on seeing a lady, held up his hands, exclaiming in amazement, "What on earth, madam, are you doing here? Do go to some safe place, if you can find one!"

She smiled and said, with some little tartness, "Young gentleman, you seem to be a little excited. Won't you please say to General Barksdale that a lady at the door wishes to see him?"

The young man assured her General Barksdale could not possibly see her just now; but she persisted. "General Barksdale is a Southern gentleman, sir, and will not refuse to see a lady who has called upon him."

. . . The General did come to the door, but actually wringing his hands in excitement and annoyance. "For God's sake, madam, go and seek some place of safety. I'll send a member of my staff to help you find one."

She again smiled gently—while old Barksdale fumed and almost swore—and then she said quietly, "General Barksdale, my cow has just been killed in my stable by a shell. She is very fat, and I don't want the Yankees to get her. If you will send someone down to butcher her, you are welcome to the meat."

The Union cannons and three detachments of volunteers finally did their work, clearing the town of the sharpshooters who were preventing the completion of the bridges. The Union columns advanced into Fredericksburg.

Civil War Voices

Robert Stiles observed the withdrawal of the last Confederates from Fredericksburg.

The last detachment was under the command of Lane Brandon . . . my classmate at Yale. . . . Brandon captured a few prisoners and learned that the advance [Union] company was commanded by Abbot, who had been his chum at Harvard Law School when the war began. He lost his head completely. He refused to retire before Abbot. . . . The enemy finding the way now clear, were coming up the street, full company front, with flags flying and bands playing, while the great shells from the siege guns were bursting over their heads and dashing their hurtling fragments after our retreating skirmishers.

Buck [Denman] was behind the corner of a house, taking sight for a last shot. Just as his fingers trembled on the trigger, a little three-year-old fair-haired baby girl toddled out of an alley accompanied by a Newfoundland dog, and gave chase to a big shell that was rolling lazily along the pavement, she clapping her hands and the dog snapping and barking furiously at the shell.

Buck's hand dropped from the trigger. He dashed it across his eyes to dispel the mist and make sure he hadn't passed over the river and wasn't seeing his own baby girl in a vision. No—there is a baby, amid the hell of shot and shell; and here come the enemy. A moment, and he has grounded his gun, dashed into the storm, swept his great arm around the baby, gained cover again, and, baby clasped to his breast and musket trailed in his left hand, is trotting after the boys up to Marye's Heights.

As Confederate Colonel E. Porter Alexander reported to General Longstreet in describing the position of his First Corps' artillery on

Marye's Heights, overlooking the town, "A chicken could not live on that field when we open on it."

Alexander was describing six hundred yards of open field that stretched between his position and the town. The left side of Lee's line was anchored on Marye's Heights and on a 1,200-foot-long stone wall at the base of the heights. The retaining wall was built alongside the main road to Richmond, which had been cut away and sunken by years of use. Shoulder high, the stone wall was an ideal position from which to defend. Lafayette McLaws had stationed his Georgia brigade, commanded by General Thomas R. R. Cobb, in the sunken road. The Confederate soldiers were packed two ranks deep behind the wall and had a clear field of fire to their front.

Around noon on December 13, 1862, a brigade of blue-coated men filed out of Fredericksburg, formed their battle lines, and charged toward the stone wall. They were cut to pieces by the Confederate artillery and fell back before the Georgians behind the wall fired a single volley. Two more brigades charged in quick succession with the same result. McLaws ordered the South Carolina brigade to join Cobb's men behind the stone wall, making the line four ranks deep. Stepping back from the wall to reload and back up to fire, the rebel defenders lay down a rapid and continuous storm of bullets.

Civil War Voices

Union General Darius N. Couch.

[T]he whole plain was covered with men, prostrate and dropping, the live men running here and there, and in front closing upon each other, and the wounded coming back. The commands seemed to be mixed up. I had never before seen fighting like that—nothing approaching it in terrible uproar and destruction. There was no cheering on the part of the men, but a stubborn determination to obey orders and do their duty. I don't think there was much feeling of success. As they charged, the artillery fire would break their formation and they would get mixed. Then they would close up, go forward, receive the withering infantry fire and . . . fight as best they could. And then the next brigade coming up in succession would do its duty, and melt like snow coming down on warm ground.

Throughout the afternoon Burnside sent wave after wave of infantry to the slaughter. Fires set in the brush killed the wounded who could not escape. Even after the sun went down, the fighting continued in the dusk until night finally brought an end to the butchery. For the Confederate defenders on Marye's Heights it was, as one Union officer put it, "a day of such savage pleasure as seldom falls to the lot of soldiers, a day on which they saw their opponents doing just what they wished them to do."

Seven Union divisions had been spent in fourteen charges on the stone wall. No Union soldier ever reached the wall; few got within fifty yards. Burnside lost 7,000 men in the attack; the Confederate defenders lost only 1,200 men.

Finally, on the fifteenth of December, a truce was established to pick up the wounded of both armies. The Union troops withdrew across the river to their original camps. Fredericksburg was reoccupied by the Confederates. (The three-year-old child rescued by Buck Denman spent the battle with the Confederates on Marye's Heights. She was returned to her mother afterward.)

The final count at Fredericksburg revealed dreadful but uneven losses on both sides. There were some 5,000 casualties for the Confederates and 12,700 Union casualties, more than 6,000 of them dead. This defeat caused Lincoln to despair. The choice of Burnside, who ignored Lincoln's warning not to attack, had proved calamitous. Many Union soldiers and officers said that McClellan would have never assaulted such a well-defended position. But Burnside clearly didn't know when to pull back from disaster.

The Army of the Potomac, having suffered the worst defeat in the history of the American army, settled into winter quarters without much hope or confidence.

Documents of the Civil War

From General Burnside's report of the battle at Fredericksburg to Major General Henry Halleck (December 1862).

General: I have the honor to offer the following reasons for moving the army of the Potomac across the Rappahannock, sooner than was anticipated by the President, Secretary of War or yourself, and for crossing at a point different from the one indicated to you at our last meeting at the President's.

During my preparations for crossing at the place I had first selected, I discovered that the enemy had thrown a large portion of his force down the river and elsewhere, thus weakening his defenses in the front, and also thought I discovered that he did not anticipate the crossing of our whole force at Fredericksburg, and I hoped by rapidly throwing the whole command over at that place to separate, by vigorous attack, the forces of the enemy on the river below from the forces behind and on the crest in the rear of the town, in which case we could fight him with great advantage in our favor. . . .

To the brave officers and soldiers who accomplished the feat of thus recrossing the river in the face of the enemy, I owe everything.

For the failure in the attack I am responsible, as the extreme gallantry, courage, and endurance shown by them was never exceeded, and would have carried the points had it been possible.

To the families and friends of the dead I can only offer my heartfelt sympathies, but for the wounded I can offer my earnest prayers for their comfortable and final recovery.

Civil War Voices

John B. Jones, a clerk in the Confederate War Department in Richmond (December 1, 1863).

God speed the day of peace! Our patriotism is mainly in the army and among the ladies of the South. The avarice and cupidity of the men at home could only be excelled by the ravenous wolves; and most of our sufferings are fully deserved. Where a people will not have mercy on one another how can they expect mercy? They depreciate the Confederate notes by charging from $20 to $40 per bbl. for flour; $3.50 per bushel of meal, $2 per lb. for butter; $20 per cord for wood, etc. When we shall have peace let the extortionists be remembered! Let an indelible stigma be branded upon them.

A portion of the people look like vagabonds. We see men and women and children in the streets in dingy and dilapidated clothes; and some seem gaunt and pale with hunger—the speculators and thieving quartermasters and commissaries only looking sleek and comfortable. If this state of things continues a year or so longer, they will have their reward. There will be governmental bankruptcy, and all their gains will turn to dust and ashes!

chapter six

1863
"The Great Task Remaining"

The winter was severe, the snow deep. The soldiers were discouraged. They knew that they had fought bravely but there had been mismanagement and inefficient generalship. Homesickness set in and became a disease.

—CHARLES COFFIN, *BOSTON JOURNAL*
JANUARY 1863

Willing to fight for Uncle Sam but not for Uncle Sambo.

—UNION NEWSPAPER HEADLINE

It is rather for us to be here dedicated to the great task remaining before us—that from these honored dead we take increased devotion to that cause for which they gave the last full measure of devotion; that we here highly resolve that these dead shall not have died in vain; that this nation, under God, shall have a new birth of freedom; and that the government of the people, by the people, for the people, shall not perish from the earth.

—PRESIDENT ABRAHAM LINCOLN
NOVEMBER 1863

✳

* Whom Did the Emancipation Proclamation
  Emancipate?

* Why Did Lincoln Replace General Burnside with
  General Hooker?

* What Was the Richmond Bread Riot?

* What Happened at Chancellorsville?

* Why Did Lincoln Banish Clement Vallandigham?

* What Was So Important About Gettysburg?

* What Were the Draft Riots?

* Was the Film *Glory* True?

* What Was a Jayhawker?

* Was the Gettysburg Address Written on the Back of
  an Envelope?

* Who Was Johnny Clem?

It was the winter of their discontent. Abraham Lincoln told a friend, "If there is a worse place than hell, I am in it."

The year opened bleakly for the Union following the defeat at Fredericksburg. Burnside's demoralized Army of the Potomac and Lee's victorious but threadbare Army of Northern Virginia wintered in camps near Fredericksburg, separated only by the Rappahannock River. It may have been a far cry from the historic Valley Forge winter of an earlier American army, but the winter of 1863 was a hard one. Men nearly froze to death and had little to do but watch the funerals of those still dying from their wounds.

Corruption and mismanagement still kept the Union troops from getting the supplies they should have received. And what they did get was not inspiring. The common ration of "salt beef" (or "salt horse"), beef preserved in pickle brine, was inedible unless soaked in water before cooking. Standard-issue clothing was also a disgrace, giving rise to a new Americanism. *Shoddy* was originally the name of the material used to produce the Union uniforms, perhaps because men were "shod" with it. But it soon came to earn its more common definition. *Harper's Monthly* called the cloth "a villainous compound, the refuse stuff and sweepings of the shop, pounded, rolled, glued, and smoothed to the external form and gloss of cloth, but no more like the genuine article than the shadow is to the substance."

The suffering was worse on the Confederate side. General Fitzhugh Lee, Robert E. Lee's nephew, reported, "General Lee was surrounded by embarrassments. . . . The troops were scantily clothed, rations for men and animals meager. The shelters were poor, and through them broke the snows, rains, and winds. He could not strike his enemy, but must watch and be patient." (Like his uncle, a West Pointer who also fought in the Mexican War, Fitzhugh Lee rose to the rank of major general in the Confederate army. At West Point in the class of 1856, Fitzhugh came before the superintendent of the academy, his uncle Robert, for "behavior not becoming an officer and a gentleman." Uncle Robert very nearly expelled his nephew.)

In the west, supplies were causing problems of another sort for

Ulysses Grant. As is always the case, plenty of people were looking to make a buck off the ravages of war. Many Yankee merchants happily delivered scarce items to the blockaded Confederacy in exchange for cotton. The Confederate government simply looked the other way, knowing that this illicit trade was helping to keep their men from starving. But Grant lost patience with the speculators who roamed through his lines, looking for growers willing to sell cotton, along with unscrupulous peddlers and army suppliers. Late in December 1862 he issued General Order No. 11, which said, "The Jews, as a class violating every regulation of trade established by the Treasury Department, are hereby expelled from the department within twenty-four hours from receipt of this order."

Grant's order, which expressed a sentiment that was neither unusual nor unpopular in nineteenth-century America, was inaccurate as well as anti-Semitic. Most traders were not Jewish, but the Shylock stereotype of the Jew as shrewd and dishonest was widely accepted. The order did not sit well with Lincoln. On January 4 he ordered Grant to rescind his decree, saying it "proscribed a whole class, some of whom are fighting in our ranks."

Grant's bigoted order was typical of an era in which Jews, for the first time in American history, were targeted as a class. In an overwhelmingly Christian country, Jews were widely blamed for the country's political and economic woes. In this time of crisis, they emerged as the scapegoat for the nation's fears, despite the fact that thousands of Jews enlisted and served, primarily in the Union cause. When Congress authorized army and navy chaplains, Jews were excluded from service. In one semicomic incident, a member of the Jewish fraternal order B'nai B'rith was arrested by a Union detective who thought the organization was a secret group aiding the Confederacy. The Rothschilds of London, the prominent Jewish investment bankers, and their American agent, August Belmont, were accused of being war profiteers who aided the Confederacy.

In the Confederacy, Jews were doubly threatened. Already under suspicion simply because of their faith, they were viewed as disloyal to the Confederacy. Typical of attitudes toward Jews was that of a

Methodist minister who called Confederate Cabinet member Judah Benjamin "a little pilfering Jew . . . one of the tribe that murdered the savior." (Judah Benjamin, like banker August Belmont, was a nonpracticing Jew. Both men married Christians and raised their children as Christians.) As Leonard Dinnerstein documents in *Anti-Semitism in America*, "It was commonly assumed that Jewish merchants hoarded merchandise and sold goods at extortionist prices."

Grant's anti-Semitism was not Lincoln's only political headache. Back in Washington, rumors of intrigues within the administration flew fast and thick, throwing the Cabinet into a full-fledged crisis. The chief cause was a rivalry between Secretary of State Seward and Treasury's Salmon Chase, both plotting to replace Lincoln as president in 1864. Chase, who thought Seward had too much influence over Lincoln, got the other Cabinet members to call for Seward's resignation. Chase had issued the first "greenbacks" in 1862, and the first dollar bill carried his own portrait, in essence printing thousands of campaign posters. But in one of Lincoln's most adroit political maneuvers, the president got both men to offer their resignations and then declined to accept them. He now held them both over a political barrel. With this crisis defused, the chief event of the new year was the formal announcement of the Emancipation Proclamation.

Documents of the Civil War

January 1, 1863. From the Emancipation Proclamation (see Appendix VI for complete text).

That on the first day of January in the year of our Lord one thousand eight hundred and sixty-three, <u>all persons held as slaves</u> within any State, or designated part of a State, the people whereof shall then be in rebellion against the United States, <u>shall be then, thenceforth and forever free,</u> and the Executive Government of the United States, including the military and naval authorities thereof, will recognize and maintain the freedom of such persons, and will do no act or acts to repress such persons, or any of them, in any efforts they may make for their actual freedom.

Whom Did the Emancipation Proclamation Emancipate?

The White House was open for business on New Year's morning, for it was a traditional day of greeting. Hundreds of office seekers, soldiers, ordinary citizens, and politicians trooped through to shake the president's hand. Finally Secretary of State Seward entered with a document to be signed. Lincoln started, then halted, flexing his hand. He explained that all the handshaking had practically paralyzed his arm, adding, "If my name ever goes into history it will be for this act, and my whole soul is in it. If my hand trembles when I sign the Proclamation, all who examine the document hereafter will say, 'He hesitated.' "

One of the longest-running misconceptions of the Civil War was that the Emancipation Proclamation freed the slaves. It clearly did not.

Lincoln's proclamation, announced the previous July after Antietam and made official on the first day of 1863, freed only those slaves in the rebellious areas of the country—areas administered by the Confederate government, where the federal government had no control. There was a strict legal interpretation for this. Lincoln believed he did not have the constitutional power to abolish slavery, which only came with the ratification of the Thirteenth Amendment (see Appendix IX for the complete text), after Lincoln's assassination.

Lincoln had opposed slavery on moral grounds throughout his political career. During the early years of the war, he had carefully walked a fine line between the abolitionists and those fighting the war only to preserve the Union, for he was faced by a stark political reality. The border states that had remained in the Union were slave states, and Lincoln could not afford to threaten that loyalty by freeing their slaves and possibly pushing those states into the Confederacy. On every occasion that his generals had announced the emancipation of the slaves in their military districts, Lincoln had overruled them.

The president was able to issue the Emancipation Proclamation as a measure to help the war effort. For the Confederacy, slavery was an essential part of the war machine; it freed men to fight while keeping food and factory production going. Every slave who escaped or refused to work meant less production and fewer fighting men available to the Confederate armies. For Union governors, emancipated blacks would fill their enlistment quotas.

For staunch abolitionists, it was a day of jubilation—although many felt that the proclamation was a half measure because it failed to free all the slaves. And while many Americans cheered the decision, it was far from popular. Many in the Union, including thousands of soldiers, were fighting because of patriotism and loyalty. They did not think they were laying down their lives for the freedom of blacks. As one Massachusetts man wrote, "The Proclamation has in measure divided the north and the army. The army hate to think they are fighting for the negro. . . . There are officers, very few I am glad to say, who say they won't fight for the negro or under the Proclamation."

Overseas, the reaction was mixed as well. The London *Star* congratulated Lincoln for "a humanitarian act." The London *Spectator* denounced the proclamation as a sham; the pro-Confederate London *Times* described it as "a very sad document," dismissing it as "the wretched makeshift of a pettifogging lawyer." But Lincoln knew that once he took this step, any hopes the Confederate states had of foreign intervention on their behalf would be eliminated. There were massive demonstrations in English cities following the proclamation. For Great Britain, Emancipation made recognition of the Confederacy politically impossible.

Documents of the Civil War

Lincoln's own state passed a resolution, "The Emancipation Proclamation Denounced" (January 7, 1863).

Resolved: That the emancipation proclamation of the President of the United States is as unwarrantable in military as in civil law; a gigantic usurpation, at once converting the war, professedly commenced by the administration for the vindication of the authority of the constitution, into the crusade for the sudden, unconditional and violent liberation of 3,000,000 Negro slaves. . . . The proclamation invites servile insurrection as an element in this emancipation crusade—a means of warfare, the inhumanity and diabolism of which are without example in civilized warfare, and which we denounce, and which the civilized world will denounce as an uneffaceable disgrace to the American people.

Civil War Voices

Confederate artilleryman Robert Stiles.

The communication was almost constant, and the vessels, many of them really beautiful little craft, with shapely hulls, nicely painted; elaborate rigging, trim sails, closed decks, and perfect-working steering apparatus. The cargoes, besides newspapers of the two sides, usually consisted on our side of tobacco and on the Federal side of coffee and sugar; yet the trade was by no means confined to these articles, and on a sunny, pleasant day the waters were fairly dotted with the fairy fleet. Many a weary hour of picket duty was thus relieved and lightened, and most of the officers seemed to wink at the infraction of military law, if such it was. A few rigidly interdicted it, but it never really ceased.

In these dark days of homesickness and hunger, boredom and death, occasional rays of humanity sometimes slipped through. The long, dreary months were broken for some soldiers by friendly contact with the enemy. Perhaps misery does love company. It seems almost unthinkable, given the furious death and destruction of the battlefield, that the common soldiers who suffered the same awful food and cold feet felt a connection to their opposite numbers. Unofficial truces became commonplace in this winter of 1863 on the banks of the Rappahannock.

On one occasion a Confederate picket shouted to his Union counterparts across the river, "Say, Yanks, there are some fools shooting across the river up above, but we won't shoot if you don't."

Pickets from both sides began a lively trading system across the water. Small sailboats floated back and forth with items to exchange. These unofficial truces extended to officers as well. One Union officer was actually invited by his Confederate peers to attend a country dance at a nearby farmhouse. He accepted, enjoyed the dance, and was returned to his lines before daylight by the Confederates.

Why Did Lincoln Replace General Burnside with General Hooker?

While Lincoln's political troubles seemed only to deepen over the question of emancipation, far more dangerous was the military situation, for the Union leadership was still a question. Though Lincoln admired Ambrose Burnside for the way he manfully had accepted the blame for defeat, after the Fredericksburg disaster the president had lost faith in Burnside as the antidote to Robert E. Lee's genius.

Although only five weeks had passed since Burnside's troops were battered, the political pressure for action from Washington was intense. Late in January, Burnside attempted to redeem himself with another crossing of the Rappahannock, again using boats to lay pontoon bridges.

But in addition to his lack of military genius, Burnside seems to have been one of those soldiers most bedeviled by poor luck. Disastrously bad weather, including rains of deluge proportions, washed out all efforts at the crossing. The winter offensive was canceled and would be remembered instead as "the Mud March."

A New York newspaperman, William Swinton, recorded the scene:

An indescribable chaos of pontoons, wagons, and artillery encumbered the road down to the river—supply wagons upset by the roadside, artillery stalled in the mud, ammunition trains mired by the way. Horses and mules dropped down dead, exhausted with the effort to move their loads through the hideous medium. One hundred and fifty dead animals, many of them buried in the liquid muck, were counted in the course of a morning's ride.... It was now no longer a question of how to go on; it was a question of how to get back.

On the other side of the river, jeering Confederates who had mockingly shouted offers to help the Union soldiers cross the river now posted signs: BURNSIDE STUCK IN THE MUD.

"The Mud March" marked Burnside's last attempt as general-in-chief, as Lincoln recalled him to Washington. This time he turned to General Joseph Hooker (1814–1879) to command the Army of the Potomac. Born in Massachusetts, Hooker was a West Pointer (1837) and veteran of the Mexican War, where he served in the same battles with forty-year-old Captain Robert E. Lee and young Lieutenant Thomas J. Jackson, all three winning promotions for gallantry at Chapultepec. But in that war Hooker made the first of numerous political mistakes by opposing his commander, Winfield Scott. At a military inquest after the war, Hooker supported political General Gideon Pillow. Scott, head of the army when the war broke out, neither forgot nor forgave, denying Hooker a commission. Hooker compounded his error by failing to repay a loan to cover gambling debts made to him by Henry Halleck in San Francisco in the 1850s; Halleck was now chief of staff. So Hooker watched the First Battle of Bull Run as a civilian, then peppered Lincoln with letters detailing how the war ought to be fought. When his comrades from Mexico arranged a meeting with Lincoln, the ambitious Hooker told him, "I was at the battle of Bull Run the other day, and it is neither vanity nor boasting in me to declare that I am a damned sight better general than you, sir, had on that field." He was commissioned a colonel.

Hooker served well during the Peninsular Campaign but carped that McClellan, whom he called "an infant among soldiers," had failed to support him. It was during that campaign that a newspaper published a picture of Hooker with a caption that was supposed to read "Fighting—Joseph Hooker." The dash was dropped and Hooker was rechristened "Fighting Joe" Hooker, a nickname he claimed not to like; still, it served him well.

The background of a more notorious aspect of his name is more controversial. When his troops were quartered in Washington near an area filled with brothels and bars, the neighborhood came to be called "Hooker's Division"—hence the common assumption that the word *hooker* as slang for *prostitute* was connected to the general. But word historians have shown that *hooker* was already in common use by the time of the Civil War, probably relating to the Corlear's Hook section of New York, an area frequented by prostitutes and sailors. The connection between "Hooker's Division" and the prostitutes of Washington, D.C., merely cemented the word in its common usage.

Well liked by his men but distrusted by the very superiors he was often undermining, Hooker had a reputation—as Ulysses Grant did—for heavy drinking. When Lincoln promoted Hooker, Postmaster General Montgomery Blair warned, "He is too great a friend of John Barleycorn." Charles Francis Adams, a grandson of John Quincy Adams's and a son of the American minister to Great Britain, shared the widely held opinion of Hooker, commenting, "During the winter when Hooker was in command, I can say from personal knowledge and experience that the Headquarters of the army was a place no self-respecting man would like to go and no woman could go. It was a combination of barroom and brothel."

Documents of the Civil War

"The Hooker Letter," written by Abraham Lincoln on January 26, 1863, to General Joseph Hooker on his appointment to command the Army of the Potomac.

I have placed you at the head of the Army of the Potomac. Of course, I have done this upon what appear to me to be sufficient reasons. And yet I think it best for you to know that there are some things in regard to which, I am not quite satisfied with you. I believe you to be a brave and skilful (sic) soldier, which, of course, I like. I also believe that you do not mix politics with your profession, in which you are right. You have confidence in yourself, which is a valuable, if not indispensable quality. You are ambitious which, within reasonable bounds, does good rather than harm. But I think that during Gen. Burnside's command of the Army, you have taken counsel of your ambition, and thwarted him as much as you could, in which you did a great wrong to the country, and to a most meritorious and honorable brother officer. I have heard, in such a way as to believe it, of your recently saying that both the Army and the Government needed a Dictator. Of course, it was not <u>for</u> this, but in spite of it, that I have given you the command. Only those generals who gain successes, can set up dictators. What I now ask of you is military success, and I will risk the dictatorship. The government will support you to the utmost of it's (sic) ability, which is neither more nor less than it has done and will do for all commanders. I much fear that the spirit which you have aided to infuse into the Army, of criticising (sic) their Commander, and witholding (sic) confidence from him, will now turn upon you. I shall assist you as far as I can, to put it down. Neither you, nor Napoleon, if he were alive again, could get any good out of an army, while such a spirit prevails in it.

And now, beware of rashness. Beware of rashness, but with energy, and sleepless vigilance, go forward, and give us victories.

Hooker proudly showed off the letter to friends, saying, "He talks to me like a father. I shall not answer this letter until I have won him a great victory." (Lincoln's original letter is in the Library of Congress.)

Bluebloods like Charles Francis Adams might have turned their noses up at Hooker, but his appointment seemed to revitalize the Army of the Potomac. One of his first moves was to emphasize the importance of the cavalry, derided and little used by McClellan, as the "eyes of the army." Hooker upgraded and expanded its role. (That decision paid dividends on June 9, 1863, when the Union cavalry met Jeb Stuart's seemingly invincible Confederate cavalry at Brandy Station. In the largest cavalry engagement ever fought in North America, more than 20,000 mounted soldiers engaged one another for more than twelve hours in a prelude to Lee's campaign into Pennsylvania. Although Stuart held the field after the day, the Union cavalry had matched him for the first time, and Stuart was seriously criticized for the first time in the war.)

Hooker also boosted another Civil War technological advance dismissed by "old school" soldiers, the hot-air balloon, used first by the French sixty years earlier. When balloonist Thaddeus Lowe gave him a detailed map of the Confederate units, Hooker became an admirer of his experiments in aerial reconnaissance. A New Hampshire shoemaker's apprentice who had been working with balloons, Lowe had convinced Lincoln of their value when he sent a telegraph message from a balloon in June 1861. Hooker was so enthusiastic that he actually went up in one of Lowe's balloons.

But the greatest problem the new commander faced was morale. Homesickness had become a major cause of desertions, which had reached a rate of two hundred a day. According to a Boston newspaperman, Charles Coffin, 2,922 officers and 80,000 men were reported absent without leave. As he reported, "It was itself a great army."

Hooker had a simple response: He instituted a policy of revolving furloughs. Soldiers were informed that if they did not return on the day of their leave they would be court-martialed. If their absence continued, their regiment would be punished by the cancellation of any more furloughs. As Coffin reported, "It touched their honor. If they did not return, none of their comrades could go home." Hooker bolstered that order with selected executions of deserters.

Hooker also saw to it that his army was well fed. He ordered deliveries of fresh bread and vegetables and started to weed out the corrupt commissary officers who were selling their supplies to civilians on both sides of the battle lines in Virginia. By early April the army was larger and healthier.

A confident Hooker, who now possessed what he claimed was "the finest army on the planet," was hatching plans for dealing with Lee. Lincoln found Hooker overconfident, if not cocky. He said, "The hen is the wisest of all the animal creation, because she never cackles until the egg is laid."

Fighting Joe Hooker would get a chance to lay his first egg at a place called Chancellorsville.

Civil War Voices

Abraham Lincoln: "I can make more brigadier generals, but I can't make more horses."

Lincoln reportedly said this after a March 9, 1863, raid by Captain John Singleton Mosby, who had been a Virginia attorney before the war. With thousands of Union troops in the vicinity, Mosby and his band of Confederate cavalry rangers slipped into Fairfax Court House, ten miles behind the Union lines, cutting telegraph wires as they went. They knocked on the door of the headquarters of General Edwin H. Stoughton, a Vermonter with a reputed fondness for liquor and ladies. Finding him asleep following a party, Mosby slapped him on the rear and informed him that he was a prisoner, along with thirty-two other Union officers and men—and fifty-eight horses.

Mosby was controversial—both during and after the war. Union officer George Armstrong Custer called Mosby's men "guerrillas," and executed six of them in 1864. Mosby retaliated in kind and Custer stopped. Mosby said he had once nearly captured General Grant himself, who later ordered that Mosby be hanged without trial if captured.

What Was the Richmond Bread Riot?

If the winter had been hard on the troops on both sides, it was not much better on the Confederate home front. By the spring of 1863 the war was devastating the Confederate economy. Between the blockade and the diversion of supplies to arm, feed, and clothe the Confederate armies, shortages of food and basic supplies had become commonplace. In the two years of war, prices had increased seven-fold. This inflation meant that few people in the Confederate cities could afford basic necessities.

Most of the Confederacy's capital was tied up in land and slaves. Although the Confederate states accounted for 30 percent of the nation's wealth, they had only 12 percent of its banking assets. The cotton embargo, which was supposed to force Europe to recognize the Confederacy, had backfired. It was simply preventing the cotton farmers from cashing in on their principal asset. Most planters were in debt. And those with cash had to dig deeply to buy Confederate bonds at 8 percent when the rate of inflation had already reached 12 percent a month by the end of 1861. A 10 percent tax on goods and farm products was obviously unpopular.

As the war continued, the southern government printed treasury notes in ever-increasing volume. In 1862 Confederate Treasury Secretary Christopher G. Memminger warned that the printing of notes was "the most dangerous of all methods of raising money.... The large quantity of money in circulation must produce depreciation and disaster." Soon Confederate treasury notes were practically worthless.

Shortages of supplies forced the inflation rate even higher. In January 1863 a Richmond newspaper printed a schedule showing that the weekly cost to feed a small family had risen from $6.55 in 1860 to $68.25 in 1863. By March 1863 flour was selling for $100 a barrel, beef was $2 a pound, apples were $25 a bushel, boots cost $50 per pair, and wood was $30 a cord. Simple necessities had become un-affordable luxuries.

Richmond was among the cities hardest hit. As the Confederate capital, its population had more than doubled since the war began, severely straining the available supplies. The surrounding country-side had seen much of the worst of the fighting. Artillery barrages, pitched battles, and plundering soldiers had stripped nearby farms. In May 1861 $1 in gold cost $1.10 in Richmond; by June 1863 it had risen to $7; by the beginning of 1864 it would reach $20.

The crisis reached a breaking point on April 2, 1863, when hun-dreds of hungry women assembled at a church in Richmond and marched to Capitol Square. Having heard rumors that the women of Salisbury, North Carolina, had rioted to get store goods at the same prices paid by the government, the women angrily confronted Vir-ginia's governor, John Letcher, and demanded relief from the high prices. Letcher, who objected to many of Jefferson Davis's military appointments and policies but nevertheless supported the Confed-erate president, offered no solution. Soon the small group of women grew into an angry mob. Shouting "Bread! Bread!" they began smashing windows in the shopping district and emptying the stores of whatever they could grab. More destitute women and a few men converged on the scene, their number swelling to more than a thou-sand. The mayor of Richmond ordered the police to fire on the mob if it did not break up.

In the midst of the riot Jefferson Davis appeared. According to the possibly mythical version of events later told by Davis's wife, Varina, the president stood on a wagon and flung money from his pockets to the crowd. Chiding them for stealing jewelry and clothing when they were crying for bread, he said, "You say you are hungry and have no money—here is all I have." Then he held up a pocket watch and told the rioters if they did not disperse within five minutes, he would order the militia to fire. With muskets leveled at them, the rioters scattered back into the side streets and neighborhoods from which they had come. Davis had the leaders of the riot arrested; a few were convicted and briefly imprisoned.

Davis tried to suppress news of the Bread Riot, but the anti-Davis newspaper, the *Examiner*, published a full account. This brief rebellion hinted at broader strains within the rebel states. The poor and working class of the Confederacy, who were bearing the brunt of the war, were tired. Although relatively few in number, Union sympathizers in the Confederacy began to join the Union armies; every Confederate state had men fighting for the Union. And among the Confederate troops, the same problems of morale that Hooker had confronted were evident. Desertions were depleting the ranks of the armies, as soldiers went home to care for their families in the face of the severe shortages and a growing sense of the hopelessness of the Confederate cause.

Civil War Voices

General Ulysses S. Grant, during the winter of 1863 spent attempting to take the strategic river port city of Vicksburg.

The long, dreary, and—for heavy and continuous rains and high water—unprecedented winter was one of great hardship to all engaged about Vicksburg. . . . Troops could scarcely find dry ground on which to pitch their tents. Malarial fever broke out among the men. Measles and smallpox also attacked them. . . . Visitors to the camps went home with dismal stories to relate. . . . Because I would not divulge my ultimate plans to visitors, they pronounced me idle, incompetent, and unfit to command men in an emergency, and clamored for my removal. . . . I took no steps to answer these complaints, but continued to do my duty, as I understood it, to the best of my ability. . . . With all the pressure brought to bear upon them, President Lincoln and General Halleck stood by me. . . . I had never met Mr. Lincoln, but his support was constant.

This was also the winter of Ulysses Grant's discontent. Since December 1862 Grant had been struggling to remove the last major Mississippi River stronghold in Confederate hands, Vicksburg, Mississippi. If this city fell, the Union would effectively control the full length of the Mississippi, severing the Confederacy. The eastern Confederate states would be cut off from the grain, cattle, and other basic supplies coming from the western states of Texas, Arkansas, and western Louisiana. In Washington, Chief of Staff General Henry Halleck said, "In my opinion, the opening of the Mississippi River will be to us more advantage than the capture of forty Richmonds."

By the end of January, Grant was operating about ten miles above Vicksburg, aided by a Union riverboat fleet under Admiral David Dixon Porter, one of the heroes of the capture of New Orleans. On a high bluff overlooking the Mississippi, Vicksburg was all but impregnable. Confederate cannon aimed at the river from the heights controlled all river traffic. Direct assault from the river was impossible because of the steep, well-defended heights. The Union boats could not control the river until Vicksburg was neutralized.

Grant ambitiously—and unrealistically—planned to dig canals connecting one part of the Mississippi to another on the Louisiana side of the river, allowing Porter's fleet to bypass Vicksburg. After weeks of fruitless digging, the idea was abandoned.

Grant then took another gamble. He would march his men south of Vicksburg on the opposite, or western, shore. Under cover of night, Porter would take the fleet past Vicksburg. The army would then be ferried across the river below the city and march north to attack it from the inland side. The plan was dangerous because it would leave Grant's army out of communication with Washington and other Union forces and without supplies. It also risked the loss of the entire riverboat fleet if Porter could not get his ships safely past Vicksburg's defenses.

On the night of April 16 Porter started his river run. But Confederates patrolling the river in small boats soon discovered the Union plan and set fire to several houses on the river's banks to illuminate the night. The batteries in Vicksburg burst into action.

Accompanied by his wife and children, Grant watched Porter's run anxiously from another boat. One steamer, the *Henry Clay*, was set afire by Confederate guns, but two and a half hours later the cannons fell silent. Albert Richardson of the *New York Tribune* wrote, "On the gunboats not a man was killed, and only eight were wounded. On the steamboats and barges nobody was even hit. Before daylight the entire fleet, save the ill-fated *Henry Clay*, was received at New Carthage by Grant's infantry with shouts of delight."

To the sailors who had run the steamers on this daring race, the general promptly gave furloughs of forty days and transportation to and from home.

On April 30 Grant began to cross the river with his army of 50,000 men. By early May he had won five small battles, scattering the Confederate troops and sending them back to Vicksburg. Jefferson Davis wired John C. Pemberton, the Confederate general in command, to "hold Vicksburg at all hazard."

In mid-May, Grant tried to assault the heights of Vicksburg, but after suffering severe losses, he changed tactics. As he wrote, "I determined upon a regular siege. . . . With the navy holding the river, the investment of Vicksburg was complete. . . . The enemy was limited in supplies of food, men and munitions of war. . . . These could not last."

What Happened at Chancellorsville?

"May God have mercy on General Lee, for I will have none," said Fighting Joe Hooker after months of planning, training, and restoring his army's morale. He was ready, eager, and, as always, self-assured.

Late in April 1863, with an army of 130,000 men, Hooker was ready to launch what he thought was a perfect strike against Lee's 60,000 troops. One part of the army would strike again at Fredericksburg, scene of Burnside's disaster; 10,000 cavalrymen would operate against Lee's lines of supply and communication to Richmond; and, finally, Hooker himself would lead the main body of

70,000 troops nine miles west of Fredericksburg and across the Rappahannock River in an effort to surprise Lee. On April 30 Hooker reached the town of Chancellorsville and made his headquarters in Chancellor House, an old plantation home surrounded by slave shanties.

But Hooker's well-laid plans did not fool Lee. Despite the show of force at Fredericksburg, Lee immediately understood the Union strategy. Seeing that Hooker was well defended in a position surrounded by a thicket of trees and brush that would come to be called "the Wilderness," Lee adjusted his plans. Sitting on broken commissary boxes that contained the hard biscuits that the Confederate Army lived on, Lee met with Stonewall Jackson in the middle of the night on May 1 in a legendary "cracker barrel" conference. The next day he sent Jackson's 26,000 men twelve miles past the Union front and around to its weakest spot while making a diversionary show of force against the center of Hooker's lines.

Jackson's march did not go unnoticed; Union outposts spotted his army on the move. Hooker dismissed the threat, thinking that Jackson was in retreat. Union General Oliver O. Howard (1830–1909)—later instrumental in the founding of the university that would be named for him—also ignored the warnings and did not consider his troops to be in danger. When Jackson's Confederates attacked Howard at 5:15 P.M. on May 2, the Union soldiers were lounging in camp, playing cards with their rifles stacked. Jackson's assault came, in the words of one Union soldier, "like a clap of thunderstorm from a cloudless sky."

Civil War Voices

Confederate soldier John O. Casler, a member of Stonewall Jackson's army.

We ran through the enemy's camps. . . . Tents were standing and camp-kettles were on fire, full of meat. I saw a big Newfoundland dog lying in one tent as if nothing had happened. We had a nice chance to plunder their camps and search the dead, but the men were afraid to stop, as they had to keep with the artillery and were near a good many officers who might whack them over the head with their swords if they saw them plundering. But the temptation was too great, and sometimes they would run their hands in some of the dead men's pockets as they hurried along, but seldom procured anything of value.

Jackson's rampaging troops swept nearly all the way to Hooker's headquarters until reserves arrived to stop them. Among these were artillery batteries under General Alfred Pleasonton. As he was about to fire at the advancing Confederates, Pleasonton received a report that the men in front of him were actually Union soldiers. He dispatched a young officer, Clifford Thomson, on horseback to check the information. Riding toward the advancing troops, who waved the Stars and Stripes and greeted him as an ally, Thomson was still uncertain until the disguised Confederates opened fire. Pleasonton needed no further report. He ordered his twenty-two guns to fire, mowing down the Confederate lines in heaps. After the advance was halted, the fortunate Thomson told his commander, "General, those people out there are rebels!"

"Thomson, I never expected to see you again," he replied. "I thought if they didn't kill you, I should; but that was no time to stop for one man."

Civil War Voices

James Power Smith, aide-de-camp of Stonewall Jackson, on his wounding by "friendly fire."

As he [Jackson] rode near to the Confederate troops, just placed in position and ignorant that he was in the front, the left company began firing . . . and two of his party fell from their saddles dead. . . . Spurring his horse across the road to his right, he was met by a second volley from the right company. . . . Under this volley . . . the general received three balls at the same instant. One penetrated the palm of his right hand. . . . A second passed around the wrist of the left arm and out through the left hand. A third ball passed through the left arm halfway from shoulder to elbow. His horse turned quickly from the fire, through the thick bushes, which swept the cap from the general's head and scratched his forehead, leaving drops of blood to stain his face.

As the wounded Stonewall Jackson was carried from the scene, the stretcher was dropped twice as its bearers were hit by Union fire. When Confederate Brigadier General Pender reported that the troops must retire, Jackson issued what would be his last command in the field: "You must hold your ground, General Pender. You must hold your ground, sir!"

That night Jackson's left arm was amputated at the shoulder and he was taken to a farmhouse a few miles away. By Sunday, May 10, he was delirious from the fever of pneumonia. He babbled orders, spoke to his imagined wife and child, then finally said, "Let us cross over the river and rest under the shade of the trees."

Union General Oliver Howard, the victim of Jackson's last assault, summed up his death: "Jackson stood head and shoulders above his confreres, and General Lee could not replace him."

Despite the loss of his most important commander, Lee masterfully directed the battle. Jeb Stuart temporarily assumed command of Jackson's troops, and the Confederates pressed the attack on the Union forces and a seemingly befuddled General Hooker. Despite having nearly twice Lee's troops, Hooker allowed Lee to dictate thoroughly the course of the battle, which ended with Hooker's ordering a retreat across the Rappahannock in one more staggering rout of numerically superior Union armies.

Civil War Voices

In the thick tangle of woods around Chancellorsville, the brush had caught fire, trapping large numbers of the wounded from both sides. One Union man, caught in this maelstrom with both comrades and enemies, later described this experience.

> I began to pull away from the burning brushwood, and got some of them out. I tell you it was hot! Them pines was full of pitch and rosin, and made the fire as hot as a furnace.
>
> I was working away, pulling out Johnnies and Yanks, when one of the wounded Johnnies . . . toddled up and began to help. . . . The underbrush crackled and roared, and the poor devils howled and shrieked when the fire got at them. . . . By and by another reb—I guess he was a straggler—came up and began to help too. . . . We were trying to rescue a young fellow in gray. The fire was all around him. The last I saw of that fellow was his face. . . . His eyes were big and blue, and his hair like raw silk surrounded by a wreath of fire. . . . I heard him scream, "O, Mother! O, God!" It left me trembling all over like a leaf.
>
> After it was over, my hands were blistered and burned so I could not open or shut them; but me and them rebs tried to shake hands.

The aftermath of the battle at Chancellorsville was traumatic: for the Union, more than 17,000 casualties; the Confederates, 13,000, a smaller number but a higher percentage of their available men. Back in Washington, Lincoln was near despair when he got the news. "My God! my God!" he cried, "What will the country say?" Again he was left to wonder if he would find a general capable of contending with Lee. Hooker, who had suffered a concussion when his headquarters was hit by an artillery barrage, later said of himself, "To tell the truth, I just lost confidence in Joe Hooker."

Others seeing Hooker's dazed condition during the battle assumed that "John Barleycorn" was involved. Responding to the charge that Hooker was drunk, his second-in-command, General Darius Couch, later said, "He probably abstained from the use of ardent spirits when it would have been far better for him to have continued in his usual habit in that respect...." This unresolved issue aside, the reason for the loss was simple. Hooker was completely mastered by Lee, whose deification throughout the South was now complete. Chancellorsville would go down as his most dazzling victory. (Chancellorsville acquired another sort of immortality as well. It is the unnamed setting of *The Red Badge of Courage,* written by Stephen Crane, who was born after the war.)

Lee followed up on this victory with an even bolder stroke. He recommended to President Davis—who agreed—that he invade the Union once more, first moving west into the Blue Ridge Mountains of Virginia, then north into what would soon become West Virginia and, finally, into Pennsylvania.

In taking the offensive, Lee had several purposes. One motive was psychological. He believed that taking the war to his enemy's lands would boost the morale of his men as well as that of the people of the Confederacy while demoralizing the Union. Another part was political. By striking a decisive blow against the Union while continuing to threaten Washington, D.C., he hoped that he could stoke the antiwar sentiments in the Union. In Richmond, Jefferson Davis was preparing to dispatch Vice-President Stephens under a flag of truce, ostensibly to discuss prisoner exchanges. But Stephens had the authority to open peace negotiations, as he had always believed in the possibility of a negotiated settlement. With a victory in Yankee territory, a peace maneuver might be possible—on Confederate terms. The Confederate leadership also still held out the thread of a hope for European recognition. If Lee could establish his army firmly on Union soil, a slim chance remained that the European powers might aid the Confederate cause. Finally, Lee hoped to find the provisions that his poorly fed and clothed men desperately needed in the rich farmlands of southern Pennsylvania's Cumberland Valley.

Hooker was ordered to shadow Lee. As Lee led his men north, so did Hooker, moving his army in a parallel line, keeping the Army of the Potomac between Lee's troops and the nation's capital. But when Hooker asked for reinforcements, he was denied. As a result, he asked to be relieved of command. His place was taken by Lincoln's next choice for the top spot, General George Gordon Meade (1815–1872).

A Philadelphia aristocrat born in Cádiz, Spain, of American parents, Meade had graduated from West Point in the class of 1834 and served for a year before resigning to become a civil engineer. In 1842 he rejoined the army and served in Mexico. During the Civil War, he advanced through a series of battlefield promotions and led his troops well at Chancellorsville but was restrained by Hooker.

Meade and Lee would soon cross paths in a small, obscure Pennsylvania town called Gettysburg.

Civil War Voices

Boston Journal correspondent Charles Coffin reported from Harrisburg, Pennsylvania, as Lee's army moved into the state.

> Harrisburg was a Bedlam when I entered it on the 15th of June. The railroad stations were crowded with an excited people—men, women, and children—with trunks, boxes, bundles; packages tied up in bed-blankets and quilts; mountains of baggage—tumbling it into the cars, rushing here and there in a frantic manner; shouting, screaming, as if the Rebels were about to dash into the town and lay it in ashes. . . .
>
> The merchants were packing up their goods. . . . At the State House, men in their shirtsleeves were packing papers into boxes. Every team, every horse and mule and handcart in the town were employed.
>
> . . . a steady stream of teams came thundering across the bridge—farmers from the Cumberland Valley, with their

household furniture piled upon the great wagons peculiar to the locality; bedding, tables, chairs, their wives and children perched on the top, kettles and pails dangling beneath; boys driving cattle and horses.

Coffin's Pennsylvania merchants and farmers had only to fear the Confederates' method of payment. While the northern press had portrayed Lee's troops as if they were Genghis Khan's hordes, the Army of Northern Virginia was under Lee's strictest orders to behave like southern gentlemen. As one commander, John B. Gordon, later told it, "The orders from General Lee for the protection of private property and persons were of the most stringent character. . . . I resolved to leave no ruins along the line of my march through Pennsylvania; no marks of a more enduring character than the tracks of my soldiers along its superb pikes."

Lee had ordered that all supplies be paid for. His call for restraint was more than just southern chivalry, as it would later be painted. As a politician as well as a general, Lee wanted to do nothing that would antagonize the Peace Democrats, who were opposed to Lincoln and the war. But the problem was that the supplies were being paid for in Confederate paper money, which was not worth "ten cents a bushel."

The free blacks of Pennsylvania had a much greater fear. There were many reports of northern blacks being captured by the Confederates and sent to the Deep South to be sold into slavery. It was an action that Lee did nothing to prevent.

Why Did Lincoln Banish Clement Vallandigham?

"Money you have expended without limits and blood poured out like water. Defeat, debt, taxation, and sepulchers—these are your only trophies," said Congressman Clement L. Vallandigham, expressing the other Union view of the war.

Born in Ohio, Clement Laird Vallandigham (1820–1871) was the son of an Ohio preacher who became a lawyer and Democratic member of the House from 1858 to 1863. Married to the daughter of a Maryland planter, he was a states' rights believer and a bitter opponent of Lincoln and the war. During the war, he became the most outspoken of the so-called Copperheads, the northern Democrats who advocated a compromise with the Confederate states. The derivation of the name is ambiguous, but it most likely referred to the copperhead, a poisonous snake. However, Copperheads also wore buttons cut from copper coins depicting the goddess of liberty; it is unclear if they wore these badges before or after the name was given to them.

By 1863 Vallandigham had become so outspoken in his virulent attacks on Lincoln and the war that he called on soldiers to desert and avoid conscription. Campaigning for the governorship of Ohio, he threatened to join the western states in the Confederacy. Ambrose Burnside, the disgraced commander of Fredericksburg who now commanded the Department of the Ohio, arrested Vallandigham and tried him in a military court on charges of treason; he was convicted and sentenced to a military prison. A writ of habeas corpus to release Vallandigham was ignored by Burnside. But other Democratic governors, along with some Republicans, protested. Lincoln had no sympathy for his antagonist and said, "Must I shoot a simpleminded soldier who deserts, while I must not touch a hair of the wily agitator who induces him to desert?" Facing a political revolt, Lincoln commuted the sentence to banishment to the Confederate lines. But Vallandigham was not much more welcome there and eventually wound up in Canada, where he continued his campaign for governor of Ohio. (He lost the election, and as the Union's fortunes changed, he lost most of his influence. After the war, he resumed his legal practice, and during a case in which he was demonstrating how a murder victim might have killed himself, he shot himself with a pistol and later died.)

What Was So Important About Gettysburg?

Legend has it that the barefooted men of General A. P. Hill's division needed shoes. On June 30 a brigade of Confederate soldiers went into the pastoral college town of Gettysburg looking for some. Actually this brigade was doing reconnaissance. They returned to report that the town was occupied by Union cavalry. General Henry Heth asked Hill for permission to take some men to town the next day to clear out the Union detachment.

At 9:00 A.M. on July 1, Heth's men reached the outskirts of Gettysburg, where they exchanged a few shots with the unmounted cavalrymen of Union General John Buford—the first shots fired on the first of three days of fighting. A skirmish between two advance units soon mushroomed as the two great armies, drawn like moths to a flame, converged on this small crossroads and railroad junction in southern Pennsylvania, the home of thirty-five hundred farmers, merchants, and students and teachers at the town's seminary. It soon became the site of the largest and certainly one of the most significant battles ever fought in the Western Hemisphere.

Arriving on the outskirts of Gettysburg after these initial skirmishes, Lee reluctantly selected the town as the site where he would finally crush the Union Army of the Potomac. Out of contact with Jeb Stuart's cavalry, which was supposed to provide intelligence about the Union, Lee did not know where the Army of the Potomac was until a shadowy figure, known through history as "the spy Harrison," a man who worked for General Longstreet but essentially disappeared after the battle, reported that General Meade was now in command of the Union army and was marching north to meet the Confederates. Although it wasn't the strategic position Lee would have ideally chosen, he knew a victory here meant a free line of march toward Washington, Lincoln's political downfall, and a negotiated end to the war.

But the arrival of Buford's cavalry just slightly before the Confederate troops reached Gettysburg and Buford's quick thinking in establishing his men on the high ground were critical. In a reversal of

Burnside's 1862 disaster at Fredericksburg, a Union army would now be holding the high ground against an attacker. Buford's dismounted cavalrymen, armed with the new generation of breech-loading carbine rifles, were able to hold the strategic ground until they were reinforced by the arrival of General John F. Reynolds.

A competent, battle-tested veteran who had turned down the command of the Army of the Potomac before it was offered to Meade, Reynolds also recognized the importance of the terrain and expertly positioned his troops. Soon after assessing the situation, however, he was shot through the head by a sharpshooter, falling dead instantly. His place was taken by General Abner Doubleday, one of the heroes of Fort Sumter.

In the meantime, Meade also dispatched General Winfield Hancock to Gettysburg. Like Buford and Reynolds, Hancock immediately recognized the strategic importance of the hills around the town.

As thousands of men poured into the area from both sides, the Confederates launched an attack. Overwhelmed for the moment, the Union armies began to reel backward through Gettysburg. On July 1, it looked as if Lee might get the decisive victory for which he hoped.

Civil War Voices

Union cannoneer Augustus Buell described the first day's action (July 1, 1863).

> Up and down the line, men reeling and falling; splinters flying from wheels and axles where bullets hit; in the rear, horses tearing and plunging, mad with wounds or terror; drivers yelling, shells bursting; shot shrieking overhead, howling about our ears or throwing up great clouds of dust where they struck; the musketry crashing ... bullets hissing, humming, and whistling everywhere; cannon roaring, all crash on crash, and peal on peal; smoke, dust, splinters, blood; wreck and carnage indescribable. . . . Every man's

shirt soaked with sweat, and many of them sopped with blood from wounds not severe enough to make such bull-dogs let go—bareheaded, sleeves rolled up, faces black-ened.

Following what seemed like the beginning of yet another rout of a Union army, Lee sent orders to General Ewell to follow up the success "if he found it practicable." It was Lee's courteous but fatally ambiguous way of giving an order, and Ewell misinterpreted it. Thinking he had been given the discretion to either attack or hold his ground, Ewell allowed his troops to rest. It was a crucial error. Military historians have long wondered what would have happened had Ewell continued; would the rout have been completed and the Battle of Gettysburg ended there?

Bemoaning Ewell's decision, Lee supposedly said later, "If I had Stonewall Jackson at Gettysburg, I would have won that fight; and a complete victory would have given us Washington and Baltimore, if not Philadelphia, and would have established the independence of the Confederacy."

As the first day ended, Doubleday's men were rather ominously camped in Gettysburg's cemetery, and the rest of the Army of the Potomac was still arriving by forced march. By midnight of Day One, General Meade had arrived. Continuing to stream into the Gettysburg vicinity, the Union troops, which would finally total close to ninety thousand, were deployed in a line taking the shape of an upside-down J or an inverted fishhook, three miles long, on the high ground outside Gettysburg.

In poor health and suffering from diarrhea that leveled so many soldiers, Lee remained confident. Against the advice of General James K. Longstreet, his most trusted aide since the death of Jackson, he believed that his troops could overcome the massed Union forces. Recognizing the strength of the Union position, Longstreet urged Lee to move south and attack Meade from the rear. But with recent history to support him, Lee felt that the Union army, demoralized by the defeats at Fredericksburg and Chancellorsville, was vulnerable and would crack once more if pressed hard enough by his seasoned fighters.

On the sultry, murky morning of July 2—Day Two at Gettysburg—Lee planned for Longstreet to deliver a smashing blow against the Union's left side while Ewell attacked the right. With the reluctant Longstreet lagging, the attack was slow in developing, and didn't come until well into the hot, humid July afternoon.

Opposing Longstreet were the forces under General Daniel Sickles, one of the Union's notorious "political generals." A Tammany Hall Democrat and former congressman, he acquired notoriety as the first person to successfully plead temporary insanity as a murder defense. Before the war, he had killed Philip Barton Key, his wife's lover and the son of Francis Scott Key, composer of "The Star-Spangled Banner," in Lafayette Park, across the street from the White House. Edwin M. Stanton, now Lincoln's war secretary, had argued this novel defense. Although far more adept in the field than many other political appointees in uniform, here Sickles moved his troops out of their assigned position, a violation of orders that almost proved catastrophic.

More Union troops continued to arrive and take up reserve positions. All day long Meade shuttled his forces around to fill weak points in the lines. The hottest spots of the day were at a small hill called Little Round Top, nearby Devil's Den, with its cropping of large boulders that provided perfect cover for Confederate riflemen, a wheat field, and a peach orchard. All became legendary scenes.

One of the key Union positions, Little Round Top had been left undefended, a fact that New York General Gouverneur K. Warren, a topographer, astutely recognized. He quickly diverted a regiment to the hill's defense, but they were soon under intense fire from Confederates in front of them as well as sharpshooters who had worked their way among the boulders of Devil's Den.

Nearby, two Alabama regiments were preparing to attack the extreme end of the Union line, which was anchored by an undermanned 20th Maine regiment under the command of Colonel Joshua L. Chamberlain. A thirty-three-year-old professor of rhetoric and languages at Maine's Bowdoin College, Chamberlain seemed an unlikely candidate for heroism. But he was one of the best examples of the seemingly ordinary man thrust into extraordinary circumstances. Directed to hold his ground at all costs, Chamberlain directed a heroic stand by the Maine soldiers. Facing another Confederate charge with his men nearly out of ammunition, he ordered a seemingly suicidal bayonet charge that repulsed the attackers, saving the Union's end from a collapse that would have allowed the Confederates to pour in behind the Union lines. Chamberlain was wounded and won the Congressional Medal of Honor for his actions.

A little later, another Union disaster was averted by a similarly extraordinary performance by the 1st Minnesota Volunteers. After long delays, Longstreet's Confederates had overwhelmed Sickles's men, and Hancock, who had demonstrated a knack for being wherever the danger was greatest, was at hand to command. Calling for reinforcements, Hancock knew he needed to hold the Confederate advance until his reserves arrived. Looking at the 262 men of the regiment, he ordered them to make another nearly suicidal charge into the Confederate lines—nearly 1,600 Alabamans supported by artillery.

As one of the Minnesota men, William Lochren, said, "Every man realized in an instant what the order meant—death or wounds to us all; the sacrifice of the regiment to gain a few minutes time and save the position, and probably the battlefield—and every man saw and accepted the necessity for the sacrifice."

The small band of men leveled their bayonets and ran straight into the face of the Confederates in an incredible act of bravery.

Civil War Voices

William Lochren of the 1st Minnesota Volunteers.

What Hancock had given us to do was done thoroughly.
The regiment had stopped the enemy, and held back its

mighty force and saved the position. But at what sacrifice! Nearly every officer was dead or lay weltering with bloody wounds, our gallant colonel and every field officer among them. Of the two hundred and sixty-two men who made the charge, two hundred and fifteen lay upon the field, stricken down by rebel bullets, forty-seven were still in line, and not a man was missing.

With these separate Confederate assaults just barely repulsed, Meade's troops still held the strong defensive position in the hills south of the town. On Day Three—July 3—Lee was determined to attack once more with the full fury of his army. Twice he thought he had nearly defeated the Union forces. With supreme confidence in his men, Lee believed that one more concerted effort would bring about a Union collapse. Again Longstreet counseled—unsuccessfully—against an attack.

At 3:00 P.M. on a fiery, red-hot day, 15,000 disciplined Confederate soldiers emerged from the high ground in the trees on Seminary Ridge and formed perfectly aligned battle ranks facing the Union position, a mile away on Cemetery Ridge. Although just three of the nine brigades in the Confederate force were commanded by Major General George E. Pickett, a thirty-eight-year-old soldier from Virginia, it was his name that would be attached to this valiant but doomed effort.

Before the attack, the massed Confederate guns had attempted to dislodge and soften up the Union troops. More than 150 cannons fired relentlessly at the Union position. In an unprecedented exchange, it was answered in almost equal measure by Union cannon.

Civil War Voices

Sergeant John W. Plummer of the 1st Minnesota Volunteers.

The first gun was the signal for a hundred more to open, at less than half a mile distance, which till then their existence was perfectly unknown to us. Such an artillery fire has never been witnessed in this war. The air seemed to be filled with the hissing, screaming, bursting missiles, and all of them really seemed to be directed at us. They knew our exact position, for before we lay down, they could with the naked eye plainly see us, but, fortunately, most of them just went far enough to clear us, while many struck in front of us and bounded over us. We lay behind a slight rise of ground, just enough, by laying close, to hide us from the view of the rebs. A good many shells and pieces struck mighty close to us, and among us, but strange to say, none of us were injured, while the troops that lay behind us had many killed and wounded. Our batteries replied, but for the first time in our experience, they were powerless to silence the rebs, and, in fact, many of our guns were silenced. So many of their horses and men were killed that they could not work their guns and drew them off the field. Caisson after caisson were blowed up, and still the rebs' fire was fierce and rapid as ever. I kept thinking surely they cannot fire much longer, their guns will get so hot they will have to stop, and they cannot afford, so far from their base, to waste such ammunition. It was awful hot where we lay, with the sun shining down on us and we so close to the ground that not a breath of air could reach us. We kept wishing and hoping they would dry up, as much to get out of the heat as the danger, for the latter we thought little of.

Plummer was describing a ferocious two-hour artillery duel that was heard a hundred and forty miles away in Pittsburgh, supposedly making it one of the loudest noises ever heard in North America. Following the barrage, the Confederates began advancing "in magnificent order, with the step of men who believed themselves invincible," according to one Union officer. "Pickett's Charge" began with thousands of Confederate infantry marching abreast in a long straight line heading uphill toward a waiting Union army.

The Union artillery, which the Confederates had thought was silenced, immediately opened fire once more, and huge gaps were blown in the Confederate lines. Under orders not to fire and not to let loose their now famous Rebel Yell, the Confederate ranks closed and kept advancing. The Union artillery switched to firing canister— tin cans packed with iron balls—and mowed great swaths through the attackers. As the lines of Confederates moved closer, it was the turn of the Union infantry to send volley after volley of rifle fire into the oncoming Confederate ranks.

Civil War Voices

Union officer Warren Goss.

> They return our fire and rush upon the wall. As they are about to close in upon us they are met by a volley. On they come.... The shock is terrible, and its full strength falls upon Webb's brigade. Our men are shot with the rebel muskets touching their breasts. A fierce encounter now takes place. Great God! The line at the stone wall gives way!

In desperate hand-to-hand fighting, the Union line was penetrated. Some Union soldiers began to break and run but were driven back by their officers' sabers. Once again, General Hancock was at the crucial spot, as were members of the 1st Minnesota Volunteers. As on Day Two, the Minnesota regiment received orders to charge the Confederates.

Civil War Voices

Lieutenant William Harmon of the 1st Minnesota Volunteers.

If men ever become devils that was one of the times. We were crazy with the excitement of the fight. We just rushed in like wild beasts. Men swore and cursed and struggled and fought, grappled in hand-to-hand fight, threw stones, clubbed their muskets, kicked, yelled and hurrahed. But it was over in no time. . . . When the line had passed, those who were not wounded threw down their arms, I remember that a Conf. officer . . . gathered himself up as our men swept by and coolly remarked, "You have done it this time."

Suddenly it was over. Almost 4,000 of the Confederate soldiers in Pickett's Charge were captured. Pickett's division alone lost 75 percent of its men. The defending Union forces, just half as numerous as the rebel attackers, suffered only 1,500 casualties—only one fifth of the number they had inflicted.

Civil War Voices

Boston correspondent Charles Coffin.

The lines waver. The soldiers of the front rank look round for their supports. They are gone—fleeing over the field, broken, shattered, thrown into confusion by the remorseless fire. . . . The lines have disappeared like a straw in a candle's flame. The ground is thick with dead, and the wounded are like the withered leaves of autumn. Thousands of Rebels throw down their arms and give themselves up as prisoners. How inspiring the moment! How thrilling the hour! It is the high-water mark of the Rebellion—a turning point of history and of human destiny.

Military historians are still fighting over this one. Was Pickett where he was supposed to be? Did some Confederate units falter? Did the whole affair make any sense at all? (With perfect hindsight, modern military experts note that 90 percent of the massed assaults against defensive positions failed in the Civil War. If Lee knew that, maybe he still thought his guys would be in the lucky 10 percent.) Pickett's Charge was the last gasp of an exhausted army. Although Jeb Stuart, out of touch for weeks, had arrived the day before, Lee's formerly reliable and daunting cavalry made no difference. In fact, the somewhat eccentric Stuart had been successfully challenged by an equally dashing young Union cavalry officer wearing a lavish uniform of his own design, George Armstrong Custer.

Civil War Voices

General James K. Longstreet on the aftermath of the Confederate failure.

> General Lee came up as our troops were falling back and
> encouraged them as well as he could; begged them to re-
> form their ranks and reorganize their forces, and assisted
> the staff officers in bringing them all together again. It was
> then he used the expression that has been mentioned so
> often: "It was all my fault; get together, and let us do the
> best we can toward saving which is left us."

Union soldier Jesse Young described the next day, Independence Day, when a heavy rain set in, turning the roads and fields into mud.

> The next morning was the Fourth of July, but it seemed at
> the time to those who were at Gettysburg a somber and
> terrible national anniversary, with the indescribable horrors
> of the field, as yet hardly mitigated by the work of mercy,
> before the eye in every direction. The army did not know
> the extent of the victory; the nation did not realize as yet

what had been done. The armies were still watching each other, although the Confederates had withdrawn from the town of Gettysburg and concentrated their troops on Seminary Ridge.

The people in the village came out of their cellars and other places of refuge, and as the day broke upon them opened their doors. They had been under a reign of terror. ... Among the troops themselves, that Fourth of July, 1863 ... was a wretched, dismal, and foreboding day, a day of uncertainty and suspense for both armies, which still faced each other. Each ... was watching to find out what the other would do. Neither Meade nor Lee, just at that time, was anxious to bring about a renewal of the fight, and the time was occupied in caring for the wounded and burying the dead.

There was no shortage of wounded and dead. In a war that had presumably lost its ability to shock, the toll of Gettysburg was appalling. More than 51,000 men—23,000 for the Union, 28,000 for the Confederates, or almost one third of the total number—were killed, wounded, or missing after the three days of fighting. Seven thousand Confederate soldiers were too badly wounded to remove and were left to the Union surgeons. Remarkably, only one civilian of Gettysburg had been killed, a young woman struck by a stray bullet.

Civil War Voices

General John D. Imboden, assigned to protect the Confederate wagons as they retreated for Virginia.

Many of the wounded in the wagons had been without food for thirty-six hours. Their torn and bloody clothing, matted and hardened, was rasping the tender, inflamed, and still oozing wounds. Very few of the wagons had even a layer of straw in them, and all were without springs. The road was rough and rocky from the heavy washing of the preceding day. The jolting was enough to have killed strong men, if long exposed to it. . . . During this one night I realized more of the horrors of war than I had in all the two preceding years.

John Buford, one of the heroes of Gettysburg, later contracted typhoid fever and died within six months of the battle, on December 16, 1863. Daniel Sickles, whose misguided troop placement had almost proven fatal to the Union, was hit in the leg during the battle. He was carried from the field, coolly smoking a cigar, and his leg was amputated a half hour later. (He donated it to a medical museum and reputedly continued to visit it.) Joshua Chamberlain of the 20th Maine was wounded twice; he was relieved and later suffered from malaria. He resumed command of his regiment about a year later at Cold Harbor, where he was again wounded and promoted on the spot by Ulysses Grant.

George Pickett, the Virginian who led a third of the troops in the fateful charge on Day Three, was bitter. After the war he said to another officer, "That old man [Lee] . . . had my division massacred."

General Meade, satisfied with what he had accomplished, reported to Lincoln that he had driven "from our soil every vestige of the presence of the intruder."

"My God, is that all?" cried Lincoln, who wanted and expected the destruction of Lee and his weakened army as they retreated across the Potomac, back into Virginia, where Lee would regroup to fight for nearly two more years.

Civil War Voices

From the diary of a young woman who lived through Grant's siege of Vicksburg.

We are utterly cut off from the world, surrounded by a circle of fire. . . . The fiery shower of shells goes on day and night.

...People do nothing but eat what they can get, sleep when they can, and dodge the shells....

I think all the dogs and cats must be killed or starved. We don't see any more pitiful animals prowling around....

The confinement is dreadful....

...This place has two large underground cisterns of good cool water, and every night in my subterranean dressing-room a tub of cold water is the nerve-calmer that sends me to sleep in spite of the roar.

One cistern I had to give up to the soldiers, who swarm about like hungry animals seeking something to devour. Poor fellows! My heart bleeds for them. They have nothing but spoiled, greasy bacon, and bread made of musty pea-flour, and but little of that. The sick ones can't bolt it. They come into the kitchen when Martha [a slave] puts the pan of corn-bread in the stove, and beg for the bowl she mixes it in. They shake up the scrapings with water, put in their bacon, and boil the mixture into a kind of soup, which is easier to swallow than pea-bread. When I happen in, they look so ashamed of their poor clothes. I know we saved the lives of two by giving a few meals....

The churches are a great resort for those who own no caves. People fancy they are not shelled so much, and they are substantial, and the pews good to sleep in.

Even as the two great armies in the East were moving toward their appointment with destiny in Gettysburg, the citizens and defenders of Vicksburg, Mississippi, were growing desperate. Soldiers whose daily rations had been cut to one small biscuit and a piece of bacon threatened mutiny. The city was cut off from outside contact and most supplies other than what could be smuggled in by night. In order to escape the relentless shelling from the Union guns, the people of Vicksburg had taken to living in caves dug into the ground.

After forty days of siege—on the same day that Pickett was sent against the Union army at Gettysburg—negotiations between Grant and the defenders of Vicksburg began. This time Grant allowed conditions, one of which was to "parole" the soldiers of Vicksburg. Not wanting to burden the navy with the task of transporting 30,000 prisoners to confinement in the North, he let them go with a pledge not to fight anymore.

Vicksburg surrendered on July 4, 1863. As the Union troops marched in, the young lady diarist reported:

> What a contrast to the suffering creatures we had seen were these stalwart, well-fed men, so splendidly set up and accoutred! Sleek horses, polished arms, bright plumes—this was the pride and panoply of war. Civilization, discipline, and order seemed to enter with the measured tramp of those marching columns; and the heart turns with throbs of added pity to the worn men in gray who were being blindly dashed against this embodiment of modern power.

The Union troops entered the fallen town and their bread and supplies were soon being distributed to the starving Confederates. Grant telegraphed the news to Lincoln, who was already savoring, if not with complete satisfaction, the outcome at Gettysburg. Upon receipt of Grant's message, the president said, "The Father of Waters again goes unvexed to the sea."

The city of Vicksburg would not officially celebrate Independence Day again until 1945, at the end of the Second World War.

Civil War Voices

According to a *New York Tribune* correspondent, Lincoln responded to complaints about rumors of Grant's drinking habits with this comment:

> By the way, can you tell me where he gets his whiskey? He has given us successes and if his whiskey does it, I

should like to send a barrel of the same brand to every general in the field.

MILESTONES IN THE CIVIL WAR: 1863

January 1 President Lincoln formally issues the *Emancipation Proclamation*, which frees the slaves only in rebel states with the exception of thirteen parishes in Louisiana, forty-eight counties in the future State of West Virginia, and seven counties in eastern Virginia. The proclamation does not apply to the Border States and their 300,000 slaves. But in England, mass rallies celebrate the emancipation, making British recognition of the Confederacy politically impossible.

January 3 The *Battle of Murfreesboro* (*Stones River*), *Tennessee*, a costly draw, concludes as Braxton Bragg retreats.

January 4 Grant is ordered to repeal his General Order No. 11, which expelled Jews from his area of operations.

January 25 Lincoln replaces the ineffective General Burnside with General Hooker as head of the Army of the Potomac.

The governor of Massachusetts receives the authority to recruit black troops. The 54th Massachusetts Volunteers is the first black regiment from the North.

The Confederate Congress calls black troops and their commanders criminals; black prisoners of war are effectively sentenced to death or enslavement if captured.

March 3 Lincoln signs the first Conscription Act. Enrollment is demanded of males between the ages of twenty and forty-five, although hired substitutes or payments of $300 can be used for an exemption.

April 17 Union gunboats begin to run past the citadel of *Vicksburg* under cover of night.

May The Bureau of Colored Troops is established by the War Department, beginning the aggressive recruitment of black soldiers.

May 1–4 *Battle of Chancellorsville.* In a major Confederate victory, Lee's army defeats Hooker's Army of the Potomac, but the South loses one of its most effective commanders when Thomas "Stonewall" Jackson is accidentally killed by Confederate troops.

May 5 Clement L. Vallandigham, the most vocal of the Copperheads (Peace Democrats opposed to Lincoln and the war), is arrested by federal troops. Jailed in Cincinnati, he is tried for treason and sentenced to two years in a military prison. On *May 19* Lincoln commutes this sentence to banishment to the Confederacy. Angry Ohio Democrats nominate Vallandigham for governor in *June.*

May 22 In the battle to take control of the Mississippi River, Grant attacks *Vicksburg, Mississippi.* Grant sustains heavy losses and begins a siege of the city designed to starve its defenders into surrendering.

June 1 Burnside closes the *Chicago Times* for publishing disloyal statements. Lincoln revokes the suspension.

June 22 Pro-Union West Virginia, formed out of several western Virginia counties, is admitted as the thirty-fifth state. Its constitution calls for the emancipation of slaves.

June 27 Lincoln relieves Hooker of command. General George G. Meade is appointed to lead the Army of the Potomac. Lee has begun an invasion into the North, and Meade must counter it.

July 1–3 *Battle of Gettysburg.* One of the most devastating battles of the war, it is the one that ultimately turns the tide against the Confederacy. The combined casualties are more than fifty thousand dead and wounded. Lee begins a retreat back to Virginia; Meade, given his severe losses, fails to pursue the straggling Confederates.

July 2 Vice-President Stephens is sent to Hampton Roads under a flag of truce to negotiate a prisoner exchange. He is also authorized to discuss an end to the war. With the Union victory at Gettysburg, Lincoln replies, "The request is inadmissible."

July 4 The surrender of *Vicksburg.* Accepting the "unconditional surrender" of this town, Grant gains complete control of the Mississippi River, effectively bisecting the Confederacy from north to south.

July 13–16 *Draft Riots.* In a violent reaction to the Conscription Act, mobs riot in *Boston; Portsmouth, New Hampshire; Rutland, Vermont; Wooster, Ohio;* and *Troy, New York.* The worst riots occur in *New York City*, where mostly Irish workingmen attack and lynch blacks. Federal troops come from Gettysburg to stop the mayhem.

July 17 The Confederate troops abandon *Jackson, Mississippi*.

July 18 In the charge made famous by the film *Glory*, the 54th Massachusetts Volunteers assault Fort Wagner in the harbor of *Charleston, South Carolina*.

August 21 Confederate raider William Quantrill attacks *Lawrence, Kansas*, looting and burning. More than 150 civilian men and boys are murdered by Quantrill's raiders.

September 19–20 *Battle of Chickamauga*. In a huge battle in Georgia, the Confederates eventually force the Union army to retreat back to *Chattanooga, Tennessee*.

October 3 Lincoln proclaims that the last Thursday in November will be set aside as a national day of Thanksgiving.

October 16 Lincoln names Grant to head the united western armies.

November 19 Lincoln delivers the Gettysburg Address.

November 23–25 Grant drives away the Confederate forces besieging *Chattanooga*. The Confederates then abandon *Knoxville*, leaving Tennessee under Union control.

December 8 In a message to Congress, Lincoln offers amnesty to Confederates who will take an oath of allegiance.

December 12 In *Richmond*, the Confederate government orders that supplies coming from the North to feed Union prisoners are to be stopped.

What Were the Draft Riots?

When Americans see outbursts like the riots that followed the Rodney King verdict in Los Angeles, we tend to shake our heads and wonder what's wrong with the modern world. Pictures of mayhem, burning, looting, and rioters smashing a man's head with a brick have us wishing for a simpler time, "the good old days." Well, it happened during the Civil War, too. In New York City. Only this time the rioters were mostly white and the victims of their violence mostly black.

"Hell no. We won't go." If you picture hippie types from the sixties chanting that slogan during the Vietnam War, think again. Imagine instead that the chanters include J. P. Morgan, soon to become America's wealthiest man; Andrew Carnegie, not far behind Morgan; future President Grover Cleveland, then a young, aspiring attorney; and the well-to-do fathers of Teddy and Franklin D. Roosevelt. They were but a few of the thousands who sat out the Civil War after the Enrollment Act was passed on March 3, 1863. They did it legally, by paying a substitute to go in their place, but thousands of others didn't want to go either, and they turned murderous in New York in the summer of 1863.

The first effective draft by the federal government, the Enrollment Act called for men ages eighteen to forty-five to register with local militia units and be available for national service. (This was not the first American draft; the Confederate draft had preceded it by a year.) Like the act in the Confederacy, the federal law exempted men in some occupations, such as telegraph operators, railroad engineers, judges, and other certain government employees. Men with mental disabilities or with certain types of dependents were also exempted. Physical disabilities that would exempt a man included imperfect vision in the right eye, the lack of front teeth and molars, and the loss of more than one finger of the right or more than two fingers of the left hand.

The states were responsible for the draft and usually relied on lottery systems. To meet the demand for troops, each state had to fill a quota based on its population. The number of volunteers from a state would be subtracted from its quota, and the difference would be made up by conscription. If a draftee volunteered, he was eligible to collect a bounty of $100 from the federal government plus additional bounties from the state and local communities. The bounties, or enlistment rewards, could exceed $500—about the average yearly wage in those days.

A draftee could also gain an exemption by paying $300 or by hiring a substitute. The obvious inequity of this solution prompted the cry of "rich man's war, but poor man's fight" on both sides. Even Lincoln, though technically too old for the draft, hired a substitute, feeling that he had to set an example. His oldest son, Robert Todd, a Harvard student, was also commissioned into the army when his civilian status proved an embarrassment.

The draft, bounties, lottery, and substitution systems all provided plenty of opportunity for draft dodgers. Physicians began to provide deferments for a fee. "Bounty jumpers" flourished. They would enlist to collect their bounty in one district, then desert and enlist elsewhere to collect another bounty. (The record for bounty jumping was held by a man who admitted enlisting and jumping thirty-two times before being caught.)

Civil War Voices

An anonymous New Yorker's response to the Conscription Act.

> Since poverty has been our crime,
> We bow to the decree.
> We are the poor who have no wealth
> To purchase liberty.

Coming on the heels of the Emancipation Proclamation, the draft law was bitterly resented. By the summer of 1863 angry protests and outbreaks of violence had taken place in nearly every Union state. The headline of one Pennsylvania newspaper read, WILLING TO FIGHT FOR UNCLE SAM BUT NOT FOR UNCLE SAMBO.

Nowhere was the opposition to the draft more vocal than in New York City, where Lincoln was despised by the powerful Democratic party, which was openly critical of his administration. The working-class Irish were particularly resentful of the policies that allowed the wealthy to buy their way out of the draft, and they were hostile toward blacks, many of whom had recently been used to replace striking Irish longshoremen.

On Saturday morning, July 11, 1863, the first draftees' names were drawn and published in the papers alongside the casualty lists from Gettysburg. The following Monday, July 13, the draft office at Third Avenue and Forty-sixth Street was attacked by a mob of men armed with clubs who set the building on fire. Instead of putting out the fire, a fire brigade, angry that their jobs no longer entitled them to an official exemption, joined the mob. This was the beginning of a four-day spree of looting and arson that ended with cold-blooded murder. New York's chief of police was attacked and left unconscious in the street. The offices of Horace Greeley were attacked with mobs screaming for the outspoken abolitionist. But singled out for deadly vengeance was the city's black population. The rioters, many of them too young for the draft but caught up in the frenzy, burned and pillaged their way down Third Avenue en route to an armory, where they swarmed inside and started grabbing the thousand rifles stored there. At the same time, another mob at Fifth Avenue and Forty-third Street attacked an orphanage where black children lived. In *The New York City Draft Riots*, Iver Bernstein catalogs some of the most grotesque acts that followed. A crippled black coachman was lynched, and his body was burned after his fingers and toes had been hacked off. A sixteen-year-old Irish boy then dragged the corpse through the street by the genitals. Small boys threw stones at the windows of black families to mark them. In a grim foretaste of the scene witnessed by television viewers when truck driver Reginald Denny was assaulted in Los Angeles in 1992, a black sailor was attacked by a mob of longshoremen. He was struck with cobblestones and stabbed; men jumped on his prostrate body as each member of the gang walked up and committed some atrocity. He died soon after.

Civil War Voices

The Brooklyn, New York, correspondent for the *Christian Recorder*, the newspaper of the African Methodist Episcopal church, published in Philadelphia.

Many men were killed and thrown into rivers, a great number hung to trees and lamp-posts, numbers shot down; no black person could show their heads but what they were hunted like wolves. These scenes continued for four days. Hundreds of our people are in station houses, in the woods, and on Blackwell's island. Over three thousand are to-day homeless and destitute, without means of support for their families. It is truly a day of distress to our race in this section. In Brooklyn we have not had any great trouble, but many of our people have been compelled to leave their houses and flee for refuge. The Irish have become so brutish, that it is unsafe for families to live near them, and while I write, there are many now in the stations and country hiding from violence. . . .

. . . In the village of Flushing, the colored people went to the Catholic priest and told him that they were peaceable men doing no harm to any one, and that the Irish had threatened to mob them, but if they did, they would burn two Irish houses for every one of theirs, and would kill two Irish men for every colored man killed by them. They were not mobbed.

During the worst of the four days of rioting and burning, Governor Horatio Seymour, a Democrat and a critic of Lincoln's, promised to get the draft repealed if only the lawlessness would stop. This plea accomplished little. For two more days, rioters controlled the city until troops sent from West Point and the Gettysburg battlefield eventually restored order.

Civil War Voices

Social reformer Mary Livermore, reporting on a tour of the Midwest as a member of the Sanitary Commission.

Women were in the field everywhere, driving the reapers, binding and shocking, and loading grain, until then an unusual sight. At first, it displeased me, and I turned away in aversion. By and by, I observed how skillfully they drove the horses around and around the wheat field, diminishing more and more its periphery at every circuit, the glittering blades of the reaper cutting wide swaths with a rapid, clicking sound that was pleasant to hear. Then I saw that when they followed the reapers, binding and shocking, although they did not keep up with the men, their work was done with more precision and nicety, and their sheaves had an artistic finish that those lacked made by the men. So I said to myself, "They are worthy women, and deserve praise: their husbands are probably too poor to hire help, and, like the 'helpmeets' God designed them to be, they have girt themselves to this work—and they are doing it superbly." Good wives! Good women!

Livermore's observation was accurate. The war had taken many men from their jobs, and the labor shortage was met in three meaningful ways. American invention began to respond with labor-saving devices that sped the Industrial Revolution in the United States. And more than 800,000 immigrants came to America during the war years, flooding the factories and farms with inexpensive labor. The dark underside of the immigration is that about one in twenty women was forced into prostitution.

And, as Livermore reported, women moved out of their traditional roles and did more of what was once exclusively men's work. The growth of the federal government during the war provided women with some of the first opportunities in government jobs. Women also took over many traditionally male jobs, such as teaching, as many young men volunteered and were later conscripted into the army. Although it was a far cry from women's liberation, it was the beginning of a social upheaval in American life that slowly culminated in universal suffrage and the gradual acceptance of women into professions and the workplace.

Documents of the Civil War

Lincoln's General Order regarding the treatment of captured black soldiers (July 30, 1863).

> It is the duty of every Government to give protection to its citizens, of whatever class, color, or condition, and especially to those who are duly organized as soldiers in the public service. The law of nations, and the usages and customs of war, as carried on by civilized powers, permit no distinction as to color in the treatment of prisoners of war as public enemies. To sell or enslave any captured person, on account of his color, and for no offense against the laws of war, is a relapse into barbarism, and a crime against the civilized of the age.
>
> The Government of the United States will give the same protection to all its soldiers; and if the enemy enslaves any one because of his color, the offense shall be punished by retaliation upon the enemy's prisoners in our possession.
>
> It is therefore ordered that for every soldier of the United States killed in violation of the laws of war, a Rebel soldier shall be executed; and for every one enslaved by the enemy or sold into Slavery, a Rebel soldier shall be placed at hard labor on public works and continued at such labor until the other shall be released and receive the treatment due a prisoner of war.

Was the Film Glory True?

To many whites in the Union who balked at conscription, the notion of fighting *for* the freedom of blacks was bad enough. Fighting *with* them was unthinkable.

But as the Union had moved to fill out the ranks of its army, Frederick Douglass, other free black leaders, and white abolitionists were proposing another solution to the shortage: allow free blacks to fight. Douglass had said, "Let the black man get upon his person the brass letters 'U.S.,' let him get an eagle on his buttons and a musket on his shoulder and there is no power on earth which can deny that he has earned the right to citizenship in the United States."

But these pleas fell on mostly deaf ears until after the Emancipation Proclamation was formally announced in January 1863. Within days John A. Andrews, the governor of Massachusetts and an ardent abolitionist, received authority from Secretary of War Stanton to form regiments that could "include persons of African descent." Andrews, who had long advocated the recruitment of black troops, chose the white officers for the new black regiment from wealthy families prominent in the abolition movement. These families could also be counted on to finance the enlistment and outfitting of the troops.

Andrews solicited the aid of Douglass and other black abolitionists in attracting the cream of the free black population for the new regiment. Two of Douglass's sons joined. Given the considerable Union opposition to allowing blacks to be soldiers, the regiment was seen as a crucial test of the fitness of black men as soldiers and citizens. Its supporters spared no expense in their effort to prove that blacks were equal to the test.

The 54th Massachusetts Volunteers was the first black regiment recruited in the North. Colonel Robert Gould Shaw, the twenty-five-year-old son of wealthy Bostonian abolitionists, was chosen to command it. On May 28 the 54th, well-equipped and well-drilled, paraded through the streets of Boston, then boarded ships bound for the coast of South Carolina. Their first combat came in a skirmish with Confederate soldiers on July 16, when the regiment repelled an attack on their camp on James Island.

Civil War Voices

Colonel Robert Gould Shaw, the commander of the 54th Massachusetts, in an address to his troops before leading the assault on Fort Wagner: "I want you to prove yourselves. The eyes of thousands will look on what you do tonight."

A few days after the draft riots, in the charge made famous by the film *Glory*, the courage of the black soldiers of the 54th Massachusetts was going to be tested by fire. The regiment was chosen to lead the assault on Battery Wagner, a fort protecting the entrance to Charleston. Shaw led the attack that carried his men onto the parapet of the fort, where he was killed by a bullet through the heart. The Confederate soldiers, aided by battery fire from Fort Sumter, beat off repeated attacks until the Union soldiers gave up and returned to their trenches. The 600 soldiers of the 54th suffered 272 casualties. The Union losses totaled 1,515 against only 174 Confederate casualties.

Shaw's body was stripped by the Confederates and tossed into a mass grave along with the bodies of the black soldiers. His family decided to leave him there; when his father learned how his son's remains had been treated, he said, "The poor benighted wretches thought they were heaping indignities upon his dead body; but the act recoils upon them. . . . We can imagine no holier place than in which he is."

For his bravery during the charge, Sergeant William H. Carney became the first black man to win the Congressional Medal of Honor. Nearly 200,000 blacks served in the Union armies; 37,000 died. They contributed significantly. Without their services, Lincoln acknowledged that the Union armies, reduced by casualties and desertions, might have been forced to negotiate a peace with the Confederacy. In other words, the war could not have been won without the services of the black soldier.

On September 6 the Confederates, recognizing the inevitable odds against them, quietly evacuated Battery Wagner, leaving Morris Island to the Union troops. One soldier called the attempts to take Battery Wagner "the most fatal and fruitless campaign of the entire war."

Civil War Voices

Walt Whitman, serving in Washington as a volunteer nurse (August 12, 1863).

I see the President almost every day, as I happen to live where he passes to or from his lodgings out of town. He never sleeps at the White House during the hot season, but has quarters at a healthy location some three miles north of the city, the Soldiers' Home, a United States military establishment. I saw him this morning about 8 1/2 coming to business, riding on Vermont avenue, near L street. He always has a company of twenty-five or thirty cavalry, with sabers drawn and held upright over their shoulders. They say this guard was against his personal wish, but he let his counselors have their way. The party makes no great show in uniform or horses. Mr. Lincoln on the saddle generally rides a good-sized, easy-going gray horse, is dress'd in plain black, somewhat rusty and dusty, wears a black stiff hat, and looks about as ordinary in attire, etc., as the commonest man. A lieutenant, with yellow straps, rides at his left, and following behind, two by two, come the cavalry men, in their yellow-striped jackets. They are generally going at a slow trot, as that is the pace set them by the one they wait upon. The sabers and accouterments clank, and the entirely unornamental cortege as it trots towards Lafayette square arouses no sensation, only some curious stranger stops and gazes. I see very plainly ABRAHAM LINCOLN's dark brown face, with the deep-cut lines, the eyes, always to me with a deep, latent sadness in the expression. We have got so that we exchange bows, and very cordial ones. Sometimes the President goes and comes in an open barouche. The cavalry always accompany him with drawn sabers. Often I notice as he goes out evenings—and sometimes in the morning, when he returns early—he turns off and halts at the large and handsome residence of the Secretary of War, on K street, and holds a conference there. If in his barouche, I can see from my window he does not alight, but sits in the vehicle, and Mr. Stanton comes out to attend him. Sometimes one of his sons, a boy of ten or twelve, accompanies him, riding at his right on a pony. Earlier in the summer I occasionally saw the President and his wife, toward the latter part of the afternoon, out in a barouche, on a pleasure ride through the city. Mrs. Lincoln was dress'd in complete black, with a long crape veil. The equipage is of the plainest kind, only two horses, and they nothing extra. They pass'd me once very close, and I saw the President in his face fully, as they were moving slowly, and his look, though abstracted, happen'd to be directed steadily in my eye. He bow'd and smiled, but far beneath his smile I noticed well the expression I have alluded to. None of the artists or pictures has caught the deep, though subtle and indirect expression of this man's face. One of the great portrait painters of two or three centuries ago is needed.

What Was a Jayhawker?

In the twentieth century, we have become far too accustomed to tales of atrocities against civilians during wartime. The televised "ethnic cleansing" from Bosnia is just one example. In spite of the romantic visions many people retain of the Civil War as a glorious, noble conflict, that sort of grim terror aimed at civilians also took place during the Civil War.

The worst of these atrocities happened in Kansas, the site of other bloody battles during the Kansas-Nebraska fighting ten years earlier. The fighting of that period only got worse during the Civil War, as

Jayhawkers—*jayhawking* meant "stealing"—and Bushwackers—an old folk term for backwoodsmen that later came to mean "sneak attack"—continued their no-holds-barred guerrilla warfare.

There were no rules. Prisoners were rarely taken. While the Confederate guerrillas commanded by a murdering psychopath named William C. Quantrill gained the most notoriety, units like the 7th Kansas Cavalry—"Jennison's Jayhawkers"—which included abolitionist John Brown, Jr., in its ranks, could be equally rapacious in their horse-stealing and murderous attacks on Missouri slaveholders who supported the Confederacy. One of Jennison's Jayhawkers (Jennison himself was a notorious horse thief) once wrote home, "About ten of us went out jayhawking . . . caught their horses and took the best ones . . . found some silverware. . . . I got the cups, two silver Ladles and two sets of spoons."

The regular Union forces in the area were commanded by Brigadier General Thomas Ewing, General William T. Sherman's brother-in-law. He was faced with trying to stop the Confederate raiders, especially those led by Quantrill. Born in Ohio, Quantrill had lived in Lawrence for a short time before the war but was forced out of town as "an undesirable." A gambler and petty thief, he joined the Confederacy after serving briefly with the Union. Although commissioned as a captain, his ruthless tactics in attacking civilian targets were regarded as counterproductive by most professional Confederate soldiers. But his superiors were willing to look the other way because his raids kept several regiments of Union soldiers preoccupied in Missouri and out of the war's main action. Quantrill soon attracted a rogues' gallery of equally cold-blooded, apparently psychopathic killers to his guerrilla force. They eventually included Frank and Jesse James and William "Bloody Bill" Anderson. Forget the Hollywood images of Robin Hood–like outlaws stealing from the rich to give to the poor. These were murderers.

Attempting to blunt the guerrillas' activities, in the summer of 1863 General Ewing began arresting women suspected of aiding Quantrill's men and jailed some of them in a dilapidated three-story building in Kansas City. On August 14, the building collapsed, killing four of the women and seriously injuring many others. One of the dead was the sister of Bloody Bill Anderson. Seeking revenge, Quantrill, Anderson, and 450 men stormed into Lawrence at daybreak on August 21 and went on a three-hour-long spree—drinking followed by burning, looting, and massacring civilians.

Quantrill had kept a hit list of men he specifically planned to kill, and he got all but one, U.S. Senator James H. Laine, who managed to hide in a cornfield in his nightshirt. Quantrill's victims were shot down coldbloodedly in front of their families or burned alive in their houses. Among the 182 men and teenage boys who were murdered, about 20 were unarmed recruits; the others were civilians. Although no women were physically harmed, many were left widowed or fatherless. More than 180 buildings were also torched in the raid. (Four of Quantrill's raiders at Lawrence were Cole, James, John, and Robert Younger. Known as the Younger Brothers, they teamed up with Frank and Jesse James after the war to rob banks, trains, and stagecoaches.)

In retaliation, on the following day, Ewing issued General Order No. 11, calling for the removal of civilians from four Missouri counties that bordered Kansas. Ten thousand people were forced from their homes in the fruitless effort to curb Quantrill. They became refugees, living on the plains after the Unionists destroyed their homes. For years to come, the area was known as the Burnt District.

Quantrill later separated from Anderson, another madman who rode into battle with the scalps of the Union dead dangling from his saddle. In a second, equally deadly, raid, Anderson took about 150 men into Centralia, Missouri, on September 27, 1864. Getting drunk, they looted the town and stopped a train, robbing its passengers. Then they shot two dozen unarmed Union soldiers. A party of 150 Union cavalry that set out after Anderson was ambushed instead. Their bodies were later discovered; the Union troopers had been shot, scalped, bayoneted, and had their noses and ears cut off. Anderson was finally killed in a fight with the Union militia in Missouri. His head was cut off and set on a telegraph pole.

William Quantrill, known as "the bloodiest man in the Annals of America," continued his raiding and eventually joined a regular Confederate army that was unsuccessfully attempting to attack St. Louis. With the end of the war in sight, Quantrill headed east, having told people that he was going to kill Abraham Lincoln. Caught in a fight with a Union patrol, he was shot and died of his wounds in a prison infirmary on June 6, 1865. Twenty-seven years old at his death, he left behind $500 to a lady friend in St. Louis, who used the money to finance a brothel.

Civil War Voices

From the *New York Herald* (October 6, 1863).

> All our theaters are open...and they are all crowded nightly. The kind of entertainment given seems to be of little account. Provided the prices are high and the place fashionable, nothing more is required. All the hotels are as crowded as the theaters; and it is noticeable that the most costly accommodations, in both hotels and theaters are the first and most eagerly taken. Our merchants report the same phenomenon in their stores: the richest silks, laces, and jewelry are the soonest sold. At least five hundred new turnouts may be seen any fine afternoon in the park; and neither Rotten Row, London, nor the Bois de Boulogne, Paris, can show a more splendid sight. Before the golden days of Indian summer are over these five hundred new equipages will be increased to a thousand. Not to keep a carriage, not to wear diamonds, not to be attired in a robe which cost a small fortune, is now equivalent to being a nobody.
>
> This war has entirely changed the American character. The lavish profusion in which the old Southern cotton aristocracy used to indulge is completely eclipsed by the dash, parade and magnificence of the new Northern shoddy aristocracy of this period. Ideas of cheapness and economy

are thrown to the winds. The individual who makes the most money—no matter how—and spends the most money—no matter for what—is considered the greatest man. To be extravagant is to be fashionable. These facts sufficiently account for the immense and brilliant audiences at the opera and the theaters; and until the final crash comes such audiences will undoubtedly continue.

The world has seen its iron age, its silver age, its golden age and its brazen age. This is the age of shoddy. The new brownstone palaces on Fifth Avenue, the new equipages at the park, the new diamonds which dazzle unaccustomed eyes, the new silks and satins which rustle over loudly, as if to demand attention, the new people who live in the palaces and ride in the carriages and wear the diamonds and silks—all are shoddy. From the devil's dust they sprang and unto the devil's dust they shall return.

There were no such excesses in the Confederacy. In October 1863 Captain W. M. Gillaspie, in charge of outfitting the army in Alabama, made this newspaper appeal to the ladies of the state:

I want all the blankets and carpets that can possibly be spared. I want them, ladies of Alabama, to shield your noble defenders against an enemy more to be dreaded than the Northern foe with musket in hand—the snows of coming winter. Do you know that thousands of our heroic soldiers of the West sleep on the cold, damp ground without tents? Perhaps not. You enjoy warm houses and comfortable beds.

If the immortal matrons and maidens of heathen Rome could shear off and twist into bowstrings the hair of their heads to arm their husbands in repelling the invader, will not the Christian women of the Confederacy give the carpets off their floors to protect against the chilly blasts of winter those who are fighting with more than Roman heroism, for their lives, liberty, and more, their honor? Suffi-

cient blankets cannot be had in time. Food and clothing
failing the army, you and your children will belong to Lincoln. To get your daily bread, you will then be permitted
to hire yourselves to your heartless enemies as servant, or
perchance to your own slaves. Think of that! Think of your
brothers, fathers, and sons, drenched with the freezing rains
of winter, and send in at once every blanket and carpet, old
or new, you can spare. They will be held as a sacred trust.
As soon as they can be gotten ready for issue, they will be
sent to the quartermaster in chief of General Johnston's
army for distribution.

As a guarantee that a proper distribution shall be made of
such as may be donated, H. H. Ware, Esq., will receive and
receipt for the same at Selma. Honorable and well-known
names will be announced to receive and receipt for the
same at Montgomery, Tuscaloosa, Demopolis, Marion and
elsewhere.

We will pay a liberal price for all that may be delivered
at this place, or to any bonded quartermaster in this State,
upon the presentation of his certified account upon form
No. 12. . . .

Ministers of the Gospel also are urgently requested to call
the attention of their congregations to this matter, Every
one, male or female, who can furnish a blanket may save a
man to the army.

The sacrifices demanded of the southern civilian were also telling
in Richmond. In his wartime journal, John B. Jones, a clerk in the
Confederate War Department, wrote,

A poor woman yesterday applied to a merchant in Carey
Street to purchase a barrel of flour. The price he demanded
was $70.

"My God!" exclaimed she, "how can I pay such prices?
I have seven children; what shall I do?"

"I don't know, madam," said he, coolly, "unless you eat your children."

Such is the power of cupidity—it transforms men into demons. And if this spirit prevails throughout the country, a just God will bring calamities upon the land which will reach these cormorants, but which, it may be feared, will involve all classes in a common ruin.

Was the Gettysburg Address Written on the Back of an Envelope?

This is what most of us learned in school. Lincoln was on a train bound for Gettysburg when he got the idea for his speech and jotted it down on the back of an envelope in his pocket.

In fact, Lincoln was on a train, traveling to Gettysburg to help dedicate a cemetery at the site of the July battle. A cemetery was necessary not only for ceremonial reasons but because the three-day battle had left a scene of death and desolation that did not disappear when the two armies pulled out. For anyone who retains a romantic notion of war, Garry Wills painted a grim picture of the scene of battle in this once-bucolic town in *Lincoln at Gettysburg*:

> The residents of Gettysburg had little reason to feel satis-
> faction with the war machine that had churned up their
> lives. General Meade may have pursued Lee in slow mo-
> tion; but he wired headquarters that "I cannot delay to pick
> up the debris of the battlefield." That debris was mainly a
> matter of rotting horseflesh and manflesh—thousands of
> fermenting bodies, with gas distended bellies, deliquescing
> in the July heat. For hygienic reasons, the five thousand
> horses (or mules) had to be consumed by fire, trading the
> smell of burning flesh for that of decaying flesh. Eight thou-
> sand human bodies were scattered over, or (barely) under,
> the ground. Suffocating teams of soldiers, Confederate pris-
> oners, and dragooned civilians slid the bodies beneath a

minimal covering, as fast as possible—crudely posting the names of the Union dead with sketchy information on boards, not stopping to figure out what units the Confederate bodies had belonged to. It was work to be done hugger-mugger or not at all, fighting clustered bluebottle flies black on the earth, shoveling and retching by turns. The buzzards themselves had not stayed to share in this labor—days of incessant shelling had scattered them far off.

Even after most bodies were lightly blanketed, the scene was repellent. A nurse shuddered at the all-too-visible "rise and swell of human bodies" in these furrows war had plowed. A soldier noticed how the earth "gave" as he walked over the shallow trenches. Householders had to plant around the bodies in their fields and gardens, or brace themselves to move the rotting corpses to another place. Soon these uneasy graves were being rifled by relatives looking for their dead—reburying other bodies they turned up, even more hastily (and less adequately) than had the first disposal crews. Three weeks after the battle, a prosperous Gettysburg banker, David Wills, reported to Pennsylvania's Governor Curtin: "In many instances arms and legs and sometimes heads protrude and my attention has been directed to several places where the hogs were actually rooting out the bodies and devouring them."

The work of creating a suitable memorial for the Union dead of Gettysburg—the Confederate dead would be left anonymously in mass graves—was a logistical nightmare, complicated by politics, of all things. The architect of the cemetery had to avoid giving the graves of one state's dead "preferential" placement. The grim reburial of the dead was complicated by many factors, including the fact that many Confederate soldiers had been wearing Union uniforms taken from dead bodies after earlier battles. (The contract to rebury the bodies went to a bidder at $1.59 per corpse.)

On November 19, 1863, the national cemetery on the Gettysburg battlefield was to be dedicated. After the three most notable poets of the day—Henry Wadsworth Longfellow, William Cullen Bryant, and John Greenleaf Whittier—all declined to contribute to the occasion, the dedication committee selected Edward Everett (1794–1865) of Massachusetts as the principal speaker. Having served as an editor, ambassador, president of Harvard University, secretary of state, and U.S. senator, Everett was a scholar and one of the most distinguished men in America. Although it might be difficult for modern audiences to imagine, he could keep a large lecture hall entranced with his dramatic, classical speeches. He had also helped Daniel Webster prepare the conclusion to his famous reply to Senator Hayne (see page 56). The committee sent an invitation to President Lincoln: "It is the desire that after the oration, you, as Chief Executive of the nation, formally set apart these grounds to their sacred use by a few appropriate remarks."

The invitation was intended only as a courtesy; this was a state, rather than a federal, ceremony, and the lines were much more carefully drawn between such events then. But Lincoln accepted the invitation because he had been seeking an appropriate opportunity to express his thoughts on the Union's purpose in the war. He was attending this event even though Mary Lincoln objected. She was nearly hysterical over the grave illness of Tad, one of the Lincolns' two surviving sons.

Lincoln prepared carefully—the speech was not composed on the back of an envelope or on the train to Gettysburg or at the spur of the moment when he rose to speak. Garry Wills notes, "These mythical accounts are badly out of character for Lincoln, who composed his speeches carefully. . . . That is the process vouched for in every other case of Lincoln's memorable public statements."

The burial site was still incomplete when the speakers arrived around noon. Almost twenty thousand spectators crowded the speakers' platform. The program included music and prayers as well as the expected two-hour oration by Everett. Lincoln's role was listed as

"Dedicatory Remarks," to be followed by a dirge and benediction. The president stood, still wearing a mourning band on his hat for his son Willie, and in his high-pitched, nearly shrill western twang delivered his 272-word address in about three minutes; he was interrupted five times by applause. His private secretary, John Hay, thought "The President, in a fine, free way, with more grace than is his wont, said his half dozen words of consecration."

Lincoln made one mistake in his speech when he said, "The world will little note, nor long remember what we say here." This raises another part of the Gettysburg Address mythology: that Lincoln himself was disappointed in the speech. He supposedly told one friend, "It is a flat failure and the people are disappointed." But Garry Wills has rebuffed that traditional interpretation as having no basis in fact. To be sure, there were some critics. "The cheek of every American must tingle with shame as he reads the silly, flat, and dishwatery utterances of . . . the President" was the reaction of the *Chicago Times*, which also objected to the content of the speech, claiming that Lincoln had subverted the very reason for the war. The correspondent for the London *Times* also complained peevishly, "Anyone more dull and commonplace it would not be easy to produce."

However, *The New York Times* was complimentary, as were the reports in *Harper's Weekly*. And many other Americans soon remarked on the compact greatness of the speech. Beyond the classic elegance of Lincoln's words and the simple beauty of his prose, the Gettysburg Address accomplished two things that may seem obvious today but were considered revolutionary when Lincoln voiced them. Without mentioning slavery, nullification, or states' rights, Lincoln firmly and permanently connected the birth of the United States of America to the Declaration of Independence, with its ringing statement that "all men are created equal," rather than the Constitution of 1783, with its implicit recognition of slavery. Further, he spoke of a *nation* instead of a *union*. In this brief, simple speech, Lincoln had made it clear once and for all that America was one nation, not merely a collection of sovereign states.

Who Was Johnny Clem?

The photograph shows a small boy dressed in a soldier's uniform, almost as if he were going to a Halloween party. His face is innocent, perfectly childlike. His hand is held inside his jacket in imitation of all those "real" generals who had tried to emulate Napoleon's famed pose. But he was not a child posing for a costume party. He was Johnny Clem. A federal nurse said that Clem "was a fair and beautiful child . . . about twelve years old, but very small for his age. He was only about thirty inches high and weighed about sixty pounds." He was a soldier, one of the youngest of the many boys under the age of eighteen in the Union army and many more on the Confederate side.

Often called a Brothers' War, the Civil War was also a Children's Crusade. No one can say with any certainty how many underage soldiers—that is, younger than eighteen, the Union army's legal limit—served in the war. One estimate is of 10,000 boys in the Union and many more on the Confederate side. In *The Boys' War*, Jim Murphy says that the number could have been anywhere from 250,000 to 420,000, basing his estimate on enlistment records that later showed between 10 and 20 percent of enlistees were underage. Recruiters eager to fill their quotas were usually not concerned with birth certificates if a tall, strapping farmboy came along to sign up. But even the short and puny made it, many of them filling out the ranks of regimental musicians, especially as drummer boys and buglers.

It was as a drummer boy that young Johnny Clem became a Civil War legend. With the fall of Vicksburg and the retreat of Lee's army from Pennsylvania, the Confederacy was encircled by Union land and naval forces. In essence, Winfield Scott's much-derided 1861 Anaconda Plan had been accomplished. The "snake" was rapidly tightening its hold. Grant continued to apply pressure by striking farther into the Deep South, and one of his principal targets was Chattanooga, Tennessee, a major southern railroad center just north of the Georgia border.

Under General William Rosecrans, Union armies had taken Chattanooga in mid-September, right after the Union Army of the Ohio under the ill-fated Ambrose Burnside captured Knoxville, in eastern Tennessee. Still confronting Rosecrans was a Confederate force led by General Braxton Bragg. On September 18, after Bragg was reinforced by troops from Gettysburg under James K. Longstreet, the two armies—each with about sixty thousand men—began to skirmish at Chickamauga Creek, just south of Chattanooga.

The battle began in full on the nineteenth and continued into the next day. Sometime during the fighting a Confederate officer came upon Johnny Clem and ordered him to surrender. John Joseph Klem (he later changed the name and spelling to John Lincoln Clem) had run away from his Ohio home to join the Union army in the spring of 1861, before he was ten years old. Turned down because of his age, he managed to tag along with the 22nd Massachusetts, which eventually adopted him as its mascot and drummer boy. Officers of the unit chipped in for his $13 monthly salary, and his fellow soldiers were said to have provided Johnny with a shortened rifle and a uniform in his size. In May 1863 he officially enlisted in the regiment and was able to receive his own pay.

On September 20, 1863, many members of the 22nd Massachusetts were captured in the Battle of Chickamauga, but Johnny escaped when the Confederate officer tried to capture him. He halted as if to surrender, then leveled and fired his sawed-off musket, killing the Confederate colonel. He feigned death among the piles of wounded and dead soldiers until he could escape back to the Union lines. Union General George H. Thomas promoted Johnny to lance corporal, and the newspapers picked up his story. Little Johnny Clem became a celebrity, known as "the drummer boy of Chickamauga" and as "Johnny Shiloh," since he was alleged to have had his drum smashed by cannon fire in the Battle of Shiloh in 1862.

In October 1863, Johnny was captured again by Confederate cavalry while detailed as a train guard. He was exchanged a short time later, but the Confederacy used his captivity to show "what sore straits the Yankees are driven, when they have to send their babies to fight us." In January 1864, General Thomas assigned Johnny to his staff as a mounted orderly, and on September 19, Johnny was discharged from the army.

During that Chickamauga battle, another famous nickname was being earned on a day of ferocious combat. Having seen his lines broken by Longstreet, the Union's Rosecrans was organizing a fall-back to Chattanooga as another Union army appeared on the brink of being routed. Fortunately for Rosecrans, another part of the field was under the command of General George Thomas, a Virginian who had remained loyal to the Union. As one witness related, "Thomas had placed himself with his back against a rock and refused to be driven from the field. Here he stayed, despite the fierce and pro-longed assaults of the enemy, repulsing every attack. And when the sun went down he was still there. Well was he called the 'Rock of Chickamauga.' "

Although Thomas eventually withdrew, his stand had saved the Union army, but once again the casualties on both sides were stag-gering. Done in by the battle, the Confederates failed to press their advantage, allowing Rosecrans time to regroup in Chattanooga, where the Confederates would now begin a siege as Grant had done at Vicksburg.

Civil War Voices

W.F.G. Shanks, a war correspondent with Rosencrans in Chatta-nooga.

Life in Chattanooga was dreary enough. The enemy daily threw a few shells from the top of Lookout Mountain into our camps, but they were too wise to attack with infantry the works which . . . encircled the city.

Bragg preferred to rely for the final reduction of the gar-rison upon his ally Famine—and a very formidable antag-onist did our men find him.

Guards had to be posted to keep the soldiers from stealing grain meant for horses. The men took to following the supply wagons that did arrive to pick up grains of corn and cracker crumbs that fell to the ground. Ironically, things were little better for the opposing Confederate troops, who were on the attack. When Jefferson Davis once arrived to review these troops, Private Sam Watkins said that the nearly naked, vermin-covered soldiers shouted, "Send us something to eat, Massa Jeff. I'm hungry."

During the siege, in mid-October, Grant was given control over the consolidated armies of the Tennessee, Ohio, and the Cumberland (the Union armies were named after rivers). On October 23 he arrived in Chattanooga to take personal command of the situation. Soon, the Union armies were able to open up a supply line, "the cracker line," into the besieged city. With a steady flow of supplies, Grant could then confront the Confederate army outside Chattanooga. Commanded by Braxton Bragg, the Confederates were greatly outnumbered, with two Union armies on either side of them and another in front of their position on a hill called Missionary Ridge.

Correspondent B. F. Taylor observed the battle for Missionary Ridge.

> From division to brigades, from brigades to regiments, the order ran. A minute, and the skirmishers deploy; a minute, and the first great drops begin to patter along the line; a minute and the musketry is in full lay like the cracking whips of a hemlock fire.
>
> Men go down here and there before your eyes. The wind lifts the smoke and drifts it away. . . . The divisions of Wood and Sheridan are wading breast-deep in the valley of death.
>
> And so through the fringe of woods went the line. Now, out into the open ground they burst. . . . The tempest that now broke upon their heads was terrible. The enemy's fire burst out of the rifle pits from base to summit of Missionary Ridge; five rebel batteries . . . opened along the crest. Grape and canister and shell sowed the ground with rugged iron

and garnished it with the wounded and dead. But steady and strong our columns moved on.

After the top of the ridge was taken, Taylor continued: "Dead rebels lay thick upon the ridge. Scabbards, broken arms, artillery horses, wrecks of gun carriages, and bloody garments strewed the scene. And—tread lightly, oh loyal-hearted! . . . boys in blue are lying there. . . . A little waif of a drummer boy . . . lies there, his pale face upward, a blue spot on his breast. Muffle his drum for the poor child and his mother!"

Bragg's army broke into retreat, pursued briefly by Grant. The Battle of Chattanooga was not as costly in numbers as other recent battles, but it marked one more ratcheting of the Union's clamp on the South.

Civil War Voices

Private D. L. Day of the 25th Massachusetts Volunteers, stationed at Newport News, Virginia (December 10, 1863).

General Foster has been ordered to Knoxville, Tenn. and General [Benjamin] Butler has superseded him to this command. I am not pleased with the change. General Foster was a splendid man and fine officer, and I would rather take my chances with a regular army officer than with an amateur. The first year of the war General Butler was the busiest and most successful general we had, but since then he has kind o' taken to niggers and trading. As a military governor he is a nonesuch, and in that role, has gained a great fame, especially in all the rebellious states. He is a lawyer and man of great executive ability, and can not only make laws but see to it that they are observed, but as a commander of troops in the field, he is not just such a man as I should pick out.

He had a review of our brigade the other day and his

style of soldiering caused considerable fun among the boys who had been used to seeing General Foster. He rode on to the field with a great dash, followed by staff enough for two major-generals. He looks very awkward on a horse and wears a soft hat; when he salutes the colors he lifts his hat by the crown clear off his head instead of simply touching the brim. The boys think he is hardly up to their ideas of a general but as they are not supposed to know anything, they will have to admire that he *is* a great general. He is full of orders and laws (regardless of army regulations) in the government of his own department, and his recent order in relation to the darkies fills two columns of newspaper print and is all the most fastidious lovers of darkies in all New England could desire. Hunter and Frémont are the merest pigmies beside Ben in their care of darkies.

From the diary of Secretary of the Navy Gideon Welles (December 31, 1863).

The year closes more satisfactorily than it commenced. . . . The War has been waged with success, although there have been in some instances errors and misfortunes. But the heart of the nation is sounder and its hopes brighter.

chapter seven

1864–1865
"All the Force Possible . . ."

On the night before Sherman entered the place, there were citizens who could enumerate their wealth by millions; at sunrise the next morning they were worth scarcely a dime. . . . Government had seized their cotton; the Negroes had possession of their lands; their slaves had become freemen; their houses were occupied by troops; Confederate bonds were waste paper; their railroads were destroyed, their banks insolvent. They had not only lost wealth, but they had lost their cause.

—CHARLES COFFIN
DECEMBER 1864

I have always thought that all men should be free; but if any should be slaves, it should be first those who desire it for themselves, and secondly those who desire it for others. Whenever I hear anyone arguing for slavery, I feel a strong impulse to see it tried on him personally.

—PRESIDENT ABRAHAM LINCOLN
MARCH 17, 1865

This is a sad business, colonel. It has happened as I told them in Richmond it would happen. The line has been stretched until it is broken.

—GENERAL ROBERT E. LEE
MARCH 1865

❋

* Who Burned Columbia and Charleston,
South Carolina?

* How Did Richmond Fall?

* What Happened at Appomattox Courthouse?

The small, undistinguished man in a battered uniform checked into Washington's famed Willard Hotel with his son on March 8, 1864. The clerk nearly turned the officer away until he unceremoniously signed the register. When Abraham Lincoln heard that the man he was expecting had arrived, he invited him to the weekly White House reception that evening. There was a tremendous crush of people eager to greet this short, rough-bearded, cigar-chewing soldier. Finally the crowd grew so thick that the man had to stand on a couch to be seen and avoid the crush of the mob flocking to meet the new lieutenant general of the Armies of the United States, Ulysses S. Grant. Lincoln had found his general.

It had taken three long, bloody years. But in March 1864 Grant was ordered to Washington to assume command of the Union armies. Promoted to lieutenant general, the puny Ohio boy who had once quit the army as a depressed failure now led more than half a million Union soldiers. Unlike the collection of bickering, chest-thumping, incompetent, or small-minded men who preceded him, Grant planned to coordinate these troops and, in his words, "concentrate all the force possible against the Confederate armies in the field." This was Lincoln's dream come true: a commander who understood that rebel armies, not Confederate cities, were the true target.

In the first three years, the war had grown progressively more brutal. Now Grant and Union commanders like William T. Sherman would elevate the grim battle statistics to new levels of horror and devastation. The year of 1863 had been the bloodiest of the war. The year to come would grow even worse.

In Richmond, the situation was past bleak. As the *Richmond Examiner* stated it on New Year's Eve, "To-day closes the gloomiest year of our struggle." With the devastating military losses of 1863 and the economy in a shambles, Jefferson Davis's beleaguered government was struggling to survive. Union sympathies in the Confederate states were being openly expressed. In Raleigh, North Carolina, the *Standard*'s pro-Union stance brought a mob of Confederate troops to the newspaper's offices. At about the same time, a group of pro-Union civilians wrecked the presses of the secessionist *Raleigh*

Journal. Several peace overtures would be made this year. All would be rejected because they denied the Union. Desertions from the Confederate ranks had become epidemic. But the Confederate armies still in the field were like wounded, cornered animals protecting their lair. The men leading those armies were still dangerous—defiant and unwilling to surrender.

Growing desperately short on manpower, the Confederacy began to extend conscription. The youngest and the oldest were soon being drafted. Mail clerks in Richmond, who had gone on strike a few months earlier over wages, were also called to join all the walking wounded from the hospitals pressed to defend the city. The manpower shortage even reached the point of the unthinkable—freeing and arming slaves who would fight.

As the new year began, the Confederacy had only two substantial armies left in the field. Robert E. Lee's Army of Northern Virginia, some 66,000 ill-clad, ill-fed troops, was now falling back toward Richmond. A second army, in Georgia under Joseph E. Johnston, was attempting to hold Atlanta, the Confederacy's other remaining key city. Two smaller Confederate forces were also active. One was in Virginia's Shenandoah Valley, protecting the last great storehouse and granary of the Confederacy. The second, in Tennessee, was under the command of General Nathan Bedford Forrest, a brilliant, unconventional, and controversial cavalry genius.

Did a Confederate General Order a Massacre of Black Troops at Fort Pillow?

Early in April, Confederate General Nathan B. Forrest sent a message to his commander, Leonidas Polk. "There is a federal force of 500 or 600 at Fort Pillow which I shall attend to in a day or two, as they have horses and supplies which we need."

Constructed early in 1862 by the notorious Gideon Pillow, the Confederate fort was an earthwork garrison with a nearby trading post on a bluff high above the Mississippi River. Fifty miles north of Memphis, Fort Pillow had been captured by Union forces in June

1862, and by 1864 it was one of the many garrisons used to defend the Union supply lines. A Union gunboat lay in the river to provide additional protection for the garrison. In the spring of 1864, the fort was manned by some 295 white troops of the 13th West Tennessee Cavalry—Unionists, some of whom had deserted from the Confederate ranks and were considered traitors by the Confederate Tennesseans. Also in the fort were 292 black soldiers of the 11th U.S. Colored Troops, mostly freed slaves sent north from Memphis; they were as despised by the Confederate troops as the "renegade Tennesseans" were. Official Richmond policy dictated that black Union soldiers were to be treated as runaway slaves who should be returned to their former masters; to the Confederate troops, the nearest trader willing to pay for black captives was just as good. The Confederate Congress had also sanctioned a policy of killing white Union officers who commanded black troops. Lieutenant General Kirby Smith, who ran a practical fiefdom in the Trans-Mississippi Department known as "Kirby Smithdom," had declared a policy of killing all captured blacks in U.S. uniform, a policy the Richmond government only weakly disputed.

At this time, Tennessee was painfully feeling the endgame of the war. The state, which had actually voted against secession in a referendum in 1861, was always divided uneasily between Unionists and Confederates. Now more Tennesseans expressed Union sympathy, eager for ending the war and rejoining the Union on generous terms. But this division led to bad blood among Tennesseans. It was made worse when the Union commander at Fort Pillow, Major William Bradford of Tennessee, reportedly began to roam the country, stealing food, cattle, and anything of value from Confederate sympathizers. Confederate Tennesseans turned for help to General Forrest, a Tennessee man himself and one of the Confederate leaders most dreaded by Union soldiers.

The wily survivor of the Confederate debacle at Fort Donelson earlier in the war, Forrest had also escaped paralysis or a potentially fatal wound at Shiloh, when he was shot near the spine. A tough, mean-spirited, unflinching character, Forrest had moved from Abraham Lincoln–style log-cabin, frontier poverty by handsomely profiting in the slave trade. This business earned him a fortune large enough to mount his own cavalry battalion once the war began. "War means fightin'," he once said, "and fightin' means killin'." On at least two occasions that applied to his fellow Confederates as well as the enemy. He once came close to a duel with Confederate General Earl Van Dorn. (Van Dorn was killed later in the war by a civilian, who said that the general had "violated the sanctity of his home" in an incident involving the man's wife.) In another fabled exploit, Forrest once argued with a subordinate in the aftermath of a battle. The man started to pull a gun and it fired, striking Forrest in the left side. Opening his penknife with his teeth, Forrest grabbed the gunman's hand before he could shoot again and sunk the knife into his stomach. The young officer died soon after. Twelve days later Forrest was in the saddle again. In another famous episode, Forrest used a small force of about four hundred men to march and ride in circles to confuse the Union soldiers; he captured more than sixteen hundred troops under Union officer Abel Streight.

Forrest was called the most remarkable soldier the war produced by both Robert E. Lee and William T. Sherman—who also called him "a devil" and put out a reward for his capture. Forrest was basically uneducated, not a schooled gentleman like many of his Confederate colleagues—many of whom he openly despised. A recent Forrest biographer, Jack Hurst, summarizes the unconventional cavalry leader as

> impatient with many of the seemingly petty rules of war and warriors, unwritten laws of cavalierish gallantry which seemed oblivious to the awesome imperatives of winning. To him, everything always had depended on final triumph, not on gentlemanly gamesmanship. . . . Like all the frontier fights he had made, this was not a game. It was a struggle for no less than survival. . . . After Shiloh, he began to wage war more nearly the way he had lived the rest of his life;

not only single-mindedly, but confident in his own counsel
and following his own rules.

And it was the notoriety of those personal rules that would blacken
Forrest's name at Fort Pillow in April 1864.

A few weeks earlier, Forrest and his men had been turned away
at Paducah, Kentucky, by Union troops who resisted his demands for
surrender. That was a first; Forrest's fearsome legend was so powerful
that his previous surrender demands had always been accepted.
When the general learned that black Union troops helped repel his
men at Paducah, it seemed only to make them angrier about their
failure and the garrison at Fort Pillow was the target of their vengeance. At 5:30 A.M. on April 12, fifteen hundred of Forrest's men
surrounded the fort. Having ridden seventy-two miles in twenty-
seven hours, the general arrived at the fort at 10:00 A.M. to position
his men where they could fire down on the Union soldiers. At 3:30
P.M., Forrest sent out a flag of truce and demanded the fort's surrender, promising to treat all the men of Fort Pillow as prisoners of war
in his note to the Union commander, Major Lionel F. Booth. A Philadelphia clerk who had joined the army in 1858, Booth had risen to
command a battalion of the 1st Alabama Siege Artillery (African Descent). But he had been killed by a sharpshooter earlier that day.
Without letting on that Booth was dead, Major William F. Bradford,
the supposedly renegade Tennessean, requested an hour to consult
with his subordinates. Bradford was playing for time, expecting reinforcements to arrive on Mississippi transports.

Forrest could see the smoke from the steamers on the river and,
suspecting the Union of delaying tactics, he gave the fort only twenty
minutes to decide. He also repositioned his men under the flag of
truce, a no-no in nineteenth-century warfare, where such niceties
were still honored. (Forrest had been accused of using this tactic on
other occasions.) At the end of that time the surrender was refused,
and Forrest ordered an assault.

The Confederates drove the Union soldiers out of the fort and
down the riverbank, where many tried to surrender. What happened

next was long debated, both during and after the war. Without question, the Union troops died in unusually large numbers, suggesting that a massacre had taken place. Witnesses on both sides recounted scenes of murder as the surrendering Union soldiers, blacks in particular, were simply shot dead. One Union witness said, "I also saw at least 25 negroes shot down, within 10 or 20 paces from the place where I stood. They had also surrendered, and were begging for mercy." He later testified hearing a Confederate officer order his men to "kill the last God damned one of them." And as one member of the Confederate 20th Tennessee put it a week later, "The slaughter was awful." He also wrote that General Forrest "ordered them shot down like dogs." The Union troops suffered 231 killed, 100 wounded, and 26 captured. The black units suffered 66 percent killed, the white units only 35 percent, most casualties occurring after the walls had been breached. The attacking Confederates suffered only 14 killed and wounded.

Bradford was captured. Two days later, as he was being transported to a prison camp, he was shot; his captors claimed he was trying to escape. A Union sympathizer said he'd been assassinated.

Civil War Voices

Confederate General Nathan Forrest, following the Fort Pillow massacre: "The river was dyed with the blood of the slaughtered for 200 yards. It is hoped that these facts will demonstrate to the northern people that negro soldiers cannot cope with Southerners."

Did Forrest actually order a massacre? It was widely thought so at the time, although his supporters point to his surrender demand promising fair treatment for all prisoners. Perhaps it is more likely that Forrest had been promising a "no quarter" threat—that is, to show no mercy or clemency to the defeated—for months. And his men took him at his word. Given the stated policy of the Richmond government toward black troops and their Union officers, and the charged feelings of hatred toward armed blacks among many Confederate soldiers, the seeds of an atrocity had been sown. Forrest simply nursed them. He may not have actually ordered the execution of the black soldiers, but he had created an atmosphere so poisonous that a massacre was likely, if not inevitable. As the commander, he must be ultimately deemed responsible for the actions of his men.

The Fort Pillow incident was not the only massacre of black captives. In North Carolina, when the Confederates recaptured coastal Plymouth in April 1864, a Union sergeant testified that blacks in uniform were shot, hanged, or had their brains beaten out by musket butts. White officers were reportedly dragged through town with ropes around their necks. Similar cases were reported elsewhere, though never on the scale of Fort Pillow. To black Union fighting men, "Remember Fort Pillow" became a rallying cry. And at Petersburg in June 1864, Charles Francis Adams reported of the black troops, "The darkies fought ferociously. If they murder prisoners, as I hear they did . . . they can hardly be blamed."

How Did Both Sides Treat Their POWs?

As the events at Fort Pillow demonstrated, the lot of Civil War prisoners had become a dangerous one. Passions and propaganda had destroyed the accepted conventions of warfare and treatment of captives, earlier codified in Europe. By 1864, vengeance replaced any gentleman's code of "civilized" warfare. If there was ever "fair play" in wartime, the brutality of the Civil War had destroyed it.

Formal prisoner exchanges, used earlier in Europe, were not instituted initially because Lincoln did not want to recognize the Confederacy. Gradually, however, informal exchanges under flags of truce became common between the two sides but did not involve large numbers of men in the early years of the war. Finally, in 1862, an agreement was made between the two armies—rather than the two governments—under which all prisoners were supposed to be exchanged within ten days of their capture. Under this agreement, a certain number of enlisted men could be exchanged for an officer; a general was worth from forty-six to sixty privates; a major general, forty privates; and finally down to two privates per noncommissioned officer.

There was also a curious system called "parole," borrowed from the European military tradition, in which released prisoners agreed not to serve a military role until they had been officially exchanged for an enemy prisoner. Using the honor system, parole worked well enough to keep the prisons relatively empty, although it created a mountain of paperwork and a bureaucratic nightmare for the two governments. That was good for soldiers, who often waited months before being told to report back for duty when they had been officially "exchanged." In 1863, after capturing Vicksburg, Grant paroled more than 31,000 Confederate soldiers rather than transporting them back to Union prison camps. He was furious when some of these parolees turned up among the Confederate captives at Chattanooga; in April, after taking overall command, he ordered an end to prisoner exchanges until the Confederates agreed to honor the one-for-one bargain.

In the wake of Fort Pillow and other reports of the killing of black prisoners, Grant also demanded that the Confederates make no distinction between white and black prisoners. It was this April 17 order, coupled with Grant's desire to place a further strain on Confederate manpower and resources, that effectively halted prisoner exchanges. Grant's decision certainly added to the woes of the hard-pressed Confederacy, but its unintended effect was a grim death sentence for thousands of Union prisoners.

Prisoner of war camps were among the most tragic and inhumane disgraces of the war. Just as the character of the war changed as the years went by and the casualties mounted, the character of the prisons grew bleaker and more cruel. Retribution on the part of sadistic and vindictive administrators on both sides eventually replaced the most basic humane treatment.

The very first Confederate prison camp was in Charleston, inside Castle Pinckney, and its Union captives were treated quite gallantly by their captors, the young cream of Charleston society. But that brand of treatment would not last long. As the battles grew in size, the number of captives on each side grew as well. The Confederate government converted the Libby & Son Ship Chandlers & Grocers, a cotton warehouse near downtown Richmond, into a prison for Union officers. The 150-by-100-foot three-story prison held twelve hundred men in eight crowded, vermin-infested drafty rooms. One of its notorious "turnkeys," Dick Turner, reportedly shot anyone who went to a window for light and air. Libby was the scene of the largest mass prison escape of the war, led by Union Colonel Thomas E. Rose of Pennsylvania, captured at Chickamauga in September 1863. Using makeshift tools, Rose planned an elaborate escape route through a tunnel from the prison's cellar. On February 7, 1864, Colonel Abel D. Streight, the officer who had surrendered his large force to Nathan Forrest's much smaller cavalry, declared the tunnel long enough and demanded, as the senior officer, to be the first to escape.

One attempt was aborted when the tunnel proved too short and the prisoners had to dig farther for two more days. While the other prisoners held a musical show to mask the escape, Rose and his men went out through the tunnel and disappeared on the streets of Richmond. The next day's roll call revealed that 109 prisoners were missing. Fifty-nine made it back to the Union lines. Two drowned while crossing streams, and the remaining forty-eight, including Rose, were recaptured. Exchanged in July, Rose went on to fight in the Battles of Franklin and Nashville, Tennessee, the last two major battles in the West. A career soldier, he resigned from the army with the rank of major in 1894. Streight rejoined his men and also fought at Franklin and Nashville.

Libby Prison and the nearby Belle Isle Prison, where thousands of Union prisoners lived in tents with little to eat, were rightfully notorious. But the conditions in these camps paled in comparison to the treatment that lay in store for the captives in a Confederate prison camp officially named Camp Sumter but better known as Andersonville.

When the concentration of war prisoners in Richmond drained the local food supply, a new prison site was selected in the heart of Georgia, near the village of Andersonville in Sumter County. The first prisoners arrived in February 1864 and found a fifteen-foot-tall stockade surrounding about sixteen acres of open land. Designed for ten thousand men, there was little in the way of housing, clothing, or medical care. The only fresh water was a thin trickle from Stockade Creek, a stream that flowed through the prison yard into nearby Sweet Water Creek. Cooking waste flowed into the stream, with the downstream serving as the prison latrine. Bloodhounds were kenneled nearby to track down runaways, who were chained or put in stocks as punishment for escaping. During the next few months, 400 new prisoners arrived each day, and in June the prison was expanded to twenty-six acres. By August it contained more than 33,000 Union prisoners, making it the fifth-largest city in the Confederacy.

Andersonville was initially overseen by John Henry Winder, the Confederate commander responsible for all prisons. His death of fatigue in February 1865 probably saved him from a postwar trial and execution. Winder was succeeded by Henry Wirz, a Swiss-born Confederate officer with a reputation for cruelty and sadism. Under Wirz, Andersonville was certainly worse than any other prison, Union or Confederate. Georgia's summer heat, disease, and inadequate food and medical care took a terrible toll: of the 45,000 prisoners at Camp Sumter, at least 13,000 died. (Some have argued that the camp at Salisbury, North Carolina, was even worse. Although smaller than Andersonville, with only 10,000 prisoners, Salisbury saw a mortality rate of 34 percent, higher than Andersonville's 29 percent.)

Civil War Voices

Henry Hernbaker, a Union soldier captured at Gettysburg and a prisoner at both Belle Isle Prison on the James River near Richmond and Andersonville.

> I was now taken to Andersonville, where I remained about seven months, and the horrors I met there it is useless for

me to attempt to describe.... The sun was scorching hot, and having nothing to protect us from its burning rays, the whole upper surface of our feet would become blistered, and then would break, leaving the flesh exposed, and having nothing to dress it or protect it in any way, gangrene would follow, and some would lose their feet, and part of a limb, and death would soon follow. And I have seen others die from want of nourishment. The amputations would average as many as six per day, and I saw not a single instance of a recovery from them. Some became victims of total blindness, occasioned by constant exposure to the heat of the sun and its action on the nervous system.

In the month of June it rained twenty-one days in succession, and there were often fifteen thousand of us in the stockade without any shelter of any kind. In the hospital the poor sick who were too feeble to help themselves, and literally swarming with lice, had all their life's blood taken away from them in that way. In fact, it is hardly possible to conceive a greater accumulation of woes to come upon mortal man than fell to the lot of the prisoners at Andersonville. Three thousand died in the month of August, and I have counted one hundred dead bodies in a row, and some of them so decomposed as to fall to pieces upon being removed. There have been as many as one hundred and fifty died in one day, and you can imagine what a foul atmosphere there would be. Day after day, I have gone the rounds of the wretched hospitals and looked upon every variety of suffering that the human frame is capable of presenting. I have heard Capt. Wirz swear that he was killing more d——d Yankees with his treatment than they were with powder and lead in the army.... Another old friend of mine ... was shot dead by the rebel guard for reaching under the dead line for a drink of water, and another was shot dead at my feet for reaching under the dead line for a piece of mouldy bread.

Another Andersonville inmate was Sergeant Lucius Barber of Illinois, captured while fighting in Georgia under Sherman.

> At 4 o'clock P.M. the Georgia Hell, which clutched in its iron grasp ten thousand Union soldiers, was seen in the distance. We were marched up to the commandant's headquarters ... where a rigid search was performed. ... This devil in human shape, Wirz, I will briefly describe. Any man gifted with any discernment would pronounce him a villain at first sight. ... As he moved around amongst us, he spit out his vile abuse in the most disgusting manner, nearly every word an oath. ... About sundown we were marched to the outside gate of Hell. ... Its huge doors swung open to admit us and we were in the presence of—I do not know what to call them. It was evident they were human beings but hunger, sickness, exposure and dirt had so transformed them that they more resembled walking skeletons, painted black. Our feelings cannot be described as we gazed on these poor human beings. ... Such squalid, filthy wretchedness, hunger, disease, nakedness and cold, I never saw before. Thirty-five thousand souls had been crowded into this pen, filling it completely. Poorly clad and worse fed, drinking filth and slime, from one hundred to three hundred of these passed into the gate of the eternal world daily.

There were nearly 13,000 graves at Andersonville, although the actual death toll was probably much higher. After the war, Clara Barton went to the prison site to bury properly and identify the dead and missing of Andersonville. What she saw enraged her, and she tried to testify at Wirz's trial. Even without her testimony, Wirz was quickly convicted. He was executed as the man responsible for Andersonville's conditions and held accountable for at least 10,000 of the Union deaths. Wirz was the only Confederate soldier executed by the Union after the war. (Long after, the controversy over the Confederate treatment of Union prisoners continued. Many defenders of Wirz and the Confederate cause faulted Grant's halting of prisoner exchanges, blaming the many prisoner deaths on food shortages alone. The Daughters of the Confederacy even erected a memorial to Wirz in 1909, saying he was "judicially murdered" by Yankees and claiming Grant was responsible for the Union deaths.)

The Confederacy did not have a monopoly on inhumane prison conditions. Point Lookout, Maryland, was the largest Union POW camp; designed for 10,000 men to live in tents, it often contained twice that number. Another Union prison, Fort Jefferson, in the Dry Tortugas near Key West, Florida, was an old fort that was converted to a military prison late in 1861. It was used not for POWs but criminals in the Union armies. Like the notorious Devil's Island, it had a merciless climate and a brutal work camp noted for disease and hard labor. (Dr. Samuel Mudd, the doctor convicted for treating the broken leg of John Wilkes Booth, Lincoln's assassin, was sentenced to life imprisonment at Fort Jefferson.)

The worst Union prison was in Elmira, in upstate New York, where 2,963 Confederate soldiers died, nearly a quarter of the 12,123 men held there. This death rate was only slightly less than Andersonville's and more than double the average death rate in the other Union prison camps. Built in May 1864, after prisoner exchanges were halted, the camp was designed to hold 5,000 men. The deaths at Elmira were caused by diseases brought on by starvation and terrible living conditions. During a bitterly cold winter, clothes sent by families for the prisoners were deliberately withheld, and hundreds of men, forced to live in tents with no blankets, froze to death. In May 1864 War Secretary Stanton ordered prisoner rations reduced to the same amount issued to Confederate soldiers. This supposedly ensured that Confederate prisoners were receiving the equivalent of the rations Union prisoners were getting. In other words, in the midst of plenty in the Union, malnourished Confederate prisoners suffered epidemics of scurvy, diarrhea, pneumonia, and smallpox. Survivors of the camp nicknamed the prison "Hellmira."

Civil War Voices

General Grant to General Meade (April 9, 1864): "Wherever Lee goes, there you will go also."

Why Were the Two Armies Back at the Wilderness near Chancellorsville?

By May, Grant was ready to attack, pressing his large army once more toward the Confederacy. His focus was not on Richmond but on the remaining Confederate armies in the field. On May 4, 1864, about 120,000 men in the Army of the Potomac crossed the Rapidan River and started what Confederate General John B. Gordon called "the seventh act in the 'On to Richmond' drama played by the armies of the Union." It was familiar terrain for Lee and his troops as well as many of the Union veterans. The Battle of Chancellorsville, Lee's greatest victory, had been fought here one year earlier.

Civil War Voices

Union Private Warren Goss, reaching Chancellorsville and the black-ened ruins of the old Chancellor House, found grim reminders of Hooker's 1863 debacle (May 1864).

> Weather-stained remnants of clothing, rusty gun-barrels and bayonets, tarnished brasses and equipments, with bleaching bones and grinning skulls, marked this memorable field. In the cavity of one of these skulls was a nest with three speck-led eggs of a field bird. In yet another was a wasp nest. Life in embryo in the skull of death.

On May 5, Lee's 66,000-man Army of Northern Virginia attacked the Army of the Potomac in the Wilderness, the same tangle of woods that had seen such terrible fighting in 1863.

Civil War Voices

Private Warren Goss, 2nd Massachusetts Artillery, described the Wilderness (May 5, 1864).

> No one could see the fight fifty feet from him. The roll and
> crackle of musketry was something terrible, even to the veterans of many battles. The lines were very near each other,
> and from the dense underbrush and the tops of trees came
> puffs of smoke, the "*ping*" of the bullets, and the yell of
> the enemy. It was a blind and bloody hunt to the death, in
> bewildering thickets, rather than a battle. . . .
> . . . The uproar of battle continued through the twilight
> hours. It was eight o'clock before the deadly crackle of musketry died gradually away, and the sad shadows of night,
> like a pall, fell over the dead in these ensanguined thickets.
> The groans and cries for water or for help from the
> wounded gave place to the sounds of the conflict, and thus
> ended the first day's fighting of the Army of the Potomac
> under Grant.

The Reverend A. M. Stewart, a Pennsylvania chaplain who had been
with the Army of the Potomac since the Peninsula Campaign, was at
the first Battle of Chancellorsville (May 5, 1864).

> What awful, what sickening scenes! No, we have ceased to
> get sick at such sights. Here a dear friend struck dead by a
> ball through the head or heart! Another fallen with leg or
> thigh broken, and looking, resignedly yet wistfully, to you
> for help away from the carnage. Another dropping his gun,
> quickly clapping his hand upon his breast, stomach or bowels, through which a Minnie has passed, and walking slowly
> to the rear to lie down and die. Still another—yea, many
> more—with bullet holes through various fleshy parts of the
> body, from which the blood is freely flowing, walking back

and remarking, with a laugh somewhat distorted with pain,
"See, the rascals have hit me!" All this beneath a canopy
of sulphur and a bedlam of sounds, like confusion confounded.

Night at length put an end to the carnage, and left the
two armies much in the same position as at the opening of
the strife.

The "work of death," as Reverend Stewart called it, began once
again the next day, Friday, May 6. When a Union attack smashed
through the Confederate line, Lee himself spurred his horse to the
front to lead a counterattack. He was pulled back by his own men,
who shouted, "Go back, General Lee! Go back!" Also in the thick
of the fighting was Confederate General James K. Longstreet. In an
eerie repeat of the day a year earlier in Chancellorsville in which
Stonewall Jackson had been wounded by his own troops, Longstreet
was also hit. Unlike Jackson, however, he survived his injury.

As the two armies battled, fire once again broke out in the woods,
adding to the soldiers' woes.

Civil War Voices

Private Warren Goss (May 6, 1864):

Flames sprang up in the woods in our front, where the fight
of the morning had taken place. With a crackling roar, like
an army of fire, it came down upon the Union line. The
wind drove the blinding smoke and suffocating heat into
our faces. This, added to the oppressive heat of the
weather, was almost unendurable. It soon became terrible.
The line of fire, with resistless march, swept the thickets
before its advance, then reaching out its tongue of flame,
ignited the breastworks of resinous logs, which soon roared
and crackled along their entire length. The men fought the
enemy and the flames at the same time. Their hair and

beards were singed and their faces blistered....

During the conflict our men had exhausted their ammunition and had been obliged to gather cartridges from the dead and wounded. Their rifles, in many instances, became so hot by constant firing, that they were unable to hold them in their hands. The fire was the most terrible enemy our men had that day, and few survivors will forget the attack of the flames.

By the end of two days of close combat in the Wilderness, the battle was at a standstill, with heavy losses on both sides. The Union had casualties of more than 17,000 (2,246 killed and 12,073 wounded); the Confederates suffered approximately half that many. But in the grim new mathematics of this war, the Union could afford such ghastly losses, while the dwindling Confederate forces would be hard-pressed in the coming war of attrition.

Acting on a hunch that Grant would next try to get between his Confederate forces and Richmond, Lee pulled his troops out of the Wilderness, concentrating them near Spotsylvania Court House, a strategic crossroad. Several more days of all-out fighting took place there as the Union army tried to seize the momentum and crush the Confederate army.

Calling this battle "the most desperate engagement in the history of modern warfare," Horace Porter, an aide to Grant, described the combat:

It was chiefly a savage hand-to-hand fight across the breastworks. Rank after rank was riddled by shot and shell and bayonet thrusts, and finally sank, a mass of torn and mutilated corpses; then fresh troops rushed madly forward to replace the dead; and so the murderous work went on. Guns were run up close to the parapet, and double charges of canister played their part in the bloody work. The fence-rails and logs in the breastworks were shattered into splinters, and trees over a foot and a half in diameter were cut completely in two. . . .

The opposing flags were in places thrust against each other, and muskets were fired with muzzle against muzzle. Skulls were crushed with clubbed muskets and men stabbed with swords and bayonets thrust between the logs in the parapet which separated the combatants. Wild cheers, savage yells, and frantic shrieks rose above the sighing of the wind and the pattering of the rain and formed a demoniacal accompaniment to the booming of the guns. . . . Even the darkness of night and the pitiless storm failed to stop the fierce contest.

At about the time of the fighting at Spotsylvania, Lee lost another key lieutenant. On May 8, 1864, Union General Philip H. Sheridan boasted that his cavalry could whip Confederate General Jeb Stuart out of his boots. Grant said, "Let him start right out and do it."

With the most powerful cavalry force the Army of the Potomac had ever mounted—more than 10,000 troopers riding four abreast—Sheridan's column stretched for thirteen miles. The confident general made no effort to conceal his movement. Alerted to Sheridan's advance, Stuart positioned his 4,500 troopers between the Union column and Richmond. At Yellow Tavern, an abandoned inn six miles north of Richmond, the two cavalry forces fought for three hours on May 11. Even though they were outnumbered, the Confederate troops stubbornly defended their position until the Union force withdrew. But during that withdrawal, a dismounted Union cavalryman shot at a large, red-bearded Confederate officer on a horse thirty feet away. Fatally wounded, Stuart died the next day. Lee had lost his greatest cavalry leader. Hearing the news, he told a staff member, "I can scarcely think of him without weeping."

Documents of the Civil War

Ulysses Grant's bulletin to the War Department.

> Headquarters in the Field
> May 11, 1864—8 A.M.

> We have now ended the sixth day of very heavy fighting. The result, to this time, is much in our favor.

> Our losses have been heavy, as well as those of the enemy. I think the loss of the enemy must be greater.

> We have taken over 5,000 prisoners by battle, whilst he has taken from us but a few, except stragglers.

> I propose to fight it out on this line, if it takes all summer.
> U. S. Grant
> Lieut. Gen. Commanding the Armies
> of the United States

Ten days later, after more than two weeks of constant fighting at Spotsylvania Court House, Grant pulled his Army of the Potomac out of its trenches and sent it marching southeast. Lee took a parallel course, always staying between Grant's army and Richmond. On June 1, their paths converged at Cold Harbor, the site of a McClellan-Lee battle in 1862. After twenty-eight days of almost constant fighting, Grant's men had suffered more than 31,000 casualties, more than his western armies had lost in three years of war. Now, just a few miles northeast of Richmond, Grant faced Lee, who had pushed the Army of Northern Virginia to Cold Harbor in time to construct a formidable defensive position.

Civil War Voices

Sergeant George Cary Eggleston of Virginia, recalling the Confederates' arrival at Cold Harbor (June 1, 1864):

By the time we reached Cold Harbor we had begun to understand what our new adversary meant, and therefore, for the first time, I think, the men in the ranks of the Army of Northern Virginia realized that the era of experimental campaigns against us was over; that Grant was not going to retreat; that he was not to be removed from command because he had failed to break Lee's resistance; and that the policy of pounding had begun, and would continue until our strength be utterly worn away, unless by some decisive blow to the army in our front, or some brilliant movement in diversion ... we should succeed in changing the character of the contest. We began to understand that Grant had taken hold of the problem of destroying the Confederate strength in the only way that the strength of such an army, so commanded, could be destroyed, and that he intended to continue the plodding work till the task should be accomplished, wasting very little time or strength in efforts to make a brilliant display of generalship in a contest of strategic wits with Lee. We at last began to understand what Grant had meant by his expression of determination to "fight it out on this line if it takes all summer."

... We had absolute faith in Lee's ability to meet and repel any assault that might be made, and to devise some means of destroying Grant. There was, therefore, no fear in the Confederate ranks, of anything that General Grant might do; but there was an appalling and well-founded fear of starvation, which indeed some of us were already suffering. ...

But what is the use of writing about the pangs of hunger? The words are utterly meaningless to persons who have never known actual starvation, and cannot be made otherwise than meaningless. Hunger to starving men is totally unrelated to the desire for food as that is commonly understood and felt. It is a great agony of the whole body and of the soul as well. It is unimaginable, all-pervading pain inflicted when the strength to endure pain is utterly gone. ...

When we reached Cold Harbor the command to which I

belonged had been marching almost continuously day and night for more than fifty hours without food and for the first time we knew what actual starvation was. It was during that march that I heard a man wish himself a woman—the only case of that kind of remark I ever heard of—and he uttered the wish half in grim jest and made haste to qualify it by adding, "or a baby."

In the next two days, Grant lost another 5,000 men. On June 3, the third day at Cold Harbor, the Union troops massed for a great push that opened at 4:30 A.M. What followed was a terrible slaughter during an assault on well-dug Confederate trenches. In approximately *one hour* of fighting, the Union suffered between 6,000 and 7,000 casualties. By noon the attack was called off. And afterward Lee's army still stood between Grant's men and Richmond.

Grant decided on a new strategy. On June 12 he packed up his men and moved them south to Petersburg, twenty miles below Richmond, the hub of every railroad connecting Richmond to the rest of the eastern Confederacy.

The six weeks of fighting since Grant moved into the Wilderness had produced unprecedented and astonishing losses. Assessing this campaign, historian Richard Wheeler writes in *Voices of the Civil War,*

> Grant's Wilderness campaign had cost him more than 50,000 casualties. During the campaign's progress, Washington sent him about 40,000 reinforcements. President Lincoln accepted the horrendous casualty reports without protest, but he must have agonized over them in private. They affected him not only as a man of conscience but as a politician. This was the Presidential election year, and his conduct of the war was the leading issue.

Civil War Voices

Susan Lee Blackford, a Virginia volunteer in the Women's Relief Society, the Confederate equivalent of the Union's Sanitary Commission (May 12, 1864).

Mrs. Spence came after me just as I was about to begin this morning and said she had just heard that the Taliaferro's factory was full of soldiers in a deplorable condition. I went down there with a bucket of rice, milk, etc. to see what I could do. I found the house filled with wounded men and not one thing provided for them. They were lying about the floor on a little straw. Some had been there since Tuesday and had not seen a surgeon. I washed and dressed the wounds of about fifty and poured water over the wounds of many more. The town is crowded with the poor creatures, and there is really no preparations for such a number. If it had not been for the ladies many of them would have starved to death. The poor creatures are very grateful, and it is a pleasure to us to help them in any way. I have been very hard at work ever since the wounded commenced coming. I went to the depot twice to see what I could do. I have had the cutting and distribution of twelve hundred yards of cotton cloth for bandages, and sent over three bushels of rolls of bandages, and as many more yesterday. I have never worked so hard in all my life and I would rather do that than anything else in the world. I hope no more wounded are sent here as I really do not think they could be sheltered. The doctors, of course, are doing much, and some are doing their full duty, but the majority are not. They have free access to the hospital stores and deem their own health demands that they drink up most of the brandy and whiskey in stock, and, being fired up most of the time, display a cruel and brutal indifference to the needs of the suffering, which is a disgrace to their profession and to humanity.

MILESTONES IN THE CIVIL WAR: JANUARY–JUNE 1864

January 2 General Patrick R. Cleburne, an Irish immigrant who had become one of the most respected Confederate commanders in the West, proposes to his high command that some slaves be freed and trained to fight in the Confederate armies. His colleagues call the idea "revolting" and believe that it would demoralize the army. (Cleburne died later that year, and the Confederacy moved to arm some slaves four months later, but the war ended before any blacks ever joined a Confederate army.)

January 19 *Arkansas*, a slaveholding state that remained in the Union, adopts a new antislavery constitution.

February 3–14 General William T. Sherman captures *Meridian, Mississippi*. Following the strategy he will use as he drives east to the Atlantic coast from the Mississippi, Sherman begins to destroy supplies, buildings, and railroads in his path.

March 10 Grant is named commander of the Union armies following his victory at *Chattanooga*.

April 12 The *Fort Pillow Massacre*. Confederate General Nathan B. Forrest captures this *Tennessee* fort on the Mississippi River. In the aftermath, hundreds of black and fifty-three white soldiers are murdered.

April 30 As the Union armies begin to tighten their grip, Jefferson Davis and his pregnant wife, Varina, suffer a personal tragedy. Joe, their five-year-old son and Davis's favorite, falls from a balcony at the Confederate White House and dies the next day. Mary Boykin Chesnut notes in her diary, "These people, the Davises, had enough to bear without calamity at home. And now it comes to the very quick." Davis later has the balcony torn down.

May 3 Grant, with 120,000 troops, begins his inexorable advance into Virginia, driving toward *Richmond* and Lee's sixty thousand troops.

May 4 General Sherman begins his march toward *Atlanta* with an army of 110,000 men.

May 5–6 *The Wilderness.* A bloody but inconclusive battle is fought in tangled woods. Brushfires kill hundreds of wounded men unable to be evacuated.

May 8–12 *Battle of Spotsylvania.* Attempting to maneuver around Lee, Grant's army is met at this *Virginia* town. Five days of fighting lead to a stalemate.

May 15 A Union force entering the *Shenandoah Valley* in *Virginia* is defeated by Confederate General Jubal Early, setting back a Union offensive.

May 31 In *Cleveland,* Radical Republicans who oppose President Lincoln nominate General John C. Frémont as their presidential candidate.

June 1–3 *Battle of Cold Harbor.* Grant assaults Lee's defenses, with devastating losses on each side. Grant recognizes his mistakes but realizes that he has the superior numbers with which to wear down the Confederate army.

June 7 In *Baltimore,* the Republicans nominate President Lincoln to run for a second term despite his national unpopularity. His new vice-presidential candidate is Andrew Johnson, a southern War Democrat who is expected to widen the ticket's appeal in the border states.

June 15–18 Lee fights off another of Grant's attacks on *Petersburg, Virginia.* Grant settles into a long siege of the city in a repeat of his successful tactic at Vicksburg a year earlier.

June 27 *Battle of Kenesaw Mountain (Georgia).* Sherman orders an assault on Confederate positions, resulting in a deadly failure; nearly two thousand Union men are casualties. The Confederates win a defensive victory.

June 30 Congress passes the Internal Revenue Act to increase taxes to finance the war.

Treasury Secretary Chase resigns, and this time the resignation is accepted by Lincoln.

How Did the North and South Pay for the War?

Hate the IRS? Thank the Civil War. Along with all the other modern developments for which the Civil War is justly famed—advances in communications, technology, medicine, and military hardware—it also created a major transformation of the federal government. Before the war, the average American might have come into contact with the federal government in only one way—getting mail delivered. There was no Social Security, no Medicare, no student loan guarantees, no lunch programs, no highway fund to build interstates, no air traffic controllers, FCC, or Education Department. And for much of the war there was no tax man. There had been plenty of death during the Civil War; now there were also taxes.

Born in New Hampshire, Salmon P. Chase (1808–1873) was a devout Episcopalian who liked to recite psalms while bathing and dressing. Moving to Ohio, he became a lawyer and eventually an ambitious politician who wanted very much to become president. During the 1830s he had taken a strong stand against slavery in speeches and writings, declaring that the Constitution was an antislavery document. He joined the Free-Soil party in 1848 and as a Free-Soiler and Democrat was elected to the Senate, where he became a vociferous abolitionist voice, opposing both the Compromise of 1850 and the Kansas-Nebraska Act. A two-term governor of Ohio, Chase was a serious contender for the 1860 Republican presidential nomination, but he was undone by his antislavery views, which would have spelled defeat in the general election. Instead, he was extremely influential in switching Ohio's convention votes to Lincoln, who rewarded Chase with a Cabinet post despite his lack of any financial experience.

Chase was a quick study. As secretary of the treasury, he was faced with the enormous task of financing the war. Initially he relied on loans from private financiers. When they proved insufficient, he was forced to issue war bonds, which were marketed by Philadelphia banker Jay Cooke, whose brother knew Chase from Ohio politics.

While Cooke earned millions in commissions (and was later accused of profiteering), his bond sales actually brought the government a great deal of money—more than a billion dollars. But as sources of private capital dried up, Chase was forced to print paper money. In 1862, when the government was spending about *$2.5 million a day* on the war, "greenbacks," backed by Treasury gold, were introduced as the first federal paper money. (Individual states had printed paper money previously.) These greenbacks were accepted only over serious congressional reservations about the practice of printing paper money. Some legislators even argued that the Constitution said that Congress could "coin money" and took that literally to mean that only coins could be produced. But Chase's arguments won the day, and the Legal Tender Act was passed in 1862. (Ironically, as the Supreme Court's Chief Justice, he later declared the act unconstitutional.) The first bills carried his picture, hundreds of thousands of little campaign posters for a presumed run at the presidency.

At the same time the Legal Tender Act was passed, Congress also passed the first Internal Revenue Act. As military defeats piled up for the Union, the money crisis deepened, ruining the government's credit rating. The Internal Revenue Act provided for taxes on just about everything. Read this and weep: The first income tax rates were *3 percent*! There were also sin taxes (on tobacco and liquor, among other items), luxury taxes (on jewelry and yachts), license taxes, inheritance taxes. You name it, and the Civil War taxed it. (The tax laws died after the war. It was only in 1913, with the ratification of the Sixteenth Amendment, that the income tax became part of the Constitution.)

Although no country bumpkin—he had been a well-paid corporate lawyer before the war—Lincoln was not familiar with finance and allowed Chase enormous leeway. Under Chase's tenure, the Treasury expanded enormously, with the number of clerks growing from 383 to more than 2,000. This rapid growth created many of the first federal jobs, mostly clerical, for women. It also gave Chase enormous patronage power. A dispute with Lincoln over one of these appointments finally led to Chase's resignation in June 1864.

If anything, Chase's Confederate counterpart had a far more difficult, if not impossible, task. Christopher G. Memminger was born in Germany and brought to South Carolina after his father's death. He became a member of Davis's Cabinet on February 21, 1861, and was faced with the dismal job of keeping the Confederacy financially solvent. In its earliest days, the Confederacy lacked the organization and mechanics to establish a system of taxes or tariffs. So it relied on the sale of bonds, which would be redeemed after the war. As a wave of patriotism swept the Confederacy after Fort Sumter, the initial reaction to these bonds was strong. Foreign buyers were also sought, and Memminger opposed the cotton embargo because he realized that it was the only source of Confederate credit with foreign lenders.

But the Confederacy's poor credit rating forced Davis's administration to turn to taxes that were considered even more onerous and unconstitutional than they were in the Union. In 1863 the Confederate states imposed an income tax along with a 10 percent "tax in kind." This essentially meant that farmers had to turn over a portion of their crops to agents of the Richmond government. Slaves, incidentally, were not taxed as property. There was a careful legal argument for this bit of reasoning, but it's worth noting that most of the lawmakers held slaves; most of the working poor, who had to turn over their crops and couldn't buy their way out of conscription, did not.

Civil War Voices

Victoria V. Clayton, the wife of Confederate General Henry D. Clayton of Alabama, wrote of the changes southern women were facing as the war closed in on them.

We were blockaded on every side, could get nothing from without, so had to make everything at home; and having been heretofore only an agricultural people, it became necessary for every home to be supplied with spinning wheels and the old-fashioned loom, in order to manufacture clothing for the members of the family. This was no small undertaking. I knew nothing about spinning and weaving cloth. I had to learn myself, and then to teach the Negroes. Fortunately for me, most of the Negroes knew how to spin thread, the first step towards clothmaking. Our work was hard and continuous. To this we did not object, but our hearts sorrowed for our loved ones in the field. . . .

We were required to give one-tenth of all that was raised, to the government. There being no educated white person on the plantation except myself, it was necessary that I should attend to the gathering and measuring of every crop and the delivery of one-tenth to the government authorities. This one-tenth we gave cheerfully and often wished we had more to give.

My duties . . . were numerous and often laborious; the family on the increase continually, and every one added increased labor and responsibility. And this was the case with the typical Southern woman.

What Was "the Crater"?

Lee's Army of Northern Virginia, now encamped in Petersburg, had taken severe losses during the Wilderness and later campaigns. But dug in behind heavy fortifications, Lee's men were in a strong position. So Grant had returned to the tactic that had worked at Vicksburg, a siege of the city. The Union little appreciated his idea. A few months earlier Grant had been hailed as a savior. At the Wilderness he may have been thinking that a breakthrough to Richmond could bring him the presidency. Now the newspapers were calling him "Butcher Grant." Nearly a year after Gettysburg, Vicksburg, and Chattanooga, the war seemed no closer to being over.

Outside besieged Petersburg, the ill-starred Ambrose Burnside had a plan literally to blow a hole in the Confederate defenses and send the Union troops streaming into the breach. The idea was brought to Burnside by Colonel Henry Pleasants, a coal mine engineer who led a Pennsylvania regiment of coal miners. Occupying a position just a hundred and fifty yards from the Confederate lines, they planned to tunnel under the Confederate fortification and blow it up. Burnside quickly approved the plan, and the regiment began work on June 25.

The tunnel was four feet wide at the bottom, two feet wide at the top, and five feet high. On July 27 the miners carried four tons of black powder to the end of the tunnel and placed it in the side galleries. Then they carried dirt back into the main tunnel and tamped it into the last thirty-four feet so that the blast would not come back out the entrance. Burnside organized an assault force to rush past the gap the explosion would create in the rebel defenses.

At 3:15 A.M. on July 30, Colonel Pleasants lit the ninety-eight-foot fuse and sprinted back out of the tunnel. When there was no explosion after forty-five minutes, two volunteers entered the tunnel and relit the fuse. In minutes, a hundred-and-seventy-foot section of the Confederate entrenchments suddenly erupted in a huge blast, throwing dirt and timbers a hundred feet up in the air. An entire Confederate regiment was buried by the debris.

Under the initial plan, the attack through the exploded breach was to be carried out by black soldiers who had already proven themselves and had been trained for this task for two weeks. But General George Meade, still uncertain that black troops were reliable and afraid of bad publicity, worried their possible failure would lead to the criticism that he didn't value their lives. At the last minute, he ordered a different division to lead the attack. Straws were drawn, and the division under James H. Ledlie, a drunkard with political connections, got the assignment. Ledlie stayed behind drinking rum as his men, completely unprepared, charged into the gaping hole instead of around it.

The Battle of the Crater proved disastrous. The Union took four thousand casualties as the disorganized men were trapped inside the crater, easy targets for the Confederate riflemen. Afterward Grant said, "It was the saddest affair I have witnessed in the war."

Burnside's checkered Civil War career was finally finished and he resigned a few months later, never to serve in the army again.

Civil War Voices

Luther Rice Mills, a Confederate soldier, described life under siege in Petersburg.

Something is about to happen. I know not what. Nearly everyone who will express an opinion says Gen'l Lee is about to evacuate Petersburg. . . . I go down the lines, I see the marks of shot and shell, I see where fell my comrades, the Crater, the grave of fifteen hundred Yankees, when I go to the rear I see little mounds of dirt, some with headboards, some with none, some with shoes protruding, some with a small pile of bones on one side near the end showing where a hand was left uncovered, in fact, everything near shows desperate fighting. And here I would rather "fight it out." If Petersburg and Richmond is evacuated—from what I have seen and heard in the army—our cause will be hopeless. It is useless to conceal the truth any longer. Many of our people at home have become so demoralized that they write to their husbands, sons and brothers that desertion now is not dishonorable. It would be impossible to keep the army from straggling to a ruinous extent if we evacuate. I have just received an order . . . to carry out on picket tonight a rifle and ten rounds of cartridges to shoot men when they desert. The men seem to think desertion is no crime and hence never shoot a deserter when he goes over—they always shoot but never hit.

Did Admiral David Farragut Really Say "Damn the Torpedoes. Full Speed Ahead!"?

Among those disgusted by the Union's failure to end the war was David G. Farragut, the veteran sailor who had fought in the War of 1812 and was responsible for the pivotal capture of New Orleans in 1862. Now, in the summer of 1864, he said, "The only comfort I have is that the Confederates are more unhappy, if possible, than we are."

Commanding the naval squadron maintaining the coastal blockade, Farragut's main target was Mobile, on Alabama's Gulf coast. It was a city that he had wanted to take as soon as New Orleans fell, but he had been directed to other assignments. One of the two functioning Confederate ports, Mobile was still used by the blockade runners who kept the Confederacy's lifeline to the world open. Opposing Farragut as commander of Mobile's naval defenses was Franklin Buchanan, who had been the Confederate commander of the *Virginia* in the famous first fight between the ironclads at Hampton Roads (see page 215). Once again, it was a case of two comrades now turned against each other, for in their early navy days, Farragut and Buchanan had battled pirates together in the West Indies. As Farragut waited, he wrote, "I am tired of watching Buchanan and wish from the bottom of my heart that Buck would come out and try his hand upon us."

Thirty miles long, Mobile Bay was protected by three Confederate forts, three gunboats, and the *Tennessee*, a ram modeled on the *Virginia* and considered the strongest ironclad ever built. The channel into the harbor, a narrow pass bristling with Confederate guns, had also been filled with obstructions and lined with mines, or "torpedoes"— beer kegs filled with powder that exploded on contact with a passing ship.

After making a personal reconnaissance of Mobile Bay, the crusty old Farragut decided to take on the seemingly impossible and put his squadron into action on August 5. With four ironclad "monitors" in the lead and fourteen wooden vessels lashed together in pairs, the fleet would run past the forts and the *Tennessee*. Aboard the *Hartford*, his flagship, Farragut was in the leading pair of wooden vessels. When the Confederate forts and ships began their incessant, heavy fire, he climbed into the *Hartford*'s rigging, and a sailor was sent to lash him on so he wouldn't fall. As the smoke of battle around him grew so thick he couldn't see, Farragut untied himself, climbed higher, and relashed himself, remaining aloft for the rest of the action.

One of the Union monitors, the *Tecumseh*, sank after striking a torpedo, and a Union naval disaster seemed imminent as the fleet was almost stationary, with ships nearly colliding with one another in a sort of seagoing traffic jam. This was Farragut's immortal moment. According to one witness, the admiral ordered, "Damn the torpedoes! Go ahead!" (not "Full speed ahead"), and the *Hartford* surged forward, passing the fort and the Confederate fleet, which had been hit hard by the Union guns, leading the Union ships through.

With one Confederate gunboat captured and two others out of action, only the *Tennessee* was still battleworthy. But Buchanan was not going to give up, and he aimed this ram at Farragut's flagship and the rest of the Union fleet. The two ships collided without extensive damage, then the entire Union fleet turned its guns on the *Tennessee*. Finally, the concentrated fire became too much for the Confederate ironclad. Buchanan, already limping from his wounds at Hampton Roads, was hit, and his leg was broken. Soon after, the *Tennessee* called for a truce.

The Union losses were 145 dead, including 93 who went down with the *Tecumseh*. During the next two days, two of the forts guarding Mobile fell, and on August 23 Fort Morgan, the Confederate stronghold overlooking the bay, was taken. While the city itself remained in Confederate hands, the port of Mobile was sealed off. Word of the victory was welcome news for Lincoln. Any Union progress was significant at this point, for his prospects for a second term were seriously in doubt.

Civil War Voices

Abraham Lincoln (August 1864).

> This morning, as for some days past, it seems exceedingly
> probable that this Administration will not be re-elected.
> Then it will be my duty to so co-operate with the President
> elect, as to save the Union between the election and the
> inauguration; as he will have secured his election on such
> ground that he cannot possibly save it afterwards.

Who Ran Against Lincoln in 1864?

Lincoln's pessimism was well founded. After nearly four years of war
that seemed no closer to being over, with battlefield casualties growing by enormous numbers, and with a war-weary nation angry and
disillusioned, Lincoln's prospects seemed worse than bleak. A prominent New York Republican, Thurlow Weed, was convinced he could
not be reelected, saying, "The people are wild for peace." Although
since Lincoln's time no sitting American president has ever lost an
election in time of war, recent history was not on Lincoln's side. No
incumbent president since Andrew Jackson had been reelected. Initially, there was some doubt that Lincoln would even receive his
party's nomination, as a Cabinet revolt coalesced around Treasury
Secretary Salmon P. Chase in March 1864. Ambitious and smart,
Chase was also self-assured. One Senate colleague said, "Chase is a
good man, but his theology is unsound. He thinks there is a fourth
person in the Trinity."

As treasury secretary, Chase had used patronage to build a strong
power base among an inner circle of Republicans who considered
themselves stronger than Lincoln. An effort early in the year to replace Lincoln with Chase on the Republican ticket failed when word
of this palace coup was made public and Chase appeared disloyal; he
resigned his treasury post a few months later.

Lincoln won the party's nomination, although his vice-president,

Hannibal Hamlin, was dropped in favor of Tennessee Governor Andrew Johnson, a Democrat who brought more regional balance to the ticket. The Republicans also decided to run as the Union party in 1864. But some disenchanted Republicans, still eager to replace Lincoln, turned to John C. Frémont, the famed general and explorer who had been the first Republican presidential nominee, in 1856, and whom Lincoln had removed from command in a controversy over freeing slaves. Frémont eventually withdrew from the race, convinced by fellow Republicans that he could only throw the election to a Democrat.

Lincoln's real opposition came from another general he had fired, George B. McClellan, who became the Democrats' nominee at their Chicago convention in August. Although running on a peace platform that included "efforts be made for a cessation of hostilities," McClellan ran with the notion of continuing the war. His essential argument was over Lincoln's conduct of the war and the issue of emancipation, not the war itself.

Lincoln's strategy was simple: Portray the Democrats as disloyal and pray for victories.

Did General Sherman Really Say "War Is Hell"?

Lincoln's presidency was probably saved on September 2, 1864, the day General William T. Sherman marched into Atlanta after four months of fighting throughout Georgia.

After capturing Atlanta, Sherman almost immediately ordered the civilian evacuation of the city. He said later, "I was resolved to make Atlanta a pure military garrison or depot, with no civil population to influence military measures."

Civil War Voices

General William T. Sherman in a letter to the mayor and town commissioners of Atlanta (September 12, 1864).

You cannot qualify war in harsher terms than I will. War is cruelty, and you cannot refine it. And those who brought war into our country deserve all the curses and maledictions a people can pour out. I know I had no hand in making this war, and I know I will make more sacrifices today than any of you to secure peace. . . . You might as well appeal against the thunderstorm as against these terrible hardships of war. They are inevitable, and the only way the people of Atlanta can hope once more to live in peace and quiet at home is to stop the war.

Sherman's most famous words were not spoken until after the war, if at all. In remarks attributed to him on June 19, 1879, in an address at Michigan Military Academy, the general said, "War is at best barbarism. . . . Its glory is all moonshine. It is only those who have neither fired a shot, nor heard the shrieks and groans of the wounded who cry aloud for blood, more vengeance, more desolation. War is hell."

What he did in Atlanta in 1864 was living proof of what he later said. With Atlanta under Sherman's control and the Confederate army under General John Hood a diminishing threat, Sherman was ready to embark on the next leg of his journey through the heartland of the Confederacy, the notorious March to the Sea. But before he left Atlanta, he planned to make sure the city would no longer contribute to the Confederate war effort.

On November 15 one of his aides wrote:

A grand and awful spectacle is presented to the beholder in this beautiful city, now in flames. By order, the chief engineer has destroyed by powder and fire all the storehouses, depot buildings and machine-shops. The heaven is one expanse of lurid fire; the air is filled with flying, burning cinders; buildings covering two hundred acres are in ruins or in flames; every instant there is the sharp detonation or the smothered booming sound of exploding shells and powder concealed in the buildings, and then the sparks and flame shoot away up into the black and red roof, scattering cinders far and wide.

These are the machine-shops where have been forged and cast the Rebel cannon, shot and shell that have carried death to many a brave defender of our nation's honor. These warehouses have been the receptacle of munitions of war, stored to be used for our destruction. The city, which next to Richmond, has furnished more material for prosecuting the war than any other in the South, exists no more as a means for injury to be used by enemies of the Union.

Civil War Voices

David P. Conyngham, a correspondent for the *New York Herald,* with Sherman's army in Atlanta.

Sherman's orders were that Atlanta should be destroyed by the rearguard of the army, and two regiments were detailed for that purpose. Although the army... did not pass through until the 16th, the first fires burst out on the night of Friday, the 11th of November, in a block of wooden tenements on Decatur Street, where eight buildings were destroyed.

Soon after, fires burst out in other parts of the city. These certainly were the work of some of the soldiers, who expected to get some booty under the cover of the fires.

... On Sunday night a kind of long streak of light, like an aurora, marked the line of march and the burning stores, depots and bridges in the train of the army.

The Michigan engineers had been detailed to destroy the depots and public buildings in Atlanta. Everything in the way of destruction was now considered legalized. The workmen tore up the rails and piled them on the smoking fires. Winship's iron foundry and machine shops were early set on fire. This valuable property was calculated to be worth about half a million of dollars. . . . Next followed a freight warehouse, in which were stored several bales of cotton. The depot, turning-tables, freight sheds, and stores around were soon a fiery mass. The heart was burning out of Atlanta.

The few people that had remained in the city fled, scared by the conflagration and the dread of violence.

. . . The Atlanta Hotel, Washington Hall, and all the square around the railroad depot, were soon in one sheet of flame. Drug stores, dry goods stores, hotels, negro marts, theatres, and grog-shops were all now feeding the fiery element. Worn-out wagons and camp equipage were piled up in the depot and added to the fury of the flames.

A store warehouse was blown up by a mine. Quartermasters ran away, leaving large stores behind, The men plunged into the houses, broke windows and doors with their muskets, dragging out armfuls of clothes, tobacco, and whiskey, which was now more welcome than the rest. The men dressed themselves in new clothes and then flung the rest in the fire.

The streets were now in one fierce sheet of flame; houses were falling on all sides, and fiery flakes of cinders were whirled about. Occasionally, shells exploded and excited men rushed through the choking atmosphere, and hurried away from the city of ruins.

At a distance the city seemed overshadowed by a cloud of black smoke, through which, now and then, darted gushing flames of fire or projectiles hurled from the burning ruin. The sun looked, through the hazy cloud, like a blood-red ball of fire; and the air for miles around felt oppressive and intolerable. . . . the "Gate City" was a thing of the past.

Sherman's capture of Atlanta, combined with Farragut's success at Mobile Bay, was the deciding factor in the election of 1864. Lincoln pronounced the fall of Atlanta a gift from God and ordered a day of thanksgiving. One-hundred-gun salutes were fired from Boston to St. Louis and New Orleans. Before Atlanta, the fate of the Lincoln-Johnson ticket was uncertain. The war's toll had the country in an evil mood and the sentiment against Lincoln grew ugly. For instance the *New York World,* long an enemy of Lincoln's complained, "In a crisis of the most appalling magnitude requiring statesmanship of the highest order, the country is asked to consider the claims of two ignorant, boorish, third-rate backwoods lawyers for the highest stations in the Government. Such nominations, in such a conjuncture, are an insult to the common sense of the people. God save the Republic!"

Lincoln won convincingly. Despite Republican fears that McClellan was still popular with enlisted men, the soldiers' vote went overwhelmingly to Lincoln (116,887 for Lincoln to 33,748 for McClellan). With 2,206,938 votes, Lincoln won 55 percent of the total popular vote to McClellan's 45 percent; Lincoln's margin in the Electoral College was even more substantial, 212–21. McClellan took only his home state of New Jersey and the border states of Delaware and Kentucky. While not a sweeping landslide, the victory seemed to be a mandate for Lincoln's emancipation policies. Other Republicans simply thought that the voters had rejected McClellan.

Which Civil War Battle Was Fought in Vermont?

Desperate men sometimes take desperate measures. And the Confederacy was growing more desperate by the day. A long way from Atlanta, a group of some of the most desperate Confederates was hatching a plot that would exact some retribution for the loss of Atlanta, although nobody would ever make a movie about the burning of St. Albans, Vermont.

During the war years, St. Albans was a quiet village on Lake

Champlain, just fifteen miles from the Canadian border. Far from the battlefields and devastation, life in small northern towns like St. Albans had changed little since the start of the war except for the publication of the casualty rolls. Vermont had been among the most responsive of the Union states in meeting Lincoln's calls for volunteers.

On October 10, 1864, three young men checked into a hotel. The spokesman signed the register as Bennett Young and explained that they had come from St. John's, Canada, for a sporting vacation. Every day or so two or three more men would arrive until there were about twenty. All were young and friendly. The Bible-quoting Young struck up a conversation about theology with a pretty girl he met. The friendliness ended, however, when the visitors threw off their overcoats to reveal Confederate uniforms. The whole event seemed to take on an almost comic air as Young melodramatically announced, "This city is now in the possession of the Confederate States of America." With rebel yells, the raiders entered the town's banks and made off with some $200,000 in gold. While the banks were being robbed, eight or nine of the young rebels, their guns drawn, herded the townspeople onto the common and stole their horses. When the gold proved too heavy to carry, they negotiated for cash with a resident. Young later administered to some of the bank officers and tellers an oath of allegiance to the Confederacy and Jefferson Davis. Then he ordered the men to set the town on fire. They had brought along bottles of "Greek fire," a phosphorus-based chemical hand grenade that was supposed to burst into flames when exposed to air. The townsmen, who had begun to fight back, started to fight the fires instead.

The raiders jumped on their stolen horses and crossed back into Canada, pursued by a Vermont militia that eventually caught up with them. But the sympathetic Canadian authorities refused to extradite the men, trying them instead in a Montreal court, which declared that they were soldiers under military orders. Confederate Secretary of State Judah Benjamin helped the raiders by sending along the military orders that would legitimize Young's raid. (These orders were carried to Canada by "Little Johnny" Surratt, a reliable courier who lived with his mother in a Washington boardinghouse known to be a haven for Confederate sympathizers.) Tried only for violating Canada's neutrality, the raiders were acquitted and freed. The Canadian government did return $88,000 of the funds found on the raiders.

This raid was not the Confederacy's only attempt to take the war to the Union. In fact, it was part of a much larger scheme hatched primarily by Judah Benjamin, who had masterminded a well-funded plan to use Confederate agents and sympathizers in Canada to stir up trouble in the Union. These saboteurs and raiders, Benjamin hoped, could raid the Union prisons, free the Confederate prisoners, and seize federal arsenals to arm both the prisoners as well as the Copperheads, who were growing more vocal in their opposition to Lincoln. At worst, the Copperheads might embarrass Lincoln; at best, they could contribute to his defeat at the polls. This strategy might even force the Union to divert troops to the Canadian border, weakening the assaults on Virginia and Georgia.

The St. Albans raid turned into a comic disaster, and all the well-laid plans to arm the Copperheads proved a far-flung fantasy. But another target was New York City. In a raid initially planned to take place on election day, the attack on the Union's largest city did not actually come off until late November, also in retaliation for the sacking of Atlanta and the destruction of the Shenandoah Valley. Eight Confederate agents planned to burn some of the city's hotels with a hundred and forty firebombs made from "Greek fire," the same substance present at St. Albans. The raiders checked into various hotels in New York. They set fire to piles of clothing and rubbish but, unfamiliar with the chemical mixture, did not produce the great blazes they wanted. Although a spectacular fire was set at Barnum's Museum, where a few animals were frightened and a seven-foot woman in the sideshow went berserk, there was little damage. By midnight, all the fires had been put out.

All of the conspirators escaped from New York by train through Albany and back into Canada. But one of them, Robert Cobb Kennedy, who had been expelled from West Point before the war, was later caught after another harebrained scheme to free some captured Confederate generals. He was tried by a military commission headed by General John A. Dix; its judgment read, "The attempt to set fire to the city of New York is one of the greatest atrocities of the age. There is nothing in the annals of barbarism which evinces greater vindictiveness. It was not a mere attempt to destroy the city, but to set fire to crowded hotels and places of public resort, in order to secure the greatest possible destruction of human life." Kennedy was executed as a spy by hanging on March 25, in Fort Lafayette in New York Harbor.

Disgusted by the absurdity of the results of his attempts to undermine the Union, Benjamin pulled the plug on the Canadian adventure.

Civil War Voices

The object of the firebombers was New York's wealthier locales. The Union's largest city, however, was not a place for only the wealthy. A report of the Council on Hygiene and Public Health revealed what life was like for most of the urban poor in America's richest city back in the "good old days."

> In some of the apartments of the tenant-houses, the rags that cover the floor in lieu of carpet reek with filth. They have become a receptacle for street mud, food of all kinds, saliva, urine, and faeces. . . . The bed clothing is often little better. . . .
>
> That the evils and abuses of the system continue undiminished is seen on every hand. Not only does filth, overcrowding, lack of privacy and domesticity, lack of ventilation and lighting, and absence of supervision and of sanitary regulation, still characterize the greater number of them; but they are built to a greater height in stories, there are more rear tenant-houses erected back to back with other

buildings, correspondingly situated on parallel streets; the courts and alleys are more greedily encroached upon and narrowed into unventilated, unlighted, damp, and well-like holes between the many-storied front and rear tenements; and more fever-breeding ... culs-de-sac are created as the demand for the humble homes of the laboring poor increases.

Major George W. Nichols with General Sherman's headquarters staff, Cobb's Plantation, Georgia (November 23, 1864).

General Sherman camped on one of the plantations of Howell Cobb. It was a coincidence that a Macon paper, containing Cobb's address to the Georgians as General Commanding, was received the same day. This plantation was the property of Cobb's wife, who was a Lamar. I do not know that Cobb ever claimed any great reputation as a man of piety or singular virtues, but I could not help contrasting the call upon his fellow-citizens to "rise and defend your liberties, homes, etc., from the step of the invader, to burn and destroy every thing in his front, and assail him on all sides," and all that, with his own conduct here, and the wretched condition of his Negroes and their quarters.

We found his granaries well filled with corn and wheat, part of which was distributed and eaten by our animals and men. A large supply of sirrup made from sorghum ... was stored in an old out-house. This was also disposed of by the soldiers and the poor decrepit Negroes which this humane, liberty-loving general left to die in this place a few days ago. Becoming alarmed, Cobb sent for, and removed, all the able-bodied mules, horses, cows and slaves. He left here some fifty old men—cripples—and women and children, with nothing scarcely covering their nakedness, with little or no food, and without means of procuring it. We found them cowering over the fireplaces of their miserable huts, where the wind whirled through the crevices between the logs, frightened at the approach of the Yankees, who, they had been told, would kill them. A more forlorn, neglected set of human beings I have never seen.

A few months later, as the Confederate Congress and the Davis administration debated freeing the slaves to serve in its armies, Howell Cobb, the former governor of Georgia, wrote, "Use all the Negroes you can get, for the purpose for which you need—but don't arm them. The day you make soldiers out of them is the beginning of the end of the revolution. If slaves will make good soldiers our whole theory of slavery is wrong."

What Did General Sherman Give President Lincoln for Christmas in 1864?

Confederates like Howell Cobb were not the only ones who thought that blacks should not be carrying weapons. Many of those in agreement were Union officers, perhaps none more outspokenly so than William Tecumseh Sherman (1820–1891). Having helped to secure Lincoln's reelection with the capture of Atlanta, Sherman had left Atlanta in cinders and was now in full swing on his infamous March to the Sea. His destination was Savannah, on the Atlantic coast. It was a rapid, devastating sweep through Georgia. Sherman was unforgiving in his goal of utterly destroying any Confederate capability to continue resistance, and he ordered the destruction of all railroads.

Largely because of this march, Sherman would emerge from the Civil War with the most clearly divided reputation of all the soldiers and statesmen of the era. Lionized in the Union, he was a Yankee Lucifer in the Confederate states, particularly in Georgia and the Carolinas, which would feel the uncompromising heat of his furious approach to war. Among the most perplexing men in the Civil War, Sherman was a complicated character of many moods. Though he professed to be apolitical, he owed his career in part to the powerful politicians in his family.

Having adopted Louisiana as his home before the war, he was a reluctant warrior, weeping after South Carolina's secession and prophetically telling a friend, "This country will be drenched in blood." But he would then exact the harshest retribution on the Confederate South. Once judged "insane" by newspapers, fellow officers, and the men he commanded, he later became a Union hero, second only to his comrade Grant. Never an opponent of slavery, he would issue a postwar order that gave the emancipated slaves one of the most generous new beginnings.

Born in Ohio, Sherman was named Tecumseh at birth after the powerful Shawnee Indian leader who had nearly defeated the United States Army. When neighbors said they thought the name was strange, his father prophetically said, "Tecumseh was a great warrior." After the elder Sherman, debt-ridden, died when the boy was nine, "Cump," as he was called, was taken in by Thomas Ewing, a neighbor and close friend of his father's. The Ewings adopted the boy as their own but decided that a proper name was needed, so a priest baptized him William. Among his new family were three brothers, Thomas, Jr., Charles, and Hugh, who would all join the Union army, and a "sister," Ellen, whom Sherman later married. His "father"—later father-in-law—would become a powerful senator from Ohio, influential in Republican circles. Sherman's blood brother, John, also went into politics, represented Ohio in the House of Representatives, and was made Ohio's senator when Salmon Chase was elevated to the treasury post by Lincoln. But unlike most other politically well-connected generals, Sherman had credentials that merited his appointment.

Sixth in his 1840 West Point class, Sherman had left the army along with many other officers when it seemed that civilian life offered greater chances for success. After a string of failures in banking, real estate, and law, Sherman was in Louisiana just before the war began, running a military academy that would later become the foundation for Louisiana State University. Among his students was the son of P. T. Beauregard, the Confederate general who would occupy Sherman's tent after the first day at Shiloh. Though he had great friendships with many who joined the Confederacy and had no moral qualms about slavery, Sherman felt, as many Union professional soldiers did, that secession was treason. He returned to Missouri when Louisiana seceded.

After Fort Sumter, Sherman was given command of some raw volunteers who were among the men routed at the First Battle of Manassas. Then he went west to Kentucky, where a combination of deep depression and strange decisions led to the widespread belief that he had gone insane. Influential family members, including his wife, Ellen, interceded with Lincoln, and Sherman was eventually transferred to join Grant in a fateful partnership. At Shiloh, Sherman took decisive command and helped stave off a Union rout, even though his own lack of caution nearly allowed the Confederate army of Albert Johnston and Beauregard to destroy a Union army. The success at Shiloh seemed to give Sherman a new vigor and sense of purpose, and he became Grant's most trusted ally and colleague. When Grant was later promoted to lead the Union armies, he gave Sherman his former command in the West. With that command Sherman made his decisive assault on Atlanta.

Once Atlanta was emptied and Lincoln's future as president assured, Sherman embarked on the campaign that made him notorious. Cutting his army off from its supply base, Sherman had ordered his men to forage for their food during the three-hundred-mile march. After the recent harvest, food was plentiful for the Union soldiers, who appropriated farm carriages and wagons and loaded them with bacon, meal, turkeys, chickens, and anything else they could lay their hands on. Sherman himself said,

> No doubt many acts of pillage, robbery and violence were committed by these parties of foragers usually called "bummers," for I have since heard of jewelry taken from women, and the plunder of articles that never reached the commissary; but . . . I never heard of any cases of murder or rape. And no army could have carried along sufficient food and forage for a march of three hundred miles, so that foraging in some shape was necessary.

Meeting only light resistance from the disintegrating Confederate cavalry, Sherman's march was soon being joined by newly liberated slaves, who latched onto the Union army. As General Henry Slocum wrote:

> It was natural that these poor creatures, seeking a place of safety, should flee to the army and endeavor to keep in sight of it. Every day as we marched on we could see, on each side of our line of march, crowds of these people coming to us through the roads and across the fields, bringing with them all their earthly goods, and many goods which were not theirs. Horses, mules, cows, dogs, old family carriages, carts, and whatever they thought might be of use to them were seized upon and brought to us. . . . At times they were almost equal in number to the army they were following.

These freed fugitives would generate considerable controversy. Moving rapidly and living off the land, Sherman did not want to be burdened by these liberated blacks, whose legal status was uncertain. Nor did he want to enlist any of them in the war effort. Like many other Union commanders, Sherman was racist and had no use for black troops in his command. His defiance of presidential and War Department orders on the subject bordered on insubordination, but his success in the field allowed him to set his own policies. While he thought that freed blacks would perform acceptably as laborers to dig trenches, build forts, haul water, and chop wood, he did not think they could fill the role of soldier. He even wrote to his brother, the senator, "I won't trust niggers to fight."

In one controversial incident, Sherman was accused of allowing the blacks following his army to be killed or recaptured by the Confederates at Ebenezer Creek, near Savannah, Georgia, on December 3. One of Sherman's officers, General Jefferson Columbus Davis, was a proslavery Union officer who had been at Fort Sumter. In Sherman's rear guard, Davis prevented nearly five hundred freed slaves from crossing a pontoon bridge over a river as they straggled behind the

Union troops. When the troops had safely crossed, Davis had the bridge pulled away, leaving these former slaves trapped and defenseless. As the Union soldiers watched and tried to help some of the blacks with makeshift rafts, many of them were killed or recaptured by Joseph Wheeler's Confederate cavalry riders. Sherman was unaware of the incident until a Union newspaper broke the story and made it an issue. (Jefferson C. Davis had already earned considerable infamy, apart from his name. In September 1862 Davis and Major General William Nelson got involved in a dispute after Nelson relieved Davis of duty. Davis cornered Nelson in a Louisville hotel and demanded an apology. When Nelson refused and slapped Davis in the face, Davis followed him into the street and shot him in the chest. Davis was never brought to trial for the murder.)

Although his attitude toward blacks would continue to cause Sherman trouble, particularly with War Secretary Stanton, the Ebenezer Creek incident was overlooked in light of his stunning successes. By early December, Sherman had reached the outskirts of Savannah. Defended by a small garrison, which slipped away without a fight, Savannah surrendered on December 22.

With the city in his hands, Sherman wired a message to Lincoln on December 22, 1864: "I beg to present you as a Christmas gift the city of Savannah, with one hundred and fifty heavy guns and plenty of ammunition; also about twenty-five thousand bales of cotton."

Sherman's message actually reached Lincoln on Christmas Eve and was widely published in newspapers throughout the North.

Converted into a base of Union operations, Savannah was spared the severe sentence carried out on Atlanta a few weeks earlier. A mercy ship from Boston carrying food and supplies was even sent to the occupied city, a far cry from the devastation visited upon Atlanta.

Newsman Charles Coffin described the radical impact of Sherman's arrival on this bastion of old southern life.

> Society in the South, and especially in Savannah has undergone a great change. The extremes of social life were

very wide apart before the war. They were no nearer the night before Sherman marched into the city. But the morning after, there was a convulsion, an upheaval, a shaking up and a settling down of all the discordant elements. The tread of that army of the West, as it moved in solid column through the streets, was like a moral earthquake, overturning aristocratic pride, privilege and power.

Old houses, with foundations laid deep and strong in the centuries, fortified by wealth, name, and influence, went down beneath the shock. The general disruption of the former relations of master and slave, and forced submission to the Union arms, produced a common level. . . .

On the night before Sherman entered the place, there were citizens who could enumerate their wealth by millions; at sunrise the next morning they were worth scarcely a dime. Their property had been in cotton, Negroes, houses, land, Confederate bonds and currency, railroad and bank stocks. Government had seized their cotton; the Negroes had possession of their lands; their slaves had become freemen; their houses were occupied by troops; Confederate bonds were waste paper; their railroads were destroyed, their banks insolvent. They had not only lost wealth, but they had lost their cause.

Toward the end of Sherman's occupation of Savannah, a fire was accidentally set in a wooden warehouse district, and the wind soon took it to the former Confederate arsenal, which exploded. Although the Union troops, the freed slaves, and the citizens of Savannah worked together, the fire left a "wilderness of chimneys."

Sherman prepared to move north and rendezvous with Grant, still stalled outside Petersburg. But first he had to move through the Carolinas, starting with South Carolina, the "birthplace" of the war and to many Union men "the cause of all our troubles."

Civil War Voices

From the wartime diary of Eliza Andrews, a girl in Georgia (December 24, 1864).

About three miles from Sparta we struck the "burnt country," as it is well named by the natives, and then I could better understand the wrath and desperation of these poor people. I almost felt as if I should like to hang a Yankee myself. There was hardly a fence left standing all the way from Sparta to Gordon. The fields were trampled down and the road was lined with carcasses of horses, hogs, and cattle that the invaders, unable either to consume or carry away with them, had wantonly shot down, to starve out the people and prevent them from making their crops. The stench in some places was unbearable; every few hundred yards we had to hold our noses or stop them with the cologne Mrs. Elzey had given us, and it proved to be a great boon. The dwellings that were standing all showed signs of pillage, and on every plantation we saw the charred remains of the ginhouse and packing screw, while here and there lone chimney stacks, "Sherman's sentinels," told of homes laid in ashes. The infamous wretches! I couldn't wonder now that these poor people should want to put a rope round the neck of every red-handed "devil of em" they could lay their hands on. Hayricks and fodder stacks were demolished, corncribs were empty, and every bale of cotton that could be found was burnt by the savages. I saw no grain of any sort except little patches they had spilled when feeding their horses and which there was not even a chicken left in the country to eat. A bag of oats might have lain anywhere along the road without danger from the beasts of the field, though I cannot say it would have been safe from the assaults of hungry men.

Crowds of soldiers were tramping over the road in both

directions; it was like traveling through the streets of a populous town all day. They were mostly on foot, and I saw numbers seated on the roadside greedily eating raw turnips, meat skins, parched corn—anything they could find, even picking up the loose grains that Sherman's horses had left.

MILESTONES IN THE CIVIL WAR: JULY–DECEMBER 1864

July 5 New York editor Horace Greeley receives a letter stating that the Confederate delegates in Canada wish to discuss peace. Lincoln sends an emissary to discuss their proposal. The attempt at a negotiated peace is aborted when the Confederates refuse to restore the Union and to accept the abolition of slavery as a condition of peace.
July 14 Confederate General Jubal Early is able to reach the outskirts of the *District of Columbia* until he is slowed by Union General Lew Wallace. Reinforcements arrive, and Early withdraws.
July 17 President Davis replaces Joseph Johnston, who has been attempting to blunt Sherman's march on Atlanta, with John B. Hood.
July 20–28 Hood switches to the offensive, attacks Sherman's army outside Atlanta, and suffers heavy losses that he cannot afford. Sherman spends the rest of August slowly circling Atlanta and cutting all rails and roads into the city.
July 30 General Burnside directs a mine attack on *Petersburg* with disastrous results. His military days are over.
August 5 Admiral David Farragut captures *Mobile, Alabama*, closing one of the last ports open to Confederate blockade runners. His famous words are "Damn the torpedoes. Full speed ahead!"
August 29 In *Chicago* the Democrats nominate General George McClellan to face Lincoln. The president is running against two generals he has fired during the war.
September 2 *Atlanta* falls to Sherman. Much of the city is burned, just as in *Gone With the Wind*. Sherman wires Lincoln, "Atlanta is ours, and fairly won." On *September 7* Sherman orders the evacuation of the civilian population of the city.
September 17 Republican candidate Frémont withdraws from the

presidential race. He states that he doesn't want to split the Republicans and allow McClellan to win.

September 19–October 19 Union General Sheridan defeats Confederate General Jubal Early, although Sheridan suffers heavy losses. Early is finally driven from the *Shenandoah Valley*, one of the last remaining supply sources for the Confederate army.

October 11 Chief Justice of the Supreme Court Roger B. Taney, author of the notorious Dred Scott decision, dies at the age of eighty-nine.

October 13 By a slim margin *Maryland* adopts a new state constitution, which abolishes slavery.

October 20 Lincoln proclaims that the last Thursday in November will be celebrated as Thanksgiving Day.

October 31 *Nevada*, which will be known as the Battle State for gaining statehood during the war, is admitted as the thirty-sixth state.

November 8 *Election day*. Lincoln wins reelection with a wide margin in the electoral college but by fewer than a half-million votes. The Republicans also make gains in the House and retain control of the Senate.

November 16 Sherman begins his infamous *March to the Sea*, from *Atlanta* to *Savannah* on Georgia's Atlantic coast.

November 29 *Sand Creek Massacre*. In *Colorado Territory*, citizens around *Denver* attack a peaceful Indian camp and kill some five hundred Arapahoe and Cheyenne men, women, and children.

November 30 *Battle of Franklin, Tennessee*. In an ill-planned assault, Hood's Army of Tennessee attacks a Union army. He loses nearly 6,300 dead and wounded from his force of 27,000. Among the Confederate dead are six generals: John B. Carter, States Rights Gist, H. B. Granbury, John Adams, O. F. Strahl, and Patrick Cleburne. Union General Schofield withdraws from Franklin and retreats to *Nashville* to join the Union forces under General George H. Thomas.

December 6 Grant asks, then pleads, then angrily commands George Thomas to attack the Confederates under Hood outside Nashville before they can slip past him and strike into Ohio.

Lincoln nominates his former rival Salmon P. Chase to be Chief Justice of the Supreme Court in a gesture of reconciliation.

December 15–17 *Battle of Nashville.* With 55,000 Union troops, George Thomas finally strikes Hood's much smaller army outside the city. Forty-five hundred of Hood's men are captured and another 1,500 are killed or wounded, leaving a force of 14,000 men.

December 22 Sherman enters *Savannah, Georgia,* unopposed.

* The phrase "In God We Trust" is added to American coins for the first time.

Who Burned Columbia and Charleston, South Carolina?

Shortly before moving on to South Carolina, Sherman said, "The truth is the whole army is burning with an insatiable desire to wreak vengeance upon South Carolina. I almost tremble at her fate, but feel she deserves all that seems in store for her."

Leaving Savannah on February 5, 1865, Sherman's 60,000 men took a direct line toward Columbia, South Carolina. They faced only token resistance from any organized Confederate troops.

Civil War Voices

Confederate cavalry officer J. P. Austin, among those trying to block Sherman:

> He [Sherman] swept on with his army of sixty thousand men, like a full developed cyclone, leaving behind him a track of desolation and ashes fifty miles wide. In front of them was terror and dismay. Bummers and foragers swarmed on his flanks, who plundered and robbed everyone who was so unfortunate as to be within their reach. . . .
>
> Poor, bleeding, suffering South Carolina! Up to that time she had felt but slightly—away from the coast—the dev-

astating effects of the war; but her time had come. The
protestations of her old men and the pleadings of her noble
women had no effect in staying the ravages of sword, flame,
and pillage.

Columbia's fate could readily be foretold from the destruction along Sherman's line of march after he left Savannah. Beautiful homes, with their tropical gardens, which had
been the pride of their owners for generations, were left in
ruins. . . . Everything that could not be carried off was destroyed. Thousands who had but recently lived in affluence
were compelled to subsist on the scrapings from the abandoned camps of soldiers. . . . Livestock of every description
that they could not take was shot down. All farm imple-
ments, with wagons and vehicles of every description, were
given to the flames.

By the night of February 15, the first of the Union soldiers had
reached the Congaree River, across from Columbia. The next day
they sighted their cannon on the statehouse across the river and fired
shells into the heart of the city. J. P. Austin watched from a mile
away: "It was from this point we saw the flames burst forth from
public buildings, stores, and beautiful homesteads. . . . Columbia was
in flames. . . . By 12 o'clock, the city was one great sea of fire."

Confederate troops evacuated Charleston on the same day that
Columbia was occupied by Sherman. As they withdrew, General Har-
dee, the Confederate commander, set fire to all the Charleston warehouses where cotton was stored. The fire spread and detonated a
store of gunpowder; at least two hundred people died. The next day
the Stars and Stripes again flew over Fort Sumter.

Years later, Sherman responded to the controversy over the origins
of the fires, and the charges that he had ordered them set. "Though
I never ordered it and never wished it, I have never shed many tears
over the event, because I believed it hastened what we all fought
for, the end of the war."

Civil War Voices

From Lincoln's Second Inaugural Address (March 4, 1865). (See Appendix VIII for the complete text.)

> With malice toward none, with charity for all, with firmness in the right as God gives us to see the right, let us strive on to finish the work we are in, to bind up the nation's wounds, to care for him who shall have borne the battle and for his widow and his orphan, to do all which may achieve and cherish a just and a lasting peace among ourselves and with all nations.

A week after Lincoln's second inauguration, Sherman's army occupied Fayetteville, North Carolina, reducing an arsenal to rubble. Organized resistance to Sherman was all but finished, and more Union troops were moving into North Carolina.

How Did Richmond Fall?

Lee knew his army was trapped. As Richard Wheeler writes in *Voices of the Civil War:* "By March, 1865, Grant had been besieging Petersburg for nine months. His lines, which began southeast of Richmond, ran through Petersburg's eastern environs and then curved around the city. Grant's current strength was about 125,000 men. Lee probably had less than half that number. Moreover, faith in the cause had weakened and desertions were rising."

With meager supplies of food, clothing, and shelter, Lee's soldiers had suffered a miserable nine months in trenches around Petersburg and Richmond, protecting the capital at the government's insistence. But by March, after consulting with Jefferson Davis, Lee determined to escape from Petersburg and move southeast, linking up with the remnants of General Joe Johnston's army, then trying to contain Sherman. As a diversion, he prepared an attack on a Union earthworks called Fort Stedman, which contained mortars and heavy cannon that were being used against Petersburg. The opposing lines there stood only a hundred and fifty yards apart.

Twelve thousand men from Lee's already thin lines were placed under the command of General John B. Gordon, who waited in the predawn of March 25, 1865. As fifty axmen cut down the obstructions in front of the fort, Gordon observed a curious incident. Frequently during the war, the pickets on either side had struck up conversations and a strange camaraderie during the long nights of guard duty. It was no different during the long siege of Petersburg. Hearing Confederates moving about, a Union picket asked what they were doing. One of Gordon's men said, "Never mind, Yank. We are just gathering a little corn."

"All right, Johnny," said the Union soldier. "I'll not shoot at you while you are drawing rations."

Then Gordon ordered the soldier to fire the shot signaling the assault. He hesitated, and Gordon gave the order again. The Confederate soldier, to satisfy his conscience, yelled, "Hello, Yank! Wake up. We are coming." With that he fired the signal and the attack was on.

Skirmishers went out and overcame the sleeping Union pickets, and Fort Stedman was captured quickly in the first rush. Some of the rebels turned the nine captured cannon around and fired on the Union lines. More than a thousand prisoners, including a surprised Union general, were initially taken.

But the expected Confederate reinforcements were delayed when their train broke down. And while the Union batteries in other nearby forts delivered heavy fire, Union infantry reinforcements quickly arrived and added to the counterattack. By 8:00 A.M. Lee, watching from the Confederate lines, ordered Gordon to withdraw from Fort Stedman. Many of the soldiers chose to surrender rather than face Union fire while racing back to their own lines. Once broken, the attack was a disaster for Lee; his army lost more than 4,000 men as prisoners and casualties while the Union forces suffered about 1,600 casualties.

Not far away, Lincoln was on a steamboat at City Point, meeting with Grant and his other commanders to discuss the certain fall of Richmond. In another clash at Five Forks, just southwest of Petersburg, on March 29, the Confederates under Gettysburg's George Pickett again took heavy casualties. Lee poignantly remarked to an aide, "This is a sad business, colonel. It has happened as I told them in Richmond it would happen. The line has been stretched until it is broken."

Civil War Voices

From a Richmond newspaper.

The numbers of Virginians reported absent from their regiments without leave, will, this morning, exceed fifty thousand. What can this mean?... News reaches us to-night that General Pickett has lost control of his troops at Five Forks, and that the Yankees are gradually moving towards Richmond. It seems that our troops have become discouraged and are easily confused. The Yankee assault on Pickett's Division has completely demoralized it, if reports are true.

Richmond Examiner editor Edward Pollard, described the evacuation of Richmond:

It was eleven o'clock in the morning when General Lee wrote a hasty telegram to the War Department, advising that the authorities should have everything in readiness to evacuate the capital at eight o'clock the coming night, unless before that time despatches should be received from him to a contrary effect.

A small slip of paper, sent up from the War Department to President Davis, as he was seated in his pew in St. Paul's Church, contained the news of the most momentous event of the war....

The report of a great misfortune soon traverses a city
without the aid of printed bulletins. But that of the evacuation of Richmond fell upon many incredulous ears....
There were but few people in the streets; no vehicles dis-
turbed the quiet of the Sabbath; the sound of the church-
going bells rose into the cloudless sky and floated on the
blue tide of the beautiful day. How was it possible to imagine that in the next twenty-four hours ... this peaceful
city, a secure possession for four years, was at last to succumb ... ?

As the day wore on, clatter and bustle in the streets de-
noted the progress of the evacuation and convinced those
who had been incredulous of its reality. The disorder increased each hour. The streets were thronged with fugitives
making their way to the railroad depots; pale women and
little shoeless children struggled in the crowd; oaths and
blasphemous shouts smote the air....

When it was finally announced by the Army that those
who had hoped for a despatch from General Lee contrary
to what he had telegraphed in the morning had ceased to
indulge in such an expectation and that the evacuation of
Richmond was a foregone conclusion, it was proposed to
maintain order in the city by two regiments of militia; to
destroy every drop of liquor in the warehouses and stores;
and to establish a patrol through the night. But the militia
ran through the fingers of their officers ... and in a short
while the whole city was plunged into mad confusion and
indescribable horrors.

It was an extraordinary night; disorder, pillage, shouts,
mad revelry.... In the now dimly lighted city could be seen
black masses of people crowded around some object of excitement ... swaying to and fro in whatever momentary passion possessed them. The gutters ran with a liquor freshet,
and the fumes filled the air. Some of the straggling soldiers
... easily managed to get hold of quantities of the liquor.

Confusion became worse confounded; the sidewalks were encumbered with broken glass; stores were entered at pleasure and stripped from top to bottom; yells of drunken men, shouts of roving pillagers, wild cries of distress filled the air and made night hideous.

But a new horror was to appear upon the scene and take possession of the community. To the rear-guard of the Confederate force ... on the north side of James River ... had been left the duty of blowing up the iron-clad vessels in the James and destroying the bridges across the river.... The work of destruction might well have ended here. But the four principal tobacco warehouses of the city were fired; the flames seized on the neighboring buildings and soon involved a wide and widening area; the conflagration passed rapidly beyond control. And in this mad fire, this wild, unnecessary destruction of their property, the citizens of Richmond had a fitting souvenir of the imprudence and recklessness of the departing Administration.

Morning broke on a scene never to be forgotten.... The smoke and glare of fire mingled with the golden beams of the rising sun.... The fire was reaching to whole blocks of buildings.... Its roar sounded in the ears; it leaped from street to street. Pillagers were busy at their vocation, and in the hot breath of the fire were figures as of demons contending for prey.

First into Richmond on Monday, April 3, were Massachusetts troops under General Godfrey Weitzel. Soon an unbroken line of blue uniforms was pouring into the capital as the sullen residents watched. The only cheers came from the blacks of Richmond. As the fires continued, the Union regiments organized the blacks into fire corps, but it was useless and the fires burned out by evening.

At about the same time, Grant rode into Petersburg and was soon joined by Lincoln and his eldest son, Robert Todd Lincoln, now a member of Grant's staff, his other son, Tad, and Admiral Porter, on whose boat Lincoln had been staying. As one of Grant's aides reported, Lincoln "dismounted in the street and came in through the front gate with long and rapid strides, his face beaming with delight. He seized General Grant's hand as the general stepped forward to greet him, and stood shaking it for some time and pouring out his thanks and congratulations. . . . I doubt whether Mr. Lincoln ever experienced a happier moment in his life."

The next day, Lincoln was taken to the fallen Confederate capital by boat.

Civil War Voices

Correspondent Charles Coffin on Lincoln's arrival in Richmond:

> No carriage was to be had, so the President, leading his son, walked to General Weitzel's headquarters—Jeff Davis's mansion. . . . The walk was long and the President halted a moment to rest.
>
> "May de good Lord bless you, President Linkum!" said an old Negro, removing his hat and bowing, with tears of joy rolling down his cheeks.
>
> The President removed his own hat and bowed in silence. It was a bow which upset the forms, laws, customs, and ceremonies of centuries of slavery.

What Happened at Appomattox Courthouse?

While Lincoln sat in Jefferson Davis's office and enjoyed his moment of glory, Grant had unfinished business with Lee. After evacuating the trenches at Petersburg, Lee's starving, ragged Army of Northern Virginia headed west, looking for a way around Grant's army. With the route south to North Carolina blocked, Lee had no choice but to continue marching his exhausted soldiers west.

Civil War Voices

Confederate General Gordon, commanding Lee's rear guard:

Fighting all day, marching all night, with exhaustion and hunger claiming their victims at every mile of the march with charges of infantry in the rear and of cavalry on the flanks, it seemed the war god had turned loose all his furies to revel in havoc. On and on, hour after hour, from hilltop to hilltop, the lines were alternately forming, fighting and retreating, making one almost continuous shifting battle. Here, in one direction, a battery of artillery became involved; there, in another, a blocked ammunition train required rescue. And thus came short but sharp little battles which made up sideshows of the main performance, while the different divisions of Lee's lionhearted army were being broken and scattered or captured.

At Sayler's Creek, Anderson's corps was broken and destroyed; and General Ewell, with almost his entire command, was captured as was General Kershaw, General Custis Lee—son of the general-in-chief—and other prominent officers. . . .

The roads and fields swarmed with the eager pursuers, and Lee now and then was forced to halt his whole army, reduced to less than 10,000 fighters in order to meet these simultaneous attacks. Various divisions along the line of march turned upon the Federals, and in each case checked them long enough for some other Confederate commands to move on. . . .

General Lee was riding everywhere and watching everything, encouraging his brave men by his calm and cheerful bearing. He was often exposed to great danger from shells and bullets. . . .

On that doleful retreat . . . it was impossible for us to bury our dead or carry with us the disabled wounded. There was no longer any room in the crowded ambulances which had escaped capture.... We could do nothing for the unfortunate sufferers who were too severely wounded to march, except leave them on the roadside with canteens of water.

Lee had ordered food to be sent by rail from Lynchburg, Virginia, to the town of Farmville, twenty-five miles away on the south side of the Appomattox River. On April 6 pursuing Union troops had taken advantage of a gap in the Confederate column and, in the Battle of Sayler's Creek, cut off and captured a third of Lee's men. General Philip H. Sheridan reported the victory to Grant, adding, "If the thing is pressed I think Lee will surrender." Hearing of the message, Lincoln wired Grant, "Let the thing be pressed."

On April 7 the remaining portion of Lee's army reached Farmville and the food in the waiting boxcars. Word soon came that Union troops were rapidly approaching. To slow their advance Lee had ordered a bridge destroyed, but the Union soldiers arrived in time to put out the flames. Lee's bedraggled, starving army held the federal soldiers off and then began marching toward Appomattox Courthouse. At ten o'clock that night, Lee received a message from Grant, requesting his surrender. Without comment, Lee passed the note to James Longstreet, who read it and, looking up, said, "Not yet."

On the evening of April 8 the final Confederate council of war was held. A last-ditch attempt to break through the Union lines was attempted, but the odds were simply too great. When told that his army could not go forward, Lee said, "There is nothing left me but to go and see General Grant, and I had rather die a thousand deaths." Finally, a flag of truce was carried between the two generals.

Civil War Voices

Confederate officer William Miller Owen:

> General Lee rode through our lines towards Appomattox
> Court House.... By a singular coincidence, the meeting of
> the generals took place in the house of [Wilmer] McLean,
> the same gentleman who in 1861 at the Battle of Bull Run,
> had tendered his house to General Beauregard for head-
> quarters. He removed from Manassas after the battle, with
> the intentions of seeking some quiet nook where the alarms
> of war could never find him; but it was his fortune to be in
> at the beginning and in at its death.

Lee, accompanied by a single aide, reached the McLean house
first. Grant, who had served with Lee in Mexico, soon arrived with
his entourage.

Civil War Voices

Grant's recollection of the meeting with Lee at Appomattox:

> When I left camp that morning I had not expected so soon
> the result that was then taking place and consequently was
> in rough garb. I was without a sword, as I usually was when
> on horseback in the fields, and wore a soldier's blouse for
> a coat, with the shoulder straps of my rank to indicate to
> the army who I was.
>
> When I went into the house, I found General Lee. We
> greeted each other, and after shaking hands took our seats.
> I had my staff with me, a good portion of whom were in
> the room during the whole of the interview.
>
> What General Lee's feelings were I do not know. As he
> was a man of much dignity, with an impassible face, it was
> impossible to say whether he felt inwardly glad that the end
> had finally come, or felt sad over the result, and was too

manly to show it. Whatever his feelings, they were entirely concealed from my observation; but my own feelings, which had been quite jubilant on the receipt of his letter, were sad and depressed. I felt like anything rather than rejoicing at the downfall of a foe who had fought so long and valiantly and had suffered so much for a cause, though that cause was, I believe, one of the worst for which people ever fought, and one for which there was the least excuse. I do not question, however, the sincerity of the great mass of those who were opposed to us.

General Lee was dressed in a full uniform which was entirely new, and was wearing a sword of considerable value. . . . In my rough traveling suit, the uniform of a private with the straps of a lieutenant general, I must have contrasted very strangely with a man so handsomely dressed, six feet high and of faultless form. . . .

We soon fell into a conversation about old army times. He remarked that he remembered me very well in the old army; . . . Our conversation grew so pleasant that I almost forgot the object of our meeting, and said that he had asked for this interview for the purpose of getting from me the terms I proposed to give his army. . . .

. . . When news of the surrender first reached our lines, our men commenced firing a salute of a hundred guns in honor of the victory. I at once sent word, however, to have it stopped. The Confederates were now our prisoners, and we did not want to exult over their downfall.

Grant told Lee that his men could keep both their horses and mules and their personal side arms. Lee remarked that this would have a "happy effect." Before taking leave, he also asked Grant for rations for his men. Grant authorized all the provisions he wanted.

Documents of the Civil War

Robert E. Lee's final order to the men of the Army of Northern Virginia.

General Order No. 9   April 10, 1865

After four years of arduous service marked by unsurpassed courage and fortitude, the Army of Northern Virginia has been compelled to yield to overwhelming numbers and resources. I need not tell the brave survivors of so many hard fought battles, who have remained steadfast to the last, that I have consented to this result from no distrust of them; but, feeling that valor and devotion could accomplish nothing that could compensate for the loss that must have attended the continuance of the contest, I have determined to avoid the useless sacrifice of those whose past services have endeared them to their countrymen.

By the terms of the agreement, officers and men can return to their homes and remain there until exchanged. You will take with you the satisfaction that proceeds from the consciousness of duty faithfully performed; and I earnestly pray that a Merciful God will extend to you his blessing and protection.

With an unceasing admiration of your constancy and devotion to your Country, and a grateful remembrance of your kind and generous consideration of myself, I bid you all an affectionate farewell.

(Sgd.) R. E. Lee
Genl.

On April 12, three days after the surrender, the remains of the Army of Northern Virginia, now reduced to some eighteen thousand men, stacked their weapons and were issued paroles, and the army that kept the power of the Union at bay was no more. Two days later, at the ruins of Fort Sumter, General Robert Anderson raised the flag that he had lowered exactly four years earlier.

That same day, April 14, a jubilant Lincoln invited Grant and his wife to join him and his wife in Washington. Mrs. Grant had been embarrassed and abused by Mrs. Lincoln, who was jealous of Grant's popularity and whose behavior had become increasingly erratic. So the Grants declined the offer to spend an evening with the Lincolns at Ford's Theater.

MILESTONES IN THE CIVIL WAR: 1865

January Although it was most likely the handiwork of Edwin Stanton, Sherman issues Field Order No. 15, authorizing the use of abandoned land in *Florida* and *South Carolina* for exclusive settlement by blacks, who are given land confiscated from or abandoned by the Confederates. Whites are barred from these areas. Each freed family would be given forty acres, clothing, seed, and farm equipment. (This short-lived experiment failed during Reconstruction, but many blacks continued to date history from "the time Tecumsey was here.")

January 11 A constitutional convention in *Missouri* adopts a resolution abolishing slavery in the state.

January 15 *Fort Fisher*, protecting *Wilmington, North Carolina*, falls to joint Union army-navy forces.

January 16 Politician Francis Preston Blair, Sr., an adviser to Lincoln, has been secretly meeting in *Richmond* with Jefferson Davis to discuss peace negotiations. Davis will only discuss peace "between the two nations." Lincoln rejects this language. Although Blair visits Davis again, nothing comes of their meetings.

January 30 Lincoln issues passes for three Confederate commissioners to cross the Union lines for peace discussions.

January 31 The House passes the Thirteenth Amendment, abolishing slavery. It goes to the states for ratification. *Illinois*, Lincoln's home state, ratifies first on *February 1*.

February 2 Lincoln goes to *Hampton Roads, Virginia*, to meet with the Confederate peace delegates.

February 17 The city of *Columbia, South Carolina*, is almost completely destroyed by fires of mysterious origin. Although Sherman's forces are blamed, they may have been set by departing Confederate soldiers.

February 22 *Wilmington, North Carolina*, the last open Confederate port, falls to Union forces.

March 2 Lee's request for negotiations is rejected by Lincoln, who demands surrender before negotiation.

March 3 Congress establishes the Bureau of Refugees, Freedmen, and Abandoned Lands (Freedmen's Bureau) to aid former slaves and white refugees.

March 4 Lincoln is inaugurated for a second term. (See Appendix VIII for the text of his inaugural address.)

March 13 In a desperate act, Jefferson Davis signs a bill allowing blacks to enlist in the Confederate army; slaves who enlist will be freed. Although blacks are seen in Confederate uniforms, few actually join.

March 27–28 In a meeting at *City Point, Virginia*, with his generals, Grant and Sherman, and Admiral David D. Porter, Lincoln calls for a speedy end to the war with as little loss of life as possible. He also advocates generous terms of surrender in order to avert further bloodshed.

April 1 In one of the last battles of the war, Sheridan routs a Confederate force at *Five Forks, Virginia*.

April 2 Lee withdraws from *Petersburg*, after ten months of fighting, including a six-month siege. He advises Davis to move the Confederate government from *Richmond*.

In *Selma, Alabama*, Nathan B. Forrest suffers one of his few defeats. Though he escapes capture, the Union troops capture 2,700 prisoners along with guns and supplies.

April 3 Union troops enter *Petersburg* and *Richmond*. Two days later, Lincoln arrives in Richmond to tour the city. Cheered by the city's former slaves, he also sits in President Davis's chair.

April 6 The last battle between the Army of Northern Virginia and the Army of the Potomac is fought at *Sayler's Creek, Virginia*.

April 9 Lee formally surrenders to Grant at *Appomattox Courthouse, Virginia*.

April 11 In his final public address, Lincoln urges conciliation.

April 14 Lincoln is assassinated in Ford's Theater by the actor John Wilkes Booth, a southern sympathizer. He dies the following morning, the first president to die by assassination. On *April 15* Andrew Johnson takes the oath of office as president.

April 18 Confederate General J. E. Johnston surrenders to General Sherman, marking the formal end of Confederate resistance.

April 26 John Wilkes Booth is shot and killed in *Bowling Green, Virginia*.

April 27 The steamship *Sultana*, with a capacity of four hundred passengers, is crowded with more than two thousand people; most aboard are Union soldiers returning from Confederate war prisons. When one of the ship's boilers explodes near *Memphis*, on the Mississippi, seventeen hundred people are killed. It remains the worst marine disaster in U.S. history.

chapter eight

Aftermath

And now, while the nation is rejoicing . . . it is suddenly plunged into the
deepest sorrow by the most brutal murder of its loved chief.
We are now continually passing paroled men from Lee's army on their
way to their homes. . . . Many found blackened ruins, instead, and kindred
and friends gone, they know not whither. Oh, how much misery treason and
rebellion have brought upon our land!

—SERGEANT LUCIUS BARBER
APRIL 21, 1865

It is hard for the old slaveholding spirit to die. But die it must.

—SOJOURNER TRUTH

I hereby repeat . . . my unmitigated hatred . . . to the perfidious,
malignant, & vile Yankee race.

—SECCESSIONIST EDMUND RUFFIN

✳

* Who Killed Abraham Lincoln?

* Was Jefferson Davis Wearing a Dress When
He Was Captured?

* What Did the Confederate Soldiers Come
Home to Find?

* What Does "Forty Acres and a Mule" Mean?

* Why Was President Andrew Johnson Impeached?

* Who Were the Carpetbaggers and Scalawags?

* Who Started the Ku Klux Klan?

* Whatever Became Of?

More than 360,000 dead Union soldiers. Another 275,000 wounded. At least 258,000 Confederate dead and 100,000 more wounded. (In *The Atlas of the Civil War*, James McPherson notes that if the same proportion of Americans were killed in a war today, the number of dead would exceed 5 million.) In the Confederate states, an estimated 50,000 civilians dead. These dry statistics of death and destruction can only hint at the vast expense of the Civil War, by far the most costly of America's wars.

Many of the "best and brightest" on both sides—those young men who would be expected to lead the country into the future—had been killed or disabled. Along with the horrific number of deaths and crippling wounds, much of the seceded South was left in smoldering ruins. The southern economy was practically nonexistent. The dollar value of the destruction was staggering. Although cotton resumed its significant position almost immediately, it was another twenty-five years before the number of livestock in the South returned to prewar levels. The regional rancor that had boiled over into war was replaced by the bitter hatred of the defeated for the victor. And for millions of free blacks—as well as their masters—the world had been turned upside down.

Across the country, however, both the jubilation over the war's end and the bitterness of defeat were obscured by the death of Abraham Lincoln.

Who Killed Abraham Lincoln?

On the evening of April 14, 1865, President Lincoln attended a performance of *Our American Cousin*, a popular comedy starring Laura Keene, at Ford's Theater in Washington. Arriving late, he was accompanied by his wife and two young guests, Major Henry Rathbone and his fiancée, Miss Clara Harris. At least a dozen other people had been invited to go with the Lincolns but had, for a variety of reasons, declined the honor. The play, which had already begun, was briefly halted to welcome the arriving president.

By this time in his life, Lincoln was well aware that he faced danger. He disliked the special guard assigned to him but still kept an envelope filled with the threatening letters he had received. The possibility of assassination was a real one, and Lincoln, who had thought powerfully about death all of his life, once told his old friend Ward Hill Lamon about a dream he'd had.

> There seemed to be a death-like stillness about me. Then I heard subdued sobs, as if a number of people were weeping. I thought I left my bed and wandered downstairs. . . . It was light in all the rooms; every object was familiar to me. . . . I kept on until I arrived at the East Room, which I entered. There I met with a sickening surprise. Before me was a catafalque, on which rested a corpse wrapped in funeral vestments. "Who is dead in the White House?" I demanded of one of the soldiers. "The President," was the answer, "he was killed by an assassin!" . . . I slept no more that night.

Around ten o'clock John Wilkes Booth, an actor familiar to the theater's staff, entered Ford's and joked with the ticket taker, who allowed him in "courtesy of the house." He had been at the theater earlier in the day, secretly boring a small hole in a wall through which he could open the door to Lincoln's box if it were locked. Booth climbed to the dress circle and, around ten-thirty, entered the president's box. Standing about four feet behind Lincoln, who was leaning forward to see one of his generals in the audience, Booth held a single-shot derringer with a barrel less than two inches long and fired a small round ball into the back of Lincoln's head. (The ball is currently in the National Museum of Health and Medicine.) Major Rathbone rose to stop the assassin, but Booth stabbed him with a dagger, then leaped twelve feet to the stage, breaking his leg in the process. He waved the dagger, shouted, *"Sic semper tyrannis!"* ("Thus shall it ever be for tyrants!"), and hurried to an exit, where a stagehand held a horse for him. Some in the audience were confused and thought the shooting was a part of the play until the screams came: "The president has been shot." They were followed by cries of "Booth!" from playgoers familiar with the actor.

The first doctor to reach Lincoln was Charles A. Leale, a twenty-three-year-old army surgeon fresh out of medical school. When Leale got to him, the president was already paralyzed, his eyes closed. Initially, Leale thought he was dead, but as the young doctor probed the wound with a finger, the president showed signs of life. Two other doctors rushed in, and the three agreed to have Lincoln removed to a house across the street. Too long to fit on the bed, Lincoln was stretched across it diagonally. Two more doctors, including Lincoln's family doctor, Robert K. Stone, reached the stricken president; they would be followed by a stream of medical men, sixteen in all, and all useless. (One of them, army doctor D. W. Bliss, would also preside over the care of President Garfield when he was shot sixteen years later. His performance in that shooting was equally lacking.) The wound was probed twice more, but nothing was done for the dying president except to keep the wound free of coagulating blood. At 6:50 A.M. Lincoln stopped breathing, then recovered briefly, only to stop again. At 7:22 he "breathed his last."

In a 1995 *American Heritage* article, "How Did Lincoln Die?," Dr. Richard A. R. Fraser argues that Lincoln's wound was not necessarily fatal. The attending doctors went against medical practice accepted even then by probing the wound, first with fingers and then a metal probe. According to Fraser, this unsterile activity certainly worsened Lincoln's chances for survival, which was a distinct possibility given that many people, including Civil War combat victims, have survived greater head wounds than the one Lincoln suffered.

Within hours Vice-President Andrew Johnson (1808–1875), a Unionist senator from Tennessee who had also been targeted for assassination, was sworn in as the seventeenth president. The only member of the Senate from a seceded state to remain loyal, he was also a War Democrat among the Radical Republicans in Lincoln's Cabinet and Congress. At the inauguration a few weeks earlier, Johnson had been noticeably drunk, the result of too much "medicinal" brandy. It would now fall to this former tailor, who had known frontier poverty just as Lincoln had, to stitch the severed nation back together.

Booth's attack was the culmination of months of planning. Born in Maryland into one of America's most prominent theatrical families, Booth was the son of Junius Brutus Booth, the leading Shakespearean actor in America. An actor himself, John Wilkes Booth was known to the Lincolns, who had seen him perform; Lincoln once invited him for a visit between the acts of a play, a request the actor ignored. A white supremacist and Confederate sympathizer who had never served in a Confederate army, Booth had earlier hatched a scheme to kidnap Lincoln, take him to Richmond, and exchange him for Confederate prisoners and a negotiated peace. By April, with no Confederate government in Richmond, Booth decided to kill the president instead. His fellow conspirators, some of them Confederate veterans and all but one Marylanders, were assigned to assassinate other Cabinet officers, including Vice-President Johnson, Secretary of State Seward, and General Grant. A brutal thug, Lewis Powell, was assigned to kill Seward at his home and nearly succeeded, but Seward was wearing a heavy neck brace that probably saved his life. Powell wounded two of Seward's sons and a secretary before escaping.

After racing out of Ford's to the waiting horse, Booth escaped over the Navy Yard Bridge. Joined by one accomplice, a half-witted and star-struck David Herold, they stopped at the house of Dr. Samuel A. Mudd, who set Booth's leg in a splint. (Mudd knew Booth but said he did not recognize him that night.) The two then went to the home of a wealthy Confederate sympathizer, who hid them for six days. On April 21 they went to Virginia, eventually reaching the farm of Richard H. Garrett in Bowling Green. While sleeping in Garrett's tobacco barn, Booth and Herold were surrounded by Union cavalry, who called for them to surrender. Herold did; Booth refused, and the barn was set afire. He was then shot and killed, with Sergeant Boston Corbett taking credit for the fatal shot. (According to Stewart Sifakis in *Who Was Who in the Civil War*, Corbett was armed with a rifle, but Booth was killed by a pistol shot. More likely, Booth killed himself to avoid capture.)

Amid the swirling rumors of a massive conspiracy hatched in Richmond, President Johnson appointed a nine-man military commission to investigate the assassination. It would make the Warren Commission, which probed John F. Kennedy's assassination, look like a textbook investigation. Focusing on the involvement of Jefferson Davis in the conspiracy, the commission—which was illegal in the sense that it should have been a civilian rather than a military proceeding—went down one blind alley after another. Obscuring the fact of two plots—first a kidnapping, then a murder—the commission seemed to be interested in vengeance, not truth. All of the conspirators tried were found guilty. Four, including Mary Surratt, the mother of John Surratt, at whose boardinghouse the plot was hatched, were convicted and executed in July. Three other defendants, including Dr. Mudd, were sentenced to life imprisonment. (Both Mudd and Mrs. Surratt have passionate defenders who continue to attest to their innocence. Mudd was later pardoned and released, but a century of investigation and scrutiny of his case has not been able to completely exonerate him.) John Surratt, Booth's chief accomplice and a member of a Confederate spy ring in Canada and upstate New York, escaped and fled to Canada. He went on to England and then the Papal States. Arrested in 1866, Surratt was brought back to the United States for trial but was freed after the jury split. He retired to Baltimore and lived until 1916.

While Jefferson Davis would be suspected of complicity, in fact, Lincoln's assassination was the work of a band of zealots—some psychopathic, others mentally unstable—in a small but tragically successful conspiracy.

Civil War Voices

Secretary of the Navy Gideon Welles (April 14–15, 1865).

The giant sufferer lay extended diagonally across the bed, which was not long enough for him. He had been stripped of his clothes. His large arms, which were occasionally ex-

posed, were of a size which one could scarce have expected
from his spare appearance. His slow, full respiration lifted
the clothes with each breath that he took. His features were
calm and striking. I had never seen him appear to better
advantage than for the first hour, perhaps, that I was there.
After that, his right eye began to swell and that part of his
face became discolored.

. . . The room was small and overcrowded. The surgeons
and members of the Cabinet were as many as should have
been in the room, but there were many more, and the hall
and other rooms in the front or main house were full. One
of these rooms was occupied by Mrs. Lincoln and her at-
tendants, with Miss Harris. Mrs. Dixon and Mrs. Kinney
came to her about twelve o'clock. About once an hour, Mrs.
Lincoln would repair to the bedside of her dying husband
and with lamentation and tears remain until overcome by
emotion.

[April 15]

. . . About 6 A.M. I experienced a feeling of faintness and
for the first time after entering the room, a little past eleven,
I left it and the house, and took a short walk in the open
air. It was a dark and gloomy morning, and rain set in before
I returned to the house, some fifteen minutes [later]. Large
groups of people were gathered every few rods, all anxious
and solicitous. Some one or more from each group stepped
forward as I passed, to inquire into the condition of the
President, and to ask if there was no hope. Intense grief
was on every countenance when I replied that the President
could survive but a short time. The colored people especially—and there were at this time more of them, perhaps,
than of white—were overwhelmed with grief.

. . . A little before seven, I went into the room where the
dying President was rapidly drawing near the closing moments. His wife soon after made her last visit to him. The
death struggle had begun. Robert, his son, stood with sev-

eral others at the head of the bed. He bore himself well, but on two occasions gave way to overpowering grief and sobbed aloud, turning his head and leaning on the shoulder of Senator Sumner.

Not everyone was similarly moved. Declining the honor of accompanying Lincoln's body back to Illinois, Zachariah Chandler, the Radical Republican senator from Michigan, said, "The Almighty continued Mr. Lincoln in office as long as he was useful."

Another view of the aftermath, from a Union soldier, Sergeant Lucius Barber (April 21, 1865).

Marched at half past five. The news came today that President Lincoln, Secretary Seward and son have been assassinated, resulting in the President's death and severely wounding the others. And now, while the nation is rejoicing with unspeakable joy at its deliverance, it is suddenly plunged into the deepest sorrow by the most brutal murder of its loved chief.

We are now continually passing paroled men from Lee's army on their way to their homes, or to where their homes were. Many found blackened ruins, instead, and kindred and friends gone, they know not whither. Oh, how much misery treason and rebellion have brought upon our land!

Was Jefferson Davis Wearing a Dress When He Was Captured?

With Lee's message to leave Richmond, Jefferson Davis and his Cabinet met hastily and left on April 2 for Danville, North Carolina, a hundred and forty miles to the south. Also aboard the train was the Confederate treasury's remaining gold, worth an estimated $500,000–$600,000, as well as Confederate banknotes, negotiable bonds, and a chest of jewels donated by southern women to purchase a warship. With the treasury gold and a grim determination somehow to keep fighting, this government on the run had no clear plan. Davis was still convinced that the war could go on and hoped to establish a new capital in Danville, while awaiting news from Lee, who had said he could fight in the hills of Virginia for another twenty years! When they reached Danville, the treasury was assigned a new value: $327,022. All attempts to sort out the discrepancies in values and dispensing of official funds were useless; rumors of Davis's absconding with the Confederate treasury persisted for years but were simply baseless. However, they provided Margaret Mitchell with a plot line in *Gone With the Wind*, and somebody certainly ended up with portions of the Confederate treasury. Maybe it *was* Rhett Butler after all!

After Lee's surrender, the escaping Cabinet began to scatter. Davis continued south until he was reunited with his wife and family. Secretary of State Judah Benjamin and War Secretary John Breckinridge, once the vice-president of the United States and the 1860 Democratic contender for president, escaped to Havana, Cuba, and then to Europe. Hoping to reach safe haven in Texas or Mexico, where large numbers of former Confederate soldiers were heading, Davis continued south, pursued by an eager federal army. Shortly after Lincoln's assassination, President Johnson and Secretary Stanton put a price on Davis's head and named him as a conspirator in the murder plot.

Civil War Voices

From Frank Moore's *The Rebellion Record* (1864), describing the capture of Jefferson Davis.

As he got to the tent door thus hastily equipped, and with this good intention of preventing an effusion of blood by an appeal in the name of a fading but not wholly faded authority, he saw a few cavalry ride up the road and deploy in front.

"Ha, Federals!" was his exclamation.

"Then you are captured," cried Mrs. Davis, with emotion.

In a moment she caught an idea—a woman's idea—and

as quickly as women in an emergency execute their design, it was done. He slept in a wrapper—a loose one. It was yet around him. This she fastened ere he was aware of it, and then bidding him adieu, urged him to go to the spring, a short distance off, where his horses and arms were. Strange as it may seem, there was not even a pistol in the tent. Davis felt that his only course was to reach his horses and arms, and complied. As he was leaving the door, followed by a servant with a water bucket, Miss Howell flung a shawl over his head. There was no time to remove it without exposure and embarrassment, and as he had not far to go, he ran the chance exactly as it was devised for him. In these two articles consisted the woman's attire of which much nonsense has been spoken and written, and under these circumstances, and in this way, was Jefferson Davis going forth to perfect his escape.

But it was too late for any effort to reach his horses, and the Confederate President was at last a prisoner in the hands of the United States.

Captured on May 10, 1865, in some woods near Irwinville, Georgia, Davis was soon to be humiliated by the report spread through the country that he had been captured while wearing women's clothing. Knowing the rumor to be false, Secretary Stanton apparently enjoyed the notion and encouraged the spread of the story, which eventually entered Civil War mythology. The northern newspapers had a field day, with editorial cartoons depicting the Confederate president in hoopskirts concealing a large Bowie knife. New York showman P. T. Barnum offered $500 for the dress Davis was supposed to be wearing. When it didn't materialize, Barnum displayed a figure of Davis wearing another dress, stolen from his wife's trunk. Over the next few years, it would be seen by thousands at Barnum's New York Museum.

Following his capture, Davis was taken with his wife and children to Macon, taunted along the way by the Union soldiers singing,

"We'll hang Jeff Davis from a sour apple tree." On May 22 Davis was imprisoned in Fort Monroe, Virginia, to begin what turned out to be seven hundred and twenty days of imprisonment without benefit of trial. Under constant watch by guards who had orders not to speak if spoken to, Davis endured a harsh imprisonment as ordered by Stanton, who was still eager to implicate him in the assassination conspiracy.

In a dungeonlike cell which had been prepared for him, he was shackled in leg irons on orders from Washington. When the officer arrived to shackle him, Davis protested, "These are no orders for a soldier! They're for a hangman. No soldier should accept them. The world will ring with this disgrace, I tell you. The war's over. The South is conquered. America is my only country. I plead against this degradation for the honor of America."

Davis's condition worsened. A light was kept burning in his cell around the clock, and he could not leave the cell area for any reason. The *New York Herald* described the former president as "literally in a living tomb . . . His life has been a cheat. His last free act was an effort to unsex himself and deceive the world. He is . . . buried alive." In the cold, damp cell, his poor health grew rapidly worse. His life was saved only by the intervention of John J. Craven, a medical officer who provided Davis with some basic comforts and kept an account of his imprisonment.

News of this degrading and humiliating treatment eventually evoked an outpouring of sympathy in both the South and North. Even the staunch Republican Unionist and abolitionist Horace Greeley, editor of the *New York Tribune*, took up Davis's case. Later in the summer, Davis was permitted to take walks in the open air. Under Craven's attention, his prison conditions and health gradually improved. Never brought to trial, Davis was finally released after two years of confinement on $100,000 bail after a surety bond was signed by, among others, Greeley and another abolitionist, Gerrit Smith. He was now homeless and penniless.

Documents of the Civil War

General Grant proclaimed the end of hostilities.

> War Department, Adj.-Gen's. Office
> Washington, D.C., June 2, 1865

Soldiers of the Armies of the United States:

By your patriotic devotion to your country in the hour of danger and alarm, your magnificent fighting, bravery, and endurance, you have maintained the supremacy of the Union and the Constitution, overthrown all armed opposition to the enforcement of the laws and of the proclamation forever abolishing Slavery—the cause and pretext of the Rebellion—and opened the way to the rightful authorities to restore order and inaugurate peace on a permanent and enduring basis on every foot of American soil. Your marches, sieges, and battles, in distance, duration, and resolution, and brilliancy of results, dim the luster of the world's past military achievements, and will be the patriot's precedent in defense of liberty and right in all time to come. In obedience to your country's call, you left your homes and families, and volunteered in her defense. Victory has crowned your patriotic hearts; and, with the gratitude of your countrymen and the highest honor a great and free nation can accord, you will soon be permitted to return to your homes and families, conscious of having discharged the highest duty of American citizens. To achieve these glorious triumphs and secure to yourselves, your fellow countrymen, and posterity, the blessings of free institutions, tens of thousands of your gallant comrades have fallen and sealed the priceless legacy with their blood. The graves of these a grateful nation bedews with tears, honors their memories, and will ever cherish and support their stricken families.

> U. S. Grant, Lt. General

MILESTONES IN THE CIVIL WAR: MAY 1865–1866

1865

May 2 President Johnson offers a reward of $100,000 for the capture of Jefferson Davis.

May 10 Johnson declares that armed resistance to the federal government is officially at an end.

May 12–13 *Battle of Palmito Ranch.* A force of 80 black and white Union infantry and cavalry attack a Confederate outpost on the banks of the Rio Grande at Palmito Ranch, twelve miles from *Brownsville, Texas.* The Confederate troops, who have done nothing to break an unofficial truce, consist of about 350 cavalrymen. In two days of back-and-forth fighting, the Confederate troops force the Union soldiers to withdraw and pursue their disorganized retreat. The Confederates suffer only five minor wounds in the skirmish, while 30 Union soldiers are killed or wounded and another 113 captured. The small battle marks the last fighting by land forces of any size during the war. A Confederate victory, its outcome has no consequence and is a futile waste of life.

May 23–24 *The Grand Review.* Some 150,000 men parade through *Washington* in the course of two days. First, the Army of the Potomac, in crisp uniforms and led by General Meade, passes by with smart discipline. In the following days, Sherman's army of Westerners, the worn veterans of the March to the Sea, are cheered wildly by thousands of spectators lining the streets.

May 25 In *Mobile, Alabama,* which had escaped serious war damage, thirty tons of munitions stored in a warehouse suddenly explode. Nearly eight city blocks are destroyed, and more than three hundred people die in the explosion and fires that follow.

May 29 Johnson proposes his plan for "Restoration." It is at odds with congressional plans for "Reconstruction." He also announces a general pardon to all persons who directly or indirectly participated in "the existing rebellion" except for a few leaders of the Confederacy.

June Congress equalizes pay, equipment, arms, and medical services for black troops.

September Congressman Thaddeus Stephens urges the confiscation of the estates of Confederate leaders and the distribution of their lands to adult freedmen in forty-acre lots.

September 5 All southern ports are reopened to foreign shipping.

November 24 *Mississippi* establishes the first of the restrictive "Black Codes," which restore many prewar rules, sharply limiting the rights and freedoms of freed blacks with respect to voting, holding property, and education, virtually reenslaving them.

December 1 The writ of habeas corpus, suspended by Lincoln, is restored.

December 4 The 39th Congress convenes. All of the Confederate states except *Mississippi* have met the president's requirements for readmission. But many are unrepentant. *Georgia* sends the former vice-president of the Confederacy, Alexander H. Stephens, to the Senate. The House simply omits the former Confederate states from the roll call. Congress then creates a Joint Committee on Reconstruction—or the Committee of Fifteen—chaired by William Fessenden of Maine and hard-liner Thaddeus Stevens of Pennsylvania.

December 18 After passing through the House of Representatives in January, the Thirteenth Amendment, which abolishes slavery, is ratified by the states and proclaimed in effect by Secretary of State Seward. (See Appendix IX for the complete text.)

December 24 Meeting in the law office of Thomas M. Jones in *Pulaski, Tennessee,* six former Confederate officers form a secret society that will become the Ku Klux Klan.

* The weekly magazine *The Nation* is first published.

* Mark Twain publishes his story "The Celebrated Jumping Frog of Calaveras County."

1866

February 19 Congress authorizes the continued existence of the Freedmen's Bureau, created a year earlier to provide medical assistance, food, and other aid to free slaves; President Johnson vetoes the act.

April 9 Congress passes a Civil Rights Act over Johnson's veto, the first significant legislation ever passed over a presidential veto. It includes many of the guarantees that will become part of the Fourteenth Amendment to the Constitution when parts of this act are questioned.

June The House passes the Fourteenth Amendment, which guarantees citizenship to all native-born or naturalized Americans, and submits it for ratification by the states. It includes a provision that excludes former Confederates from holding office. Ratification of the amendment is a condition for readmission to the Union. It is not declared ratified by the states until July 28, 1868. (See Appendix X for the complete text.)

July 10 Over Johnson's veto, Congress passes a new Freedmen's Bureau Bill, continuing the department for two more years.

July 30 In *New Orleans*, forty-eight blacks are killed when whites riot over the proposed introduction of black voting rights. Similar violence takes place elsewhere in the South.

November In the first postwar elections, Radical Republicans sweep to majorities in both houses, further weakening President Johnson.

Civil War Voices

Southern journalist Charles H. Smith:

> It were in the ded of winter, thru snow and thru sleet, over creeks without bridges and bridges without floors, thru a deserted and deserlate land wher no rooster was left to krow, no pigs to squeel, no dog to bark, wher the rooins of happy hoams adorned the way, and ghostly chimniz stood

up like Sherman's sentinels a gardin the rooins he had
made.

What Did the Confederate Soldiers Come Home to Find?

Like those who defend America's use of the atomic bomb on Japan
to save millions of lives in World War II, William T. Sherman always
maintained that the devastatingly destructive war he had waged on
the Confederacy shortened the war and saved soldiers' lives. His
good intentions went unappreciated by the victims of his ruthless-
ness. For many along his path, after Sherman's troops departed, there
was literally nothing left on which to support a family. Houses were
looted. Those animals not taken by the Union troops were killed.
Under Sherman's "scorched earth" policy, any item that could be
useful for farming or manufacturing was destroyed, the grim "justice"
for what Sherman viewed as treason.

In the aftermath of the war, the entire Confederacy, save sections
west of the Mississippi that had been spared the massive battles, was
devastated—physically, economically, even spiritually. The postwar
South was probably worse off than Europe after either of the world
wars of this century. Because of Sherman's notorious destruction of
the southern railroads, many of Lee's defeated soldiers had to walk
home from Virginia. Many found that their homes had been burned.
In some cases, entire towns and even whole counties had been evacuated. Common were the stories of farmers—many women and children included—who hitched themselves to plows in the absence of
field animals.

But there was remarkable resilience as well. In *The Long Surrender*,
Burke Davis notes that energetic Southerners, joined by newcomers
drawn south by the promise of economic revival, began a dramatic
rebuilding. The rails to Atlanta were restored only a month after
Sherman left. In fact, many southern railroad companies, led by former generals like P. T. Beauregard, had trains rolling by the time
Jefferson Davis was freed. (Still, they would be burdened for decades
by unfair rates and restrictive tariffs set by Northerners, who con-

trolled the vast majority of railways and the legislatures that set rates.) In devastated Charleston, workmen were busy rebuilding "when guns were being stacked at Appomattox." Memphis, Chattanooga, and Birmingham all witnessed an extraordinary resurgence in heavy manufacturing, with Birmingham becoming a steel center. (Birmingham's revival was led by General Josiah Gorgas, the genius behind the Confederate wartime ordnance, who kept his armies supplied with ammunition despite facing overwhelming odds.) Above all, the world was waiting for southern cotton. The industry revived almost overnight.

What Does "Forty Acres and a Mule" Mean?

As bleak as things were for the defeated Confederates, it was even more difficult and uncertain for the four million freed slaves, accustomed to centuries of a system that had not remotely prepared them for emancipation. For these millions displaced by the war's end, there was a true emergency. To address that crisis, Congress had established the Bureau of Refugees, Freedmen, and Abandoned Lands in March 1865. A temporary federal agency under the War Department, it was to assist the emancipated slaves in making the transition to freedom. A year later, when Johnson vetoed a congressional bill continuing the work of the bureau, the veto was overridden by Congress, and the department was repeatedly extended.

Headed by General Oliver Otis Howard, an abolitionist who had lost his arm during the Peninsula Campaign and later marched with Sherman to the sea, the Freedmen's Bureau, as it was known, distributed trainloads of food and clothing provided by the federal government. The supplies went both to freed slaves and southern white refugees. During a relatively brief existence, the agency spent a staggering $17 million in direct aid. Much of it went to constructing hospitals for the former slaves and providing medical care. More than four thousand schools were built for black children, and most of the major Negro colleges in the United States, including Fisk and the Hampton Institute, were founded with the bureau's assistance. (General Howard helped establish what would become a black university in Washington, D.C., which was named for him. He served as Howard's third president, from 1869 to 1874.)

But the bureau was more than a friendly welfare agency. It held far-reaching powers that proved ripe for corruption. It regulated wages and working conditions, handled legal affairs, and oversaw the confiscation and redistribution of lands, one of its most controversial roles. Fulfilling Radical Republican Congressman Thaddeus Stevens's call for "forty acres and a mule" for every freed black adult, the bureau was responsible for more than 800,000 acres of land that had been abandoned or confiscated from former Confederates, including Brierfield, the plantation of Jefferson Davis. The charges of corruption later leveled against the agency hurt its reputation, but Howard was cleared of any wrongdoing after a congressional investigation.

To white Southerners, the bureau came to represent the worst of what they thought the North had in store for them. Some blacks were settled on public lands under the Southern Homestead Act of 1867, but the hopes of massive land redistribution in the South never materialized. Property rights were still considered sacred by Congress, and much of the land meant for the former slaves ended up in the hands of speculators, lumber companies, railroads, and large plantation owners. When President Johnson announced the restoration of abandoned lands to pardoned Southerners, most free blacks had little choice but to participate in sharecropping arrangements. Working as tenant farmers who paid for seed and supplies with a portion of their crops, the sharecroppers learned that slavery had merely been replaced by this economic arithmetic that never quite added up in their favor, a new form of bondage from which there was equally little hope of escape.

As a guardian of voting rights, the Freedmen's Bureau also failed to measure up. In the worst postwar voting catastrophe, forty-eight blacks died when the police of New Orleans put down a peaceful demonstration in favor of the black vote. General Philip Sheridan, the military governor of Louisiana, had been away during this "Riot of New Orleans" and called the killings a massacre. In other southern states, "Black Codes" designed to restore the prewar condition of blacks were quickly introduced. To many of the former slaves, it must have seemed that the war had changed very little.

Civil War Voices

Former slave Sojourner Truth, an abolitionist and women's rights pioneer, wrote about postwar racial attitudes in the nation's capital (October 1, 1865).

A few weeks ago I was in company with my friend Josephine S. Griffing, when the conductor of a streetcar refused to stop his car for me, although [I was] closely following Josephine and holding on to the iron rail. They dragged me a number of yards before she succeeded in stopping them. She reported the conductor to the president of the City Railway, who dismissed him at once, and told me to take the number of the car whenever I was mistreated by a conductor or driver. On the 13th I had occasion to go for blackberry wine, and other necessities for the patients in the Freedmen's Hospital where I have been doing and advising for a number of months. I thought now I would get a ride without trouble as I was in company of another friend, Laura S. Haviland of Michigan. As I ascended the platform of the car, the conductor pushed me, saying "Go back—get off here." I told him I was not going off, then "I'll put you off" said he furiously, clenching my right arm with both hands, using such violence that he seemed about to succeed, when Mrs. Haviland told him he was not going to put me off. "Does she belong to you?" said he in a hurried angry tone. She replied, "She does not belong to me, but she belongs to humanity." The number of the car was noted, and the conductor dismissed at once upon report to the president, who advised arrest for assault and battery as my shoulder was sprained by his effort to put me off. Accordingly I had him arrested and the case tried before Justice Thompson. My shoulder was very lame and swollen, but it is better. It is hard for the old slaveholding spirit to die. But die it must.

Why Was President Andrew Johnson Impeached?

The death of Abraham Lincoln was tragic in ways beyond the obvious. Before his death, Lincoln had proposed a series of lenient postwar measures designed to restore peace and the status of the seceded states. His "10 percent" plan allowed a state to be recognized if 10 percent of the voting population agreed to abide by federal regulations and support the Constitution. With Lincoln's death, the weaker President Johnson was at the mercy of a Congress under the control of the Radical Republicans, led by hard-liners like Pennsylvania's Thaddeus Stevens, an uncompromising and longtime supporter of rights for blacks, and Charles Sumner, the abolitionist Massachusetts senator beaten on the Senate floor before the war, who would hold out for a far more punitive Reconstruction plan. The Reconstruction Acts of 1867 were passed on March 7 over Johnson's veto. The congressional plan, which held that the southern states had committed "suicide," set a harsh agenda for their return to the Union.

Under the first of the Reconstruction Acts, the South was divided into five military districts with a U.S. Army general in charge of each. In other words, the South was essentially under martial law. Generals John Schofield, Daniel Sickles, John Pope, Edward Ord, and Philip H. Sheridan now held nearly dictatorial powers over their military districts. To preserve order and carry out the dictates of Congress, some 200,000 U.S. soldiers were stationed throughout the South. These military commanders removed thousands of civil officials from their jobs and actively registered black voters. Former slaves were now in a position to dominate their former masters.

The Radicals in Congress knew that the vote of large numbers of blacks could lead to their domination of the area for years to come, and they pressed for a Civil Rights Act that would prevent the states from restricting black voting rights. When the Radical Republicans swept the congressional elections of 1866, as the balance of power had shifted totally from the White House to Congress. In one blatant attempt to consolidate that power, Congress passed the Tenure of Office Act in March 1867, which said a president could not remove any official, including his own Cabinet members, without Senate approval. To challenge this blatant assault on constitutional "checks and balances," Johnson asked for the resignation of War Secretary Edwin Stanton, a Democrat but an ally of the Radicals. The president was promptly impeached by the House—the equivalent of a grand jury indictment—on eleven charges. Tried before a Republican Senate with Chief Justice Salmon Chase presiding, Johnson was acquitted by a single vote on May 18, 1868. Those Republicans who voted for his acquittal paid with their political careers. Four days later, Ulysses S. Grant was nominated by the Republicans to run for the presidency. Johnson served out the balance of his term as a political cripple.

Who Were the Carpetbaggers and Scalawags?

The era of congressional Reconstruction was something of a mixed blessing. Well-meaning Northerners who came south to assist blacks and open schools established some of the leading black colleges in the South. And for the first time in American history blacks gained political power, providing the votes that elected Grant and putting the first blacks in Congress. Mississippi sent two black men to the Senate. Blanche Bruce was a former slave, and Hiram Revels, a clergyman who had recruited black soldiers during the war, took Jefferson Davis's old seat in the Senate. (After his term in Congress, Revels became president of Alcorn Agricultural & Mining College.) The simple fact that men who had a few years earlier been in chains were now voting and sitting in the Congress was an astonishing, revolutionary event, albeit a short-lived one.

The flip side of this achievement was the corruption that came along at the same time. For the most part uneducated, the blacks of the South were not prepared for democracy and ripe for exploitation. The worst of the exploitation often came at the hands of unscrupulous Northerners who had rushed south to fill the political void. Depicted in the contemporary press as men rushing in with their belongings in a "carpet bag," they were soon derided as "carpetbaggers" and became a popular symbol of all that was hated about the North. The traditional view that *all* of these men were corrupt vandals was soundly refuted as postwar propaganda by Eric Foner in his massive study of the period, *Reconstruction*. Foner gives convincing evidence that many so-called carpetbaggers moved to the South out of a desire to improve the lot of blacks; others were simply entrepreneurs who helped reestablish the economy of the South.

Despised even worse than carpetbaggers were those former Confederates who were seen as jumping on the northern bandwagon. These white southern politicians who joined the Republican party after the war and advocated the acceptance of and compliance with congressional Reconstruction were labeled "scalawags"—traitorous opportunists who had deserted their countrymen. Again, Foner had dismissed that idea as reflecting postwar animosity more than political reality. Many so-called scalawags, he shows, believed that acceptance of the Reconstruction Acts was the fastest way to return to home rule. But they merited the disdain and attacks of those Confederates still loyal to "the Cause." Perhaps the most hated of the scalawags was General James Longstreet, Robert Lee's "Warhorse," who compounded his sin of embracing Republicanism by suggesting that "Marse Robert" had made mistakes at Gettysburg.

Who Started the Ku Klux Klan?

One of the most far-reaching consequences of the presence of the carpetbaggers, scalawags, and federal troops in the former Confederacy was a violent backlash that formally began on Christmas Eve, 1865. As Reconstruction seemed to threaten to destroy any semblance of control over their lives, some Southerners began to form secret organizations whose single purpose was to intimidate and terrorize blacks and Republicans. With such names as Pale Faces, the Sons of Midnight, and the Knights of the White Camellia, these secret societies adopted what now sound like silly names and rules. But there was nothing amusing about their aims and tactics. Their principal goal was simple: to combat black political power and maintain white supremacy.

On December 24, 1865, a group of six Confederate veterans met in Pulaski, Tennessee, and formed another such group, giving it the name Ku Klux Klan, derived from the Greek *kuklos* ("circle"). In a Nashville convention in 1867, the KKK adopted a declaration, stating its intention was to "protect the weak, the innocent and the defenseless . . . to relieve the injured and oppressed." At this meeting, Nathan Bedford Forrest, the most feared Confederate general of the war and called "the Fort Pillow Butcher" in the northern press, became the first Grand Wizard of the Empire.

Operating as a quasi-military organization, the Klan, which had also adopted the name Invisible Empire, grew quickly with the addition of more former rebel soldiers. They initially used parlor tricks and Halloween pranks: pretending they were dead rebel soldiers from the battlefield, they appeared to drink whole buckets of water that went down a hose. At first the goal was simply to frighten blacks when they rode up in the middle of the night. Wearing white robes and pointed hoods, these "ghosts of dead rebel soldiers" gradually escalated the violence. Beatings, lynchings, burning, and other acts of terrorism, rape, and murder replaced the simple night-riding in sheets. The pretense of secrecy fell away, and most southern townspeople knew exactly who was under those robes. "Klaverns" often included the most powerful men in a town—the bankers, newspaper publishers, and business owners. A sort of Rotary Club of racist terror.

The Klan's political terrorism achieved its intended effect. Added to the "Black Codes," the threat of Klan violence kept blacks away from the polls. In fact, Forrest officially called for the Klan to disband in 1869. Although it had achieved its aims, Forrest had also apparently undergone a religious conversion. At the time of his death in 1877, he was supposedly quite repentant. According to his biographer Jack Hurst, the "Devil" Forrest disavowed the Klan and called for the social and political advancement of blacks in his last two years. But the klaverns continued to flourish. In 1872 President Grant ordered the disbanding of illegal organizations, and hundreds of Klansmen were arrested. With the demise of congressional Reconstruction in 1877 and the restoration of prewar white supremacy through much of the defeated Confederacy, the Klan lost much of its urgency, and its role diminished. However, the name, rituals, and tactics reemerged in 1915, when it was reorganized with Jews and Catholics joining blacks as its targets. After World War I, when the perceived threat of socialism grew, the Klan rapidly expanded, becoming more politically influential than ever. Over the next few decades, its membership reached as many as three million. In 1924 a resolution denouncing the Klan was defeated at the Democratic national convention, and membership was a key to holding local political power. Many prominent Southerners continued to join the Klan. In one of the most notorious instances of Klan power, Hugo Black, appointed to the Supreme Court by Franklin Roosevelt, admitted that he had been a member of the Klan.

Whatever Became Of?

A summary of the postwar lives of some of the famous and infamous, obscure, notable, and otherwise curious personalities of the Civil War period.

Charles Francis Adams, Sr. Descended from one of America's greatest families, Adams remained in England as America's ambassador until 1868. There he began negotiations regarding Britain's repaying the United States for the losses inflicted on Union merchant ships' commerce by Confederate raiders built in England. Adams played a vital role in the treaty and arbitration that settled these socalled *Alabama* claims. Never a Lincoln supporter, Adams eulogized

Secretary of State Seward in 1873 as Lincoln's superior "in native intellectual power, in the extent of acquirement, in breadth of philosophical experience, and in the force of moral discipline," making Seward the genius and driving force behind the Union success and Lincoln merely a figurehead. In defense of Lincoln, Secretary of the Navy Gideon Welles responded with a book, *Lincoln and Seward*, which presented the case for Lincoln's role as the principal figure in the Civil War government. Adams died in 1886. One of his sons, Charles Francis Adams, Jr., had a distinguished Civil War career; another, Henry Brooks Adams, served as his father's unofficial secretary in London and also wrote secretly for *The New York Times*. He became a prominent historian and is best known for his book *The Education of Henry Adams*.

Louisa May Alcott Having served as a Union nurse until she fell ill with typhoid, Alcott in 1863 released under her name a collection of wartime letters, published as *Hospital Sketches*. Although her request to go south to teach the "contrabands"—slaves held by the Union army—was denied, she wrote a story called "My Contraband," published in *The Atlantic Monthly* as "Brothers," in which a captured Confederate soldier is nursed by a contraband servant who is the soldier's mulatto half brother. After the war, Alcott published the first of her most famous books, the autobiographical stories of the March family, *Little Women* (1868), *Little Men* (1871), and *Jo's Boys* (1886). Alcott died in 1888.

Robert Anderson Although he had hoped to prevent the war, Anderson had been there at its beginning in his defense of Fort Sumter. When his health failed, he was relieved of field command and given various duties until he retired from the regular army in 1863. After the recapture of Charleston, Anderson took part in the ceremony in which he raised the flag he had lowered four years earlier. Anderson died in 1871.

Clara Barton When her wartime freelance nursing duties ended, Barton spent extensive time and energy at Johnson's request in locating missing soldiers. In that role, she visited the notorious Confederate prison at Andersonville and gathered considerable evidence of the inhumane treatment common there. But, apparently for sexist reasons, she was prevented by federal prosecutors from testifying at the trial of Andersonville's commander, Henry Wirz. Following the Civil War, she went to tend the sick in the Franco-Prussian War and eventually returned to the United States to found the American Red Cross in 1881, serving as its president until 1904.

Pierre G. T. Beauregard Anderson's opposite at Sumter, the Confederate "Hero of Fort Sumter" had undergone a serious falling-out with Jefferson Davis as the war went on. Blamed for the defeat at Shiloh, Beauregard had taken sick leave without permission and was relieved of his command by Davis. Although he continued to serve and lead efficiently, his difficult relationship with Davis limited his effectiveness. Offered commands after the war in the Egyptian and Romanian armies, he returned to Louisiana, went into railroading, and was tarnished for his supervising a corrupt Louisiana lottery. Beauregard died in 1893.

Judah Benjamin Known as "the brain of the Confederacy," the secretary of state escaped after the fall of Richmond, going first to Cuba, then London. After a period of living from hand to mouth, he began a new law career in London, eventually gaining fame and fortune. Befriended by British politician Benjamin Disraeli, he began an international law practice and wrote a definitive text that became a standard law treatise used well into the twentieth century. He died in honor and wealth in 1884.

Ambrose Everett Burnside Bedeviled by one disaster after another during the war, Burnside was relieved of command after the deadly Crater explosion at Petersburg. After the war, he served as governor of Rhode Island from 1866 to 1869 and then as U.S. senator from 1875 to 1881. He is best remembered for introducing the word *sideburns* into the language.

Benjamin Butler The notorious "Beast" of the Union, tarnished by charges of wartime corruption, Butler joined the Radical Republicans in Congress and was a leader in the impeachment movement against President Johnson. In 1868 he was given a bloodstained nightshirt belonging to A. P. Huggins, a Northerner who claimed that klansmen in Mississippi had whipped him bloody in the night. During a House speech, Butler waved "the bloody shirt" as he spoke, and from 1868 through the end of the nineteenth century, waving "the bloody shirt"—which also came to be associated with Lincoln's assassination—was standard propaganda for Republicans who wanted voters to punish Democrats and the South for the Civil War.

Christopher "Kit" Carson The fabled "mountain man" who led John C. Frémont's famed expeditions to the Rockies, Carson served with the Union in the Southwest, becoming a brigadier general. He was troubled by the harsh government treatment of Indians both during and after the war, and as an Indian agent Carson was one of a handful of whites who attempted to deal honestly and humanely with the tribes.

"Albert J. Cashier" Unremarkable in any way for serving with the Illinois Volunteers, this pensioned soldier was involved in an automobile accident in 1911. Only at the hospital did it become apparent that "Albert Cashier" was a woman, whose fellow veterans said they never suspected the truth.

Joshua L. Chamberlain One of the heroes of Gettysburg, Chamberlain received the Congressional Medal of Honor. Twice wounded and suffering from malaria, he was given lighter duties but returned to his command. At Petersburg he was wounded again and carried to the rear, where General Grant promoted him to brigadier general on the spot. He took part in the siege of Petersburg and the Appomattox campaign and was given the honor of commanding the troops that formally accepted the surrender of the Confederate army. Returning to Maine, he later served as governor of the state and as president of Bowdoin College.

Salmon P. Chase Although Lincoln had refused to acknowledge an earlier resignation by Chase, he had accepted the secretary's letter of resignation submitted in 1864. After Chief Justice Roger Taney died in October 1864, to the surprise of many, Lincoln appointed Chase to replace him. As Chief Justice, Chase presided over the 1868 impeachment trial of Andrew Johnson. His support of Johnson angered his former Radical Republican colleagues in the Senate. As a result, Chase's name was put forward as a possible Democratic candidate in 1868.

Johnny Clem President Grant appointed "the drummer boy of Chickamauga" to West Point, but he failed on several occasions to pass the entrance exam. In 1871 Grant made him a second lieutenant, and Johnny began a second army term that continued until he retired in 1915 with the rank of brigadier general. He was the last Civil War veteran on the army rolls at the time of his retirement. He died at the age of eighty-five in San Antonio, Texas, and is buried in Arlington National Cemetery.

Samuel Langhorne Clemens ("Mark Twain") This great American writer's Civil War career was distinguished by a brief stint with a band of Missouri Volunteers, which later took shape in a short story, "The Private History of a Campaign That Failed." He also befriended a sick and poor Ulysses Grant, becoming the publisher of his *Memoirs*, a huge financial and critical success.

Boston Corbett The man who claimed to have killed John Wilkes Booth was arrested for disobeying orders not to shoot at Booth. But War Secretary Stanton ordered his release and praised him. Corbett became a hero in the North, received a share of the reward money, and even lectured. Increasingly eccentric over the years, he was given a job as doorkeeper in the Kansas house of representatives but was later determined to be insane. Committed to an asylum, he escaped, turned up briefly at a friend's house, then disappeared without a further trace.

George Armstrong Custer Had it not been for his later exploits, the name Custer might have been forgotten or relegated to a Civil War footnote as a rather vain and eccentric cavalry commander. Graduating at the bottom of his West Point class of 1861, Custer joined the cavalry, serving under McClellan. At Gettysburg, wearing a lavish uniform of his own design, he helped defeat Jeb Stuart, although he had actually ignored orders in doing so. At the Grand Review, Custer was notable for "losing control" of his horse and riding past the reviewing stand twice. Remaining in the army after the war, he rose to lieutenant colonel of the Seventh Cavalry and met his grim fate at Little Big Horn on the nation's centennial in 1876. (His younger brother, Thomas Custer, the only soldier to win two Congressional Medals of Honor during the war, joined his brother's regiment and also died at Little Big Horn.)

Jefferson Davis After a period of exile in Canada, the impoverished former president finally accepted a job with an insurance company. On Christmas Day, 1868, President Andrew Johnson issued a proclamation of general amnesty for the Confederate leaders. Contentious and unapologetic, Davis spent years compiling his version of events. In *The Rise and Fall of the Confederate Government* (1881), he refought the war, castigating officers like Beauregard and Johnston for the Confederacy's defeat. He lived out his remaining years in Mississippi, never seeking the restoration of his citizenship. (It was finally restored by President Jimmy Carter.)

Abner Doubleday One of the oldest American myths has been Doubleday's supposed invention of baseball. A West Pointer who fired the first Union shots at Fort Sumter, Doubleday served through much of the war but was eventually removed from command. Born in Ballston Spa, in upstate New York, he had organized baseball games at nearby Cooperstown, leading to the notion that credited him with devising the baseball diamond and basic rules while a student at West Point. This led to the decision to place the Hall of Fame at Cooperstown, where Doubleday Field was named in his honor. (The design of the baseball diamond has since been attributed to Alexander J. Cartwright of New York City's Knickerbocker Baseball Club.) Doubleday wrote three books about the war, and his descendants founded the publishing company of Doubleday.

Frederick Douglass One of the most influential men of his century, Douglass rose from slavery to become a counselor to presidents. Having worked hard to raise regiments of black soldiers and then to ensure their equal treatment, he remained loyal to Grant after the war. He was given a number of government posts, culminating with an appointment as minister to Haiti. Following the death of his wife, he stirred more controversy with his marriage to a white woman, a longtime supporter. In response, the uncompromising Douglass said, "My first wife was the color of my mother. My second wife is the color of my father."

Sarah Edmonds Another of the many women who fought in disguise, Edmonds was an expert at disguises and became a Union spy. Before the war, she had disguised herself as a boy and then a man to obtain work. Having deserted to protect her secret, she revealed her true identity after the war in an attempt to clear the desertion charges and gain a pension.

John T. Ford Arrested in the aftermath of Lincoln's death, the owner of Ford's Theatre was jailed for twenty-nine days. The government threatened to confiscate the theater, which he then sold for $10,000. A year before his death in 1894, Ford learned of another tragedy in the theater bearing his name: a wall collapsed, killing twenty-eight War Department workers.

John Charles Frémont Few men have risen so high to fall so fast and far. Famous before the war as the great explorer of the West, Frémont was the first presidential candidate of the Republican party. The son-in-law of a powerful senator, he later lost his Civil War command for disobeying Lincoln over the emancipation of slaves in his military area. He briefly challenged the president in 1864 but gave up the race. In a postwar railroading career, "the Pathfinder" lost a fortune and, tainted by charges of corruption, died poor and powerless.

James A. Garfield Having served in the defense of Chattanooga alongside George H. Thomas—"the Rock of Chickamauga"—Garfield resigned his commission to take a seat in Congress from Ohio. During his years in the House, he sat on the commission that awarded the disputed election of 1876 to Rutherford B. Hayes. Elected president in 1880, Garfield was shot by a disgruntled office seeker in July 1881 but lingered until his death in September.

William Lloyd Garrison With the outbreak of the Civil War, Garrison had predicted the victory of the North and the end of slavery, and he ceased to advocate disunion. Lincoln's Emancipation Proclamation removed the last difference between the two men, and Lincoln paid public tribute to Garrison's long and uncompromising struggle to abolish slavery. In 1865 Garrison discontinued the *Liberator* and advocated dissolving the antislavery societies. He then became prominent in reform campaigns to achieve suffrage for American women and justice for American Indians; to establish Prohibition and eliminate the consumption of tobacco in the United States; and to promote free trade and abolish customhouses on a world scale. He died in New York on May 24, 1879.

Ulysses S. Grant Having negotiated the postwar intrigues of Washington, D.C., Grant was nominated by the Republicans for president and won a close election in 1868, his margin of victory supplied by 300,000 black votes. His two terms were plagued by economic catastrophes, including a stock market crash, created by manipulators who hoodwinked Grant, and widespread corruption culminating in the Crédit Mobilier scandal that rocked Congress. His failure as a businessman, including the demise of a brokerage firm, Grant & Ward, left him nearly bankrupt, and it was only the publication of his successful *Personal Memoirs of U. S. Grant*, dictated while he was dying of throat cancer, that kept his family afloat. Grant finally succumbed to the cancer at Mount McGregor, near Saratoga Springs, in upstate New York in July 1885.

Horace Greeley The influential editor of the *New York Tribune* took a gentle approach to Reconstruction, earning him the anger of the Republican party. He went so far as to sign Jefferson Davis's bail bond because he believed that the Confederate president's long imprisonment was a violation of Davis's constitutional rights. In 1870 Greeley, the great prophet of western expansion and exploration, founded a town in northern Colorado as a farm cooperative and temperance colony, which was later named for him. Disenchanted with the corruption of Grant's administration, he ran for president in 1872 and was endorsed by the Democrats but lost to Grant. Having lost his newspaper, the election, and his wife, he died three weeks after the election in New York, on November 29, 1872.

Winfield Scott Hancock Named for America's top military hero, Hancock became a hero himself, particularly in his performance at Gettysburg. Unlike Grant, he was unable to translate his wartime heroics into political victory, and he lost a close election in 1880 to James Garfield.

Benjamin Harrison A lawyer and politician in Indiana before the war, he served with Sherman in the March to the Sea and reentered politics after the war. In 1888 he became one of six Union officers to win the presidency.

Rutherford B. Hayes Born in Ohio, this future president rose to brevet general of volunteers. He served in Congress and then as Ohio's governor before he launched a presidential run in 1876. In the most controversial election in U.S. history, Hayes lost the general election to Samuel Tilden, but a special commission gave him the election after promises to remove federal troops from the South were made to southern legislators in exchange for their switching electoral votes to Hayes. He served a single term but had little support from his own party because of his deal with the southern Democrats.

James Butler Hickock Born in Illinois, Hickock moved to the Kansas Territory as a teenager and was elected a constable at age nineteen. After serving as a stagecoach driver, he became a Union scout during the war, later claiming to have killed fifty Confederates with fifty shots from his special rifle. After the war, he was appointed a deputy marshal in Kansas and fought in several of the major Indian battles. In 1869 he became marshal of Hays, Kansas, and for a brief period in 1871 he was marshal of Abilene, becoming famous as Wild Bill Hickock while touring the country with Buffalo Bill Cody's Wild West Show. He was shot and killed in a saloon in Deadwood, Dakota Territory, in 1876.

Joseph Eggleston Johnston   Along with Beauregard, with whom he commanded at 1st Bull Run, Johnston was the Confederate general who was most at odds with President Davis. Their acrimonious relationship, which began with petty considerations over rank and military etiquette, ended when Johnston was criticized by Davis for his tactics in trying to defend Atlanta from Sherman's advance. Relieved at Atlanta, Johnston was later recalled while Sherman marched through the South, but the Confederacy's military situation was by then hopeless. After the war, Johnston wrote his *Narrative of Military Operations During the Civil War* (1874), a highly critical account of Davis's handling of the Confederate military as well as other Confederate generals. He represented his native Virginia in the House of Representatives and was then appointed commissioner of Railroads by President Grover Cleveland. While serving as an honorary pallbearer at the funeral services of his wartime rival Sherman, Johnston removed his hat, as other mourners did. Advised to put it back on to avoid the chill, he said, "If I were in his place and he standing here in mine he would not put on his hat." He caught pneumonia and died a few weeks later.

Robert E. Lee   Poor and homeless in the war's aftermath, the great Confederate commander also suffered from a serious heart condition. Declining the offers of businessmen who wanted to capitalize on his name, Lee chose to take a position at tiny Washington College in Lexington, Virginia. At the time, he said, "I have a self-imposed task which I must accomplish. I have led the young men of the South in battle. I have seen them die in the field; I shall devote my remaining energies to training young men to do their duty in life." As the most popular figure in the Confederacy, Lee was supposedly sought to lead the Klan but declined; while he was a visible supporter of a submissive approach to the new political realities, some historians have argued that his views on the Klan were ambiguous. Avoiding any public posture, Lee focused his energy on building the college (later renamed Washington and Lee). In the spring of 1870 Lee made his only tour of the postwar South, received everywhere he went by adoring crowds, who had elevated him to the heroic pantheon alongside George Washington. Four months later he died; his supposed last words, "Strike the tent," are probably part of the myth built around him. Although Lee had signed the oath of allegiance necessary for the restoration of citizenship, in 1865, it was never officially recorded. Secretary of State Seward had given Lee's document to a friend as a souvenir. After the discovery of the missing oath in 1970, Lee's citizenship was finally restored by an act of Congress in 1975.

Mary Todd Lincoln Few figures of the Civil War era are more tragic than Lincoln's wife. The daughter of a wealthy Kentucky family, she was devoted to her husband during his life and had to survive the political insults hurled at him. She had created a few problems for Lincoln with her intemperate spending as first lady. In serious financial trouble after his death, she was forced to sell furniture and jewels. The death of her third son, Tad, in 1871 from tuberculosis sent her into a serious decline. Mary Lincoln's surviving son, Robert Todd, had her committed to an asylum after a humiliating arrest and public hearing in 1875. She won her release and spent her last years in the home of her sister, wearing black, until her death in 1882.

Robert Todd Lincoln Never close to his father, the only surviving son of Abraham Lincoln became the keeper of the Lincoln myth, supervising the publication of all his father's papers and official biographies. He was ruthless toward his mother, his attempts to have her judged insane apparently motivated by greed. Although he failed to become a major power broker in Republican politics, he did become secretary of war under President Garfield, and there were often murmurings of a Robert Lincoln candidacy for president. In 1889 he became ambassador to England under President Benjamin Harrison and later became one of America's wealthiest men as president of the Pullman Car Company, which manufactured almost all the railroad cars in America. Curiously, he witnessed two other presidential assassinations, being present at the shootings of both Garfield and McKinley. He died in Vermont at his estate, Hildene, in 1926.

Arthur MacArthur, Jr. Born in Massachusetts, this future general joined the 24th Wisconsin at age seventeen, becoming its adjutant. He saw action at Perryville, Stones River, Chickamauga, and Chattanooga. The father of General Douglas MacArthur, of World War II fame, he earned the Medal of Honor for his heroism at Missionary Ridge, an award later given to his son; they are the only father-son winners of the Congressional Medal of Honor.

George B. McClellan Broken by his failure early in the war, the man called Young Napoleon challenged Lincoln as a Democrat in 1864 but was done in by the improving fortunes of the Union armies. He later served as New Jersey's governor and died in 1885.

William McKinley Born in Ohio, McKinley served throughout the war, fighting at South Mountain, Antietam, and in Sheridan's Shenandoah Valley campaign. After his discharge, he went into law and politics, backing his comrade in arms Rutherford B. Hayes, who became president. In 1896 McKinley was elected president, the last of six Union officers to reach the White House, but was assassinated in 1901, elevating Theodore Roosevelt to the presidency. Roosevelt was the nephew of James D. Bulloch, who had purchased many of the Confederate commerce raiders used against the Union blockade.

John Singleton Mosby One of the most skillful Confederate partisan fighters, this young lawyer opted not to surrender his command at the end of the war. He simply disbanded the unit and let its members make their own peace. Only thirty-one years old, Mosby reopened his law practice in Virginia, the heart of the territory in which his rangers had operated during the war. Despite his efforts in the war, Mosby earned the enmity of many Southerners when he became a scalawag, supporting Ulysses S. Grant for the White House. Though he was ostracized, Mosby and Grant became friends. In the summer of 1878, Mosby accepted an appointment as consul to Hong Kong, where he discovered corruption and embezzlement of government funds in the Asian consulate. His investigation brought on reform in the Far East region foreign service. When Mosby was replaced in 1885, he found that Grant, on the day before his death, had secured him a job with the Southern Pacific Railroad in San Francisco.

In California, Mosby became friends with a young boy named George S. Patton, with whom he would ride while relating Civil War stories. In 1897 he lost an eye and sustained a fractured skull when kicked by a horse. Four years later he lost his job with the railroad, and President McKinley found him a job with the Department of the Interior, enforcing federal fencing laws in Omaha. Mosby was as relentless in his job as he had been in the war, and local politicians had him recalled. He was sent to Alabama to chase trespassers on government-owned land before taking a job in the Justice Department, a post he kept until 1910. He died in Washington on Memorial Day, 1916.

Allan Pinkerton During the Maryland campaign, Pinkerton continued to serve McClellan; after the Battle of Antietam, he posed for a photograph with his two former colleagues from the Illinois Central, Lincoln and McClellan. But when his chief was removed from command, Pinkerton returned to his agency in Chicago and confined his military work to investigating frauds in the army supply departments and among government contractors. After the war, he continued to build his agency and became associated with the interests of "big money." He was hated in much of the South for his role in the killing of Jesse James's eight-year-old half brother and the wounding of his mother in a bombing of the family home. His actions against the Younger Brothers did not help his reputation in the region. In the North, he was hated by the labor movement for his union-busting activities and support of the railroads and major industrial corporations. His nickname, "the Eye," eventually came to identify all members of his profession.

Henry Riggs Rathbone The young officer who was with Lincoln at his assassination later married his fiancée, Miss Harris. Appointed a diplomat by Grover Cleveland, Rathbone went to Germany with his family but became mentally ill and murdered his wife. He was committed to a German insane asylum and died there in 1911.

William Henry Seward One of those targeted in the plot to kill Lincoln, Seward was attacked by Lewis Powell, who seriously wounded the secretary of state. Although he was very nearly president, Seward thought he could control Lincoln behind the scenes, but Lincoln soon mastered him. Seward's name might well be forgotten if not for his most notable postwar achievement, known then as Seward's Folly: In 1867 he purchased Alaska from Russia. Seward served out his Cabinet post during Johnson's term, then retired from public life.

Philip H. Sheridan One of Grant's most important lieutenants, Sheridan was long detested for his destruction of the Shenandoah Valley. He headed the Reconstruction government in Louisiana and Texas and was later involved in the creation of Yellowstone Park.

William T. Sherman After the devastating March to the Sea, Sherman's name acquired a kind of infamy. Once nearly declared insane and having contemplated suicide at a low point in the war, Sherman was one of the greatest Union heroes. Uncompromising in war, he got into political hot water when the terms of surrender he offered to a Confederate general were considered too lenient, the beginning of a bitter feud with Washington's politicians. After the war, he replaced Grant as commander of the army and was a sought-after candidate who disdained politics and politicians, perhaps because of his run-ins with them during the war. His famous response to one invitation to run, often misquoted, was, "I will not accept if nominated and will not serve if elected." His other famous attribution, "War is Hell," has also never been documented, but Sherman said that he had said it many times before, during, and after the war. He published his memoirs, which were nearly as successful as Grant's. He died in 1891.

Henry Morton Stanley Born John Rowlands in Wales, he came to America as a cabin boy and adopted the name of a New Orleans benefactor. While in training as a plantation manager, he joined the "Dixie Gays" in 1861 and fought at Shiloh with an ancient musket until he was captured. A Union officer released him after he took an oath of allegiance. He was then sent to join a Union artillery unit but fell sick and was discharged. After a trip to Cuba in search of his adopted father, who had died, Stanley returned to New York and enlisted in the Union navy, from which he later deserted. After the war, he became a journalist and African explorer famed for his trek to locate the British physician whom he discovered on the shores of Lake Tanganyika and greeted, "Dr. Livingstone, I presume."

Edwin Stanton After the battle with President Johnson over the Tenure of Office Act, Stanton left the Cabinet he had served so ably as war secretary. Grant named him to the Supreme Court, but he died before being sworn in.

Alexander Stephens The Confederate vice-president, a ninety-pound thorn in Jefferson Davis's side throughout the war, was jailed briefly. He was returned to the Senate but denied his seat. One of his lasting contributions was to write a highly successful book, in which he gave the Civil War the name most Southerners used for generations, *A Constitutional View of the War Between the States*. He later returned to the House and was governor of Georgia when he died in 1883.

Sojourner Truth The outspoken evangelist who was a pioneer of both abolition and women's rights, Truth advocated a "black State" in the West. A celebrated lecturer despite being illiterate, she had a charismatic style that drew large crowds until she gave up speaking tours in 1875.

Harriet Tubman The onetime slave who had returned to the Confederacy repeatedly to bring out hundreds of other slaves, including her own parents, served as a cook, nurse, laundress, spy, and scout during the war, operating on the South Carolina coast. After the war, she maintained a home for aged freedmen in Auburn, New York, until her death in 1913.

Lew Wallace After a somewhat mixed military performance during the war, General Wallace's greatest claim to fame might have been that he presided over the trials of the Lincoln conspirators and Henry Wirz, the Andersonville prison camp superintendent. Wirz was tried by a military court, convicted, and hanged, the only Confederate officer to be executed after the war. Wallace turned to writing fiction after his military days and is famous as the author of *Ben-Hur: A Tale of the Christ* (1880), one of the most popular novels of the nineteenth century and the source of two films.

Gideon Welles Along with Seward the only Cabinet member to serve through both of Lincoln's terms, Welles was a journalist in Connecticut before the war. Having served in the Navy Department, he was given the assignment of building a navy that could sustain the blockade, a mammoth effort he carried out with remarkable efficiency. After Lincoln's death, he became outspoken in his defense of Lincoln, contradicting a view that Seward had held the power and made the decisions.

Joseph Wheeler Having survived the war and vainly battled Sherman during the March to the Sea, this Georgia-born West Pointer was captured and held in prison after the war. He later served in Congress, and when the war with Spain broke out, he rejoined the federal army as a general of volunteers; occasionally he would forgetfully urge his troops to fight the Yankees. This service entitled him to be buried in Arlington, one of few Confederate soldiers so honored.

Afterword

One day in June 1865 Edmund Ruffin, the rabid secessionist who, according to legend, had fired the first shot at Fort Sumter, put a rifle in his mouth and fired another shot. Fewer people probably took notice of this one, which ended his life. However, Ruffin's final entry in his diary gave some sense of the profound emotional state then coursing through the defeated Confederacy. He wrote:

> I hereby declare my unmitigated hatred to Yankee rule—
> to all political, social & business connection with Yankees—
> and to the Yankee race. Would that I could impress these
> sentiments, in their full force, on every living southerner,
> & bequeath them to everyone yet to be born! May such
> sentiments be held universally in the outraged & down-
> trodden South, although in silence & stillness, until the now
> far-distant day shall arrive for just retribution for Yankee
> usurpation, oppression, & atrocious outrages—& for deliv-
> erance & vengeance for the now ruined, subjugated & en-
> slaved Southern States! . . .
>   I hereby repeat . . . my unmitigated hatred . . . to the per-
> fidious, malignant, & vile Yankee race.

Call it "Yankee rule," as Ruffin did, or "Reconstruction," as schoolbooks would, federal control over the defeated Confederacy lasted for twelve years, until 1877. In that year Republican presidential candidate Rutherford B. Hayes agreed to withdraw the federal troops and return the former states of the Confederacy to home rule in exchange for their support in his contested race with Democrat Samuel Tilden. In this election Tilden, the New York governor with a reputation as a reformer and corruption fighter, had eked out a close win in the popular vote (51 to 48 percent) but didn't have sufficient electoral college votes. In what amounted to an outright theft of votes, an election commission gave Hayes the votes of electors in several southern states and put him in the White House. In *Reunion and Reaction*, C. Vann Woodward writes that Hayes's deal with the southern Democrats—the Compromise of 1877—"marked the abandonment of principles and a return to the traditional ways of expediency and concession. . . . It wrote an end to Reconstruction and recognized a new regime in the South. More profoundly than Constitutional amendments and wordy statutes it shaped the fortunes of four million freedmen and their progeny for generations to come." In other words, congressional Reconstruction and any hope of meaningful political or economic power for blacks were demolished.

To put it bluntly, not too many people seemed to care. By 1877 the nation was expanding, pushing out, casting its eyes on the rest of the world, focusing on new endeavors, including cleaning up the last of the unfinished business with the Native Americans. The Civil War was far from forgotten, but the country wanted to direct its vast energies to the present and the future. There were railroads to build, oil to be discovered, foreign territory to conquer. In this sense, the Civil War era gave way to the most rampantly corrupt period in American history, called the Gilded Age by Mark Twain and the era of the Robber Barons, those great captains of industry whose power exceeded that of any politician. In many cases, they openly and unapologetically bought and paid for elected officials.

Still, for generations to come, the Civil War shaped the country's political landscape like the great glaciers that had once carved America. In its outright hatred of the Republicanism of Lincoln, Seward, and Grant, the white South would remain solidly Democratic—and conservative—for decades. It was not until the 1968 election of Richard M. Nixon that many conservative white southern Democrats finally swung to a Republican president. That trend accelerated in the 1980s during the Reagan administration, as congressional "Boll Weevil" Democrats from the South eagerly adopted Reagan's conservative policies, giving the Republican president easy majorities in Congress with which to pass his "supply side" policies. By the early 1990s the pendulum had swung completely, as the South turned more solidly Republican and many white southern Democrats switched party affiliation.

On the other side of this racial-political equation, for most of the next century blacks would scratch and claw for the rights they had supposedly won after the Civil War. The openly repressive system of slavery had been replaced by the new "legal" repression known as "Jim Crow," or segregation. In every aspect of life—eating, sleeping, traveling, learning—black Americans were treated as second-class citizens in a country that had gone to war over their fate. The Supreme Court's famed 1896 *Plessy* v. *Ferguson* decision that "separate but equal" facilities were legal under the Constitution was a death blow to any hope of equality for blacks. Those who did struggle past the legal obstacles—poll taxes, literacy laws, property laws, and the like—remained loyal to the Republican camp for decades. Not until the Depression, when Franklin D. Roosevelt's New Deal policies won their loyalty, did blacks flock to the Democratic party. And segregation would have the full force of law for decades more, until 1954, when the Supreme Court reversed the "separate but equal" rule in the landmark *Brown* v. *Board of Education* case. The Civil Rights movement of the fifties and sixties—the bus boycotts, the Freedom Riders, the lunch counter sit-ins, the murders of civil rights workers—slowly and painfully won black Americans the basic rights to which they had always been entitled.

Time has faded the memories of battles and the dreams of glory, particularly in the former Confederacy, where "the Lost Cause" remained a powerful idea well into this century. For years the cry "The South shall rise again" expressed the visceral feeling that the former Confederacy remained separate and different from the rest of America and that its proper place would eventually be restored. In that cry, the bitterness of the Civil War defeat was passed from one generation to the next.

Today, however, it seems that those memories have been replaced by amnesia and indifference. The last vestige of any memory of the Confederacy seems to hang with the Confederate flags that decorate college fraternity house windows. Where former battlefields once stood silent witness to the war's grim passage, preservationists now fight off shopping malls and real estate developments. The battle over the American history theme park that Disney planned for the Manassas battlefield was only the most highly publicized of these struggles.

The emigration of millions of Northerners from the Rust Belt to the Sun Belt during the past twenty-five years did much to alter the character of the South. Television and shopping malls have joined to create a homogeneous American culture that has erased meaningful regional differences. The fast-food eateries and shops of Birmingham, Alabama, are the same as those in Lowell, Massachusetts. Perhaps most sadly of all, for many Americans throughout the country the Civil War has slid into the "black hole" that our schools call American History.

Yet the Civil War remains at the core of our greatest national problem: the great racial divide that grew from slavery and was a large part of the reason for the war. It was certainly not settled by the Union victory. By many measures we remain two countries, separate and unequal. But those countries are no longer North and South but black and white. This is written in the wake of a march of hundreds of thousands of black men to a rally led by a man who is an outspoken anti-Semite and who has called white people "devils." The blatant racism of a white Los Angeles police detective was one of the turning points in a notorious murder trial that split much of the country along racial lines. At the same time, a black woman who hosts a television talk show is one of the country's most influential, widely respected, and wealthy people. Every kid with a basketball on a playground wants to "be like Mike." (Jordan, that is, not Tyson.) A retired black general is one of the country's most popular figures and was unhesitatingly considered a contender for the presidency.

Nearly one hundred and fifty years ago, Abraham Lincoln predicted that blacks and whites could never live "together on terms of social and political equality." Has his grim prophecy come true?

How far we've come. How far we have to go.

Appendix I

Excerpted from the Declaration of Independence
Approved by the Continental Congress on July 4, 1776

When in the course of human events, it becomes necessary for one people to dissolve the political bands which have connected them with another, and to assume among the powers of the earth, the separate and equal station to which the the laws of Nature and Nature's God entitle them, a decent respect to the opinions of mankind requires that they should declare the causes which impel them to the separation.

We hold these truths to be self-evident, that all men are created equal, that they are endowed by their Creator with certain unalienable rights, that among these are life, liberty and the pursuit of happiness. That to secure these rights, governments are instituted among men, deriving their just powers from the consent of the governed. That whenever any form of government becomes destructive to these ends, it is the right of the people to alter or to abolish it, and to institute new government, laying its foundation on such principles and organizing powers in such form, as to them shall seem most likely to effect their safety and happiness. Prudence, indeed, will dictate that governments long established should not be changed for light and transient causes; and accordingly all experience hath shown, that mankind are more disposed to suffer, while evils are sufferable, than to right themselves by abolishing the forms to which they are accustomed. But when a long train of abuses and usurpations, pursuing invariably the same object evinces a design to reduce them under absolute despotism, it is their right, it is their duty, to throw off such governments, and to provide new guards for their future security— Such has been the patient sufferance of these colonies; and such is now the necessity which constrains them to alter their former systems of government. The history of the present King of Great Britain is a history of repeated injuries and usurpations, all having in direct object the establishment of an absolute tyranny over these states. To prove this, let facts be submitted to a candid world.

[A long indictment against King George III follows.]

... We, therefore, the representatives of the united states of America, in General Congress assembled, appealing to the Supreme Judge of the world for the rectitude of our intentions, do, in the name, and by the authority of the good people of these colonies, solemnly publish and declare that these united colonies are, and of right ought to be free and independent states; that they are absolved from allegiance to the British Crown, and that all political connection between them and the state of Great Britain, is and ought to be totally dissolved; and that as free and independent states, they have full power to levy war, conclude peace, contract alliances, establish commerce, and to do all other acts and things which independent states may of right do. And for the support of this Declaration, with a firm reliance on the protection of Divine Providence, we mutually pledge to each other our lives, our fortunes and our sacred honor.

Appendix II

American Presidents from Andrew Jackson to William McKinley

PRESIDENT	PARTY	TERM OF OFFICE
Andrew Jackson (1767–1845) Vice-President(s): John C. Calhoun; Martin Van Buren	Democratic	1829–1837
Martin Van Buren (1782–1862) Richard M. Johnson	Democratic	1837–1841
William Henry Harrison (1773–1841) John Tyler (Harrison died one month after his inauguration.)	Whig	1841
John Tyler (1790–1862) (No vice-president appointed)	Whig	1841–1845
James K. Polk (1795–1849) George M. Dallas	Democratic	1845–1849
Zachary Taylor (1784–1850) Millard Fillmore (Taylor died of cholera.)	Whig	1849–1850

Millard Fillmore (1800–   Whig   1850–1853
1874)
    (No vice-president appointed)

Franklin Pierce (1804–   Democratic   1853–1857
1869)
    William R. D. King

James Buchanan (1791–   Democratic   1857–1861
1868)
    John C. Breckinridge

Abraham Lincoln (1809–   Republican   1861–1865
1865)
    Hannibal Hamlin (first term)
    Andrew Johnson (second term)

Andrew Johnson (1808–   Democrat   1865–1869
1875)
    (No vice-president appointed)

*Ulysses S. Grant (1822–   Republican   1869–1877
1885)
    Schuyler Colfax (first term)
    Henry Wilson (second term)

*Rutherford B. Hayes   Republican   1877–1881
(1822–1893)
    William A. Wheeler

*James A. Garfield (1831–   Republican   1881
1881)
    Chester A. Arthur
    (Garfield was assassinated on July 2, 1881, by Charles
    Guiteau, a frustrated office seeker.)

*Chester A. Arthur (1830–   Republican   1881–1885
1886)
    (No vice-president appointed)

Grover Cleveland (1837–   Democrat   1885–1889;
1908)
    Thomas A. Hendricks   1893–1897
A lawyer and politician from Buffalo, New York, Cleveland was the only post–Civil War president who avoided service in 1864 by paying for a substitute when drafted. He also became the only president to be reelected after a defeat.

*Benjamin Harrison (1833–   Republican   1889–1893
1901)
    Levi P. Morton

*William McKinley (1843–   Republican   1897–1901
1901)
    Theodore Roosevelt
(After McKinley was assassinated by the anarchist Leon Czolgosz in September 1901, Roosevelt [1858–1919] became the youngest American president to date. He was the first president who was not of military age during the Civil War, although it had an impact on his life. His father had paid a substitute when he was drafted, later a source of personal shame for Teddy, and the Roosevelt household's loyalties had been divided during the war because his mother was from a Deep South family with prominent Confederate members.)

*A veteran of the Civil War

APPENDIX III

From Dred Scott *v.* Sandford
Chief Justice Roger B. Taney
for the Supreme Court
March 6, 1857

... There are two leading questions presented by the record:

1. Had the Circuit Court of the United States jurisdiction to hear and determine the case between these parties? And,

2. If it had jurisdiction, is the judgment it has given erroneous or not?

The plaintiff in error was, with his wife and children, held as slaves by the defendant, in the State of Missouri, and he brought this action in the Circuit Court of the United States for that district, to assert the title of himself and his family to freedom.

The declaration is ... that he and the defendant are citizens of different States; that is, that he is a citizen of Missouri, and the defendant a citizen of New York.

The defendant pleaded in abatement to the jurisdiction of the court, that the plaintiff was not a citizen of the State of Missouri, as alleged in his declaration, being a Negro of African descent whose ancestors were of pure African blood, and who were brought into the country and sold as slaves. ...

Before we speak of the pleas in bar, it will be proper to dispose of the questions which have arisen on the plea in abatement.

That plea denies the right of the plaintiff to sue in a court of the United States, for the reason therein stated.

If the question raised by it is legally before us, and the court should be of the opinion that the facts stated in it disqualify the plaintiff from becoming a citizen, in the sense in which the word is used in the Constitution of the United States, then the judgment of the Circuit Court is erroneous, and must be reversed. . . .

The question to be decided is, whether the plaintiff is not entitled to sue as a citizen in a court of the United States. . . .

The question is simply this: Can a Negro, whose ancestors were imported into this country, and sold as slaves, become a member of the political community formed and brought into existence by the Constitution of the United States, and as such become entitled to all the rights, and privileges, and immunities, guaranteed by that instrument to the citizen? One of which rights is the privilege of suing in a Court of the United States in the case specified by the Constitution.

It will be observed, that the plea applies to that class of persons only whose ancestors were Negroes of the African race, and imported into this country, and sold and held as slaves. The only matter in issue before the court, therefore, is whether the descendants of such slaves, when they shall be emancipated, or who are born of parents who had become free before their birth, are citizens of a State, in the sense in which the word is used in the Constitution of the United States. . . .

The words "people of the United States" and "citizens" are synonymous terms, and mean the same thing. They both describe the political body who, according to our republican institutions, form the sovereignty, and who hold the power and conduct the government through their representatives. They are what we familiarly call the "sovereign people," and every citizen is one of these people, and a constituent member of this sovereignty. The question before us is, whether the class of persons described in the plea in abatement compose a portion of this people, and are constituent members of this sovereignty? We think they are not, and that they are not included, and were not intended to be included, under the word "citizens" in the Constitution, and can, therefore, claim none of the rights and privileges which that instrument provides for and secures to the citizens of the United States. On the contrary, they were at that time considered as a subordinate and inferior class of beings, who had been subjugated by the dominant race, and whether emancipated or not, yet remained subject to their authority, and had no rights or privileges but such as those who held the power and the government might choose to grant them. . . .

In discussing this question, we must not confound the rights of citizenship which a State may confer within its own limits, and the rights of citizenship as a member of the Union. It does not by any means follow, because he has all the rights and privileges of a citizen of a State, that he must be a citizen of the United States. He may have all the rights and privileges of the citizen of a State, and yet not be entitled to the rights and privileges of a citizen of any other State. For, previous to the adoption of the Constitution of the United States, every State had the undoubted right to confer on whomsoever it pleased the character of a citizen, and to endow him with all its rights. But this character, of course, was confined to the boundaries of the State, and gave him no other rights or privileges in other States beyond those secured to him by the laws of nations and the comity of States. Nor have the several States surrendered the power of conferring these rights and privileges by adopting the Constitution of the United States. Each State may still confer them upon an alien, or anyone it thinks proper, or upon any class or description of persons; yet he would not be a citizen in the sense in which that word is used in the Constitution of the United States, nor entitled to sue as such in one of its courts, nor to the privileges and immunities of a citizen in the other States. The rights which he would acquire would be restricted to the State which gave them. . . .

It is very clear, therefore, that no State can, by any Act or law of its own, passed since the adoption of the Constitution, introduce a new member into the political community created by the Constitution of the United States. It cannot make him a member of this community by making him a member of its own. And for the reason it cannot introduce any person, or description of persons, who were not intended to be embraced in this new political family, which the Constitution brought into existence, but were intended to be excluded from it.

The question then arises, whether the provisions of the Constitution, in relation to the personal rights and privileges to which the citizen of a State should be entitled, embraced the Negro African race, at that time in this country, or who might afterwards be imported, who had then or should afterwards be made free in any State; and to put it in the power of a single State to make him a citizen of the United States, and endue him with the full rights of citizenship in every other State without their consent. Does the Constitution of the United States act upon him whenever he shall be made free under the laws of a State, and raised there to the rank of a citizen, and immediately clothe him with all the privileges of a citizen in every other State, and in its own court?

The court think the affirmative of these propositions cannot be maintained. And if it cannot, the plaintiff in error could not be a citizen of the State of Missouri, within the meaning of the Constitution of the United States, and consequently, was not entitled to sue in its courts.

It is true, every person, and every class and description of persons, who were at the time of the adoption of the Constitution recognized as citizens of this new political body; but none other; it was formed by them, and for them and their posterity, but for no one else. And the personal rights and privileges guaranteed to citizens of this new sovereignty were intended to embrace those only who were then members of the several state communities, or who should afterwards, by birthright or otherwise, become members, according to the principles on which it was founded. . . .

It becomes necessary, therefore, to determine who were citizens of the several States when the Constitution was adopted. And in order to do this, we must recur to the governments and institutions of the thirteen Colonies, when they separated from Great Britain and formed new sovereignties. . . . We must inquire who, at that time, were recognized as the people or citizens of a State. . . .

In the opinion of the court, the legislation and histories of the times, and the language used in the Declaration of Independence, show, that neither the class of persons who had been imported as slaves, nor their descendants, whether they had become free or not, were then acknowledged as a part of the people, nor intended to be included in the general words used in that memorable instrument.

It is difficult at this day to realize the state of public opinion in relation to that unfortunate race, which prevailed in civilized and enlightened portions of the world at the time of the Declaration of Independence, and when the Constitution of the United States was framed and adopted. . . .

They had for more than a century before been regarded as beings of an inferior order and altogether unfit to associate with the white race, either in social or political relations; and so far inferior that they had no rights which the white man was bound to respect; and that the Negro might justly and lawfully be reduced to slavery for his benefit. He was bought and sold and treated as an ordinary article of merchandise and traffic whenever a profit could be made by it. This opinion was at that time fixed and universal in the civilized portion of the white race. It was regarded as an axiom in morals as well as in politics, which no one thought of disputing, or supposed to be open to dispute; and men in every grade and position in society daily and habitually acted upon it in their private pursuits, as well as in matters of public concern, without doubting the correctness of this opinion.

. . . A Negro of the African race was regarded . . . as an article of property and held and bought and sold as such in every one of the thirteen Colonies which united in the Declaration of Independence and afterward formed the Constitution of the United States. The slaves were more or less numerous in the different Colonies as slave labor was found more or less profitable. But no one seems to have doubted the correctness of the prevailing opinion of the time.

The legislation of the different Colonies furnished positive and indisputable proof of this fact. . . .

The language of the Declaration of Independence is equally conclusive:

It begins by declaring that "When in the course of human events, it becomes necessary for one people to dissolve the political bands which have connected them with another, and to assume, among the powers of the earth the separate and equal station to which the laws of nature and nature's God entitle them, a decent respect for the opinions of mankind requires that they should declare the causes which impel them to separation."

It then proceeds to say: "We hold these truths to be self-evident: that all men are created equal; that they are endowed by their Creator with certain inalienable rights; that among these are life, liberty and the pursuit of happiness; that to secure these rights, governments are instituted, deriving their just powers from the consent of the governed."

The general words above quoted would seem to embrace the whole human family, and if they were used in a similar instrument at this day would be so understood. But it is too clear for dispute that the enslaved African race were not intended to be included and formed no part of the people who framed and adopted this declaration; for if the language, as understood in that day, would embrace them, the conduct of the distinguished men who framed the Declaration of Independence would have been utterly and flagrantly inconsistent with the principles they alerted; and instead of the sympathy of mankind, to which they so confidently appealed, they would have deserved and received universal rebuke and reprobation.

Yet the men who framed this declaration were great men—high in literary acquirements—high in their sense of honor, and incapable of asserting principles inconsistent with those on which they were acting. They perfectly understood the meaning of the language they used and how it would be understood by others; and they knew that it would not in any part of the civilized world be supposed to embrace the Negro race, which by common consent, had been excluded from civilized governments and the family of nations and doomed to slavery. They spoke and acted according to then established doctrine and principles and in the ordinary language of the day and no one misunderstood them. The unhappy black race were separated from the white by indelible marks, and laws long before established, and were never thought of or spoken of except as property and when the claims of the owner to the profit of the trade were supposed to need protection.

This state of public opinion had undergone no change when the Constitution was adopted, as is equally evident from its provisions and language.

The brief preamble sets forth by whom it was formed, for what purposes, and for whose benefit and protection. It declares that it is formed by the *people* of the United States; that is to say, by those who were members of the different political communities in the several states; and its great object is declared to be to secure the blessing of liberty to themselves and their posterity. It speaks in general terms of the *people* of the United States, and of *citizens* of the several states, when it is providing for the exercise of the powers granted or the privileges secured to the citizen. It does not define what description of persons are intended to be included under these terms, or who shall be regarded as a citizen and one of the people. It uses them as terms so well understood that no further description or definition was necessary. . . .

But there are two clauses in the Constitution which point directly and specifically to the Negro race as a separate class of persons, and show clearly that they were not regarded as a portion of the people or citizens of the Government then formed.

One of these clauses reserves to each of the thirteen States the right to import slaves until the year 1808, if it thinks it proper. And the importation which it thus sanctions was unquestionably of persons of the race of which we are speaking, as the traffic in slaves in the United States had always been confined to them. And by the other provision the States pledge themselves to each other to maintain the right of property of the master, by delivering up to him any slave who may have escaped from his service and be found within their respective territories. . . . And these two provisions show, conclusively, that neither the description of persons therein referred to, nor their descendants, were embraced in any of the other provisions of the Constitution; for certainly these two clauses were not intended to confer on them or their posterity the blessings of liberty, or any of the personal rights so carefully provided for the citizen. . . .

Indeed, when we look to the condition of this race in the several States at the time, it is impossible to believe that these rights and privileges were intended to be extended to them. . . .

The legislation of the States therefore shows, in a manner not to be mistaken, the inferior and subject condition of that race at the time the Constitution was adopted, and long afterwards, throughout the thirteen States by which that instrument was framed; and it is hardly consistent with the respect due to these States, to suppose that they were regarded at that time, as fellow-citizens and members of the sovereignty, a class of beings whom they had thus stigmatized. . . . More especially, it cannot be believed that the large slave-holding States regarded them as included in the word "citizens," or would have consented to a constitution which might compel them to receive them in that character from another State. For if they were so received, and entitled to the privileges and immunities of citizens, it would exempt them from the operation of special laws and from the police regulations which they considered to be necessary for their own safety. . . . And all of this would be done in the face of the subject race of the same color, both free and slaves, inevitably producing discontent and insubordination among them, and endangering the peace and safety of the State. . . .

. . . Upon a full and careful consideration of the subject, the court is of the opinion that, upon the facts stated in the plea in abatement, Dred Scott was not a citizen of Missouri within the meaning of the Constitution of the United States, and not entitled as such to sue in its courts; and, consequently, that the Circuit Court had no jurisdiction of the case, and that the judgment on the plea in abatement is erroneous. . . .

We proceed, therefore, to inquire whether the facts relied on by the plaintiff entitled him to his freedom. . . .

In considering this part of the controversy, two questions arise: 1st. Was he, together with his family, free in Missouri by reason of the stay in the territory of the United States hereinbefore mentioned? And 2nd, If they were not, is Scott himself free by reason of his removal to Rock Island, in the State of Illinois, as stated in the above admissions?

We proceed to examine the first question.

The Act of Congress, upon which the plaintiff relies [the Missouri Compromise], declares that slavery and involuntary servitude, except as a punishment for crime, shall be forever prohibited in all that part of the territory ceded by France, under the name of Louisiana, which lies north of thirty-six degrees thirty minutes north latitude, and not included within the limits of Missouri. And the difficulty which meets us at the threshold of this part of the inquiry is, whether Congress was authorized to pass this law under any of the powers granted to it by the Constitution; for if the authority is not given by that instrument, it is the duty of this court to declare it void and inoperative, and incapable of conferring freedom upon any one who is held as a slave under the laws of any one of the States.

The counsel for the plaintiff has laid much stress upon that article in the Constitution which confers on Congress the power "to dispose of and make all needful rules and regulations respecting the territory or other property belonging to the United States;" but, in the judgment of the court, that provision has no bearing on the present controversy, and the power there given, whatever it may be, is confined, and was intended to be confined, to the territory which at that time belonged to, or was claimed by, the United States, and was within their boundaries as settled by the treaty with Great Britain, and can have no influence upon a territory acquired from a foreign Government. It was a special provision for a known and particular territory, and to meet a present emergency, and nothing more. . . .

If this clause is construed to extend to territory acquired by the present Government from a foreign nation, outside of the limits of any charter from the British government to a colony, it would be difficult to say, why it was deemed necessary to give the Government the power to sell any vacant lands belonging to the sovereignty which might be found within it; and if this was necessary, why the grant of this power should precede the power to legislate over it and establish a Government there and still more difficult to say, why it was deemed necessary so specially and particularly to grant the power to make needful rules and regulations in relation to any personal or movable property it may acquire there. For the words, *other property* necessarily, by every known rule of interpretation, must mean property of a different description from territory or land. And the difficulty would perhaps be insurmountable in endeavoring to account for the last member of the sentence, which provides that "nothing in this Constitution shall be so construed as to prejudice the claims of the United States or any particular State," or to say how any particular State could have claims in or to a territory ceded by a foreign Government, or to account for associating this provision with the preceding provisions of the clause, with which it would appear to have no connection. . . .

But the power of Congress over the person or property of a citizen can never be a mere discretionary power under our Constitution and form of Government. The powers of the Government and the rights and privileges of the citizens are regulated and plainly defined by the Constitution itself. And when the Territory becomes a part of the United States, the Federal Government enters into possession in the character impressed upon it by those who created it. It enters upon it with its powers over the citizen strictly defined, and limited by the Constitution, from which it derives its own existence, and by virtue of which alone it continues to exist and act as a government and sovereignty. It has no power of any kind beyond it; and it cannot, when it enters a Territory of the United States, put off its character, and assume discretionary or despotic powers which the Constitution has denied to it. It cannot create for itself a new character separated from the citizens of the United States, and the duty it owes them under the provisions of the Constitution. The Territory being a part of the United States, the Government and the citizen both enter it under the authority of the Constitution, with their respective rights defined and marked out; and the Federal Government can exercise no power over his person or property, beyond what that instrument confers, nor lawfully deny any right which it has reserved. . . .

The rights of private property have been guarded with equal care.

Thus the rights of property are united with the rights of person, and placed on the same ground by the fifth amendment to the Constitution. . . . An Act of Congress which deprives a person of the United States of his liberty or property merely because he came himself or brought his property into a particular Territory of the United States, and who had committed no offense against the laws, could hardly be dignified with the name of due process of law. . . .

And this prohibition is not confined to the States, but the words are general, and extend to the whole territory over which the Constitution gives it power to legislate, including those portions of it remaining under territorial government, as well as that covered by States. It is a total absence of power everywhere within the dominion of the United States, and places the citizens of a territory, so far as these rights are concerned, on the same footing with citizens of the States, and guards them as firmly and plainly against any inroads which the general government might attempt, under the plea of implied or incidental powers. And if Congress itself cannot do this—if it is beyond the powers conferred on the Federal Government—it will be admitted, we presume, that it could not authorize a territorial government to exercise them. It could confer no power on any local government, established by its authority, to violate the provisions of the Constitution.

It seems, however, to be supposed, that there is a difference between property in a slave and other property, and that different rules may be applied to it in expounding the Constitution of the United States.

But . . . if the Constitution recognizes the right of property of the master in a slave, and makes no distinction between that description of property and other property owned by a citizen, no tribunal, acting under the authority of the United States, whether it be legislative, executive, or judicial, has a right to draw such a distinction, or deny to it the benefit of the provisions and guarantees which have been provided for the protection of private property against the encroachments of the Government.

Now . . . the right of property in a slave is distinctly and expressly affirmed in the Constitution. The right to traffic in it, like an ordinary article of merchandise and property, was guaranteed to the citizens of the United States, in every State that might desire it, for twenty years. And the Government in express terms is pledged to protect it in all future time, if the slave escapes from his owner.... And no word can be found in the Constitution which gives Congress a greater power over slave property, or which entitles property of that kind to less protection than property of any other description. The only power conferred is the power coupled with the duty of guarding and protecting the owner in his rights.

Upon these considerations, it is the opinion of the court that the Act of Congress [the Missouri Compromise] which prohibited a citizen from holding and owning property of this kind in the territory of the United States north of the line therein mentioned, is not warranted by the Constitution, and is therefore void; and that neither Dred Scott himself, nor any of his family, were made free by being carried into this territory; even if they had been carried there by the owner, with the intention of becoming a permanent resident....

Upon the whole, therefore, it is the judgment of this court, that it appears by the record before us that the plaintiff in error is not a citizen of Missouri, in the sense in which that word is used in the Constitution; and that the Circuit Court of the United States, for that reason, had no jurisdiction in the case, and could give no judgment in it.

Its judgment for the defendant must, consequently, be reversed, and a mandate issued directing the suit to be dismissed for want of jurisdiction.

Excerpted from Abraham Lincoln's First Inaugural Address
March 4, 1861
(Emphasis added)

. . . I hold that, in contemplation of universal law and of the Constitution, the Union of these States is perpetual. Perpetuity is implied, if not expressed, in the fundamental law of all national governments. *It is safe to assert that no government proper ever had a provision in its organic law for its own termination.*

Continue to execute all the express provisions of our National Constitution, and the Union will endure forever—it being impossible to destroy it except by some action not provided for in the instrument itself.

Again, if the United States be not a government proper, but an association of States in the nature of contract merely, can it, as a contract, be peaceably unmade by less than all the parties who made it? One party to a contract may violate it—break it, so to speak; but does it not require all to lawfully rescind it?

Descending from these general principles, we find the proposition that, in legal contemplation the Union is perpetual confirmed by the history of the Union itself. The Union is much older than the Constitution. It was formed, in fact, by the Articles of Association in 1774. It was matured and continued by the Declaration of Independence in 1776. It was further matured, and the faith of all the then thirteen States expressly plighted and engaged that it should be perpetual, by the Articles of Confederation in 1778. And, finally, in 1787 one of the declared objects for ordaining and establishing the Constitution was "to form a more perfect Union."

But if the destruction of the Union by one or by a part of only one of the States be lawfully possible the Union is less perfect than before the Constitution, having lost the vital element of perpetuity. . . .

It follows from these views that no State upon its own mere motion can lawfully get out of the Union; that resolves and ordinances to that effect are legally void; and that acts of violence, within any State or States, against the authority of the United States, are insurrectionary or revolutionary, according to circumstances.

I therefore consider that, in view of the Constitution and the laws, the Union is unbroken; and to the extent of my ability I shall take care, as the Constitution itself expressly enjoins upon me, that the laws of the Union be faithfully executed in all the States. Doing this I deem to be only a simple duty on my part; and I shall perform it so far as practicable, unless my rightful masters, the American people, shall withhold the requisite means, or in some authoritative manner direct the contrary. I trust this will not be regarded as a menace, but only as the declared purpose of the Union that it will constitutionally defend and maintain itself. . . .

If by the mere force of numbers a majority should deprive a minority of any clearly written constitutional right, it might, in a moral point of view, justify revolution—certainly it would if such a right were a vital one. But such is not our case. All the vital rights of minorities and of individuals are so plainly assured to them by affirmations and negations, guarantees and prohibitions, in the Constitution, that controversies never arise concerning them. But no organic law can ever be framed with a provision specifically applicable to every question which may occur in practical administration.

No foresight can anticipate, nor any document of reasonable length contain, express provisions for all possible questions. Shall fugitives from labor be surrendered by national or State authority? The Constitution does not expressly say. *Must* Congress protect slavery in the Territories? The Constitution does not expressly say.

From questions of this class spring all our constitutional controversies, and we divide upon them into majorities and minorities. If the minority will not acquiesce, the majority must, or the government must cease. There is no other alternative; for continuing the government is acquiescence of one side or the other.

If a minority in such case will secede rather than acquiesce, they make a precedent which in turn will divide and ruin them; for a minority of their own will secede from them whenever a majority refuses to be controlled by such a minority. *For instance, why may not any portion of a new confederacy a year or two hence arbitrarily secede again, precisely as portions of the present Union now claim to secede from it? All who cherish disunion sentiments are now being educated to the exact temper of doing this.*

Is there such perfect identity of interests among the States to compose a new Union, as to produce harmony only, and prevent renewed secession?

Plainly, the central idea of secession is the essence of anarchy. A majority held in restraint by constitutional checks and limitations, and always changing easily with deliberate changes of popular opinions and sentiments, is the only true sovereign of a free people. Whoever rejects it does, of necessity, fly to anarchy or to despotism. Unanimity is impossible; the rule of a minority, as a permanent arrangement, is wholly inadmissible; so that, rejecting the majority principle, anarchy or despotism in some form is all that is left.

This country, with its institutions, belongs to the people who inhabit it. Whenever they shall grow weary of the existing government, they can exercise their constitutional right of amending it, or their revolutionary right to dismember or overthrow it. I cannot be ignorant of the fact that many worthy and patriotic citizens are desirous of having the National Constitution amended. While I make no recommendation of amendments, I fully recognize the rightful authority of the people over the whole subject, to be exercised in either of the modes prescribed in the instrument itself; and I should, under existing circumstances, favor rather than oppose a fair opportunity being afforded the people to act upon it. . . .

Why should there not be a patient confidence in the ultimate justice of the people? Is there any better or equal hope in the world? . . .

By the frame of the government under which we live, this same people have wisely given their public servants but little power for mischief; and have, with equal wisdom, provided for the return of that little to their own hands at very short intervals. While the people retain their virtue and vigilance, no administration, by any extreme of wickedness or folly, can very seriously injure the government in the short space of four years.

In your hands, my dissatisfied fellow countrymen, and not in mine, is the momentous issue of civil war. The government will not assail you. You can have no conflict without being yourselves the aggressors. You have no oath registered in heaven to destroy the government, while I shall have the most solemn one to "preserve, protect and defend it."

I am loath to close. We are not enemies, but friends. We must not be enemies. Though passion may have strained, it must not break our bonds of affection. The mystic chords of memory, stretching from every battlefield and patriot grave to every living heart and hearthstone all over this broad land, will yet swell the chorus of the Union when again touched, as surely they will be, by the better angels of our nature.

APPENDIX V

1861: "The Blue and the Gray" at a Glance

On paper, the Confederate situation seemed hopeless when the war started. In numerical terms, it is astonishing that the Confederacy was able to prolong the war for four long years. The Union was larger, wealthier, more modern and mechanized, better educated. The Confederacy was still a rural, farming society dependent on slave labor, much as it had been in Thomas Jefferson's day. An estimated 20 percent of white Southerners could not read or write. Outside of its cotton, tobacco, and other crops, it produced little of what it needed. Everything else had to be brought in, either from the North or from Europe.

As London banker Baron Rothschild succinctly put it early in the war, the North would win because it had "the larger purse."

Of course, wars are not fought on paper as the British in America, the United States in Vietnam, and the Soviet Union in Afghanistan all learned. That is why, even against these seemingly impossible odds, the Confederacy was able to survive as long as it did in the most costly of American wars.

23 UNION STATES

(*denotes a Border State—the four slave states that
remained in the Union)

California	Minnesota
Connecticut	*Missouri
*Delaware	New Hampshire
Illinois	New Jersey
Indiana	New York
Iowa	Ohio
Kansas	Oregon
*Kentucky	Pennsylvania
Maine	Rhode Island
*Maryland	Vermont
Massachusetts	Wisconsin
Michigan	

The states of West Virginia (1863) and Nevada (1864) were added
during the war. The rest of the country was organized into seven
large territories—Arizona, Colorado, New Mexico, Utah, Nebraska,
Washington, and Indian Territory.

Population: more than 22 million; 4 million men of combat age. Two
million men joined the Union army and navy.

Economy:

* 100,00 factories, including nearly all the country's weapons factories and ship makers, employing more than 1 million workers
  with a high degree of literacy.
* 20,000 miles of railroad (70 percent of the U.S. total) and 96 percent of all the country's railroad equipment.
* the vast majority of coal mines and canals for efficient transportation of the coal that powered the mills and ships in the new
  Steam Age.
* $189 million in bank deposits (81 percent of the U.S. total).
* $56 million in gold.

11 CONFEDERATE STATES

Alabama	North Carolina
Arkansas	South Carolina
Florida	Tennessee
Georgia	Texas
Louisiana	Virginia
Mississippi	

The combined area of the Confederacy was approximately 750,000 square miles, larger than all of Western Europe, with 3,500 miles of coastline.

Population: 9 million (5.5 million whites; 3.5 million slaves); approximately 1,140,000 men of combat age, of whom 850,000 fought for the Condederacy.

Economy:
* 20,000 factories, employing about 100,000 workers. (The city of Lowell, Massachusetts, had more spindles turning thread in 1860 than did the entire Confederacy.)
* 9,000 miles of railroad.
* $47 million in bank deposits.
* $37 million in gold.

The Union states outproduced the Confederate states in every agricultural category except for one—cotton, raised by slave labor. But a misguided strategy of withholding cotton from European markets only hurt the Confederate economy. By the time that policy was abandoned, the Union naval blockade had become more efficient, further crippling the ability of the Confederacy to sell its most important cash crop.

The Emancipation Proclamation
January 1, 1863

Whereas, on the twenty-second day of September, in the year of our Lord one thousand eight hundred and sixty-two, a proclamation issued by the President of the United States containing among other things the following to wit:

"That on the first day of January in the year of our Lord one thousand eight hundred and sixty-three, all persons held as slaves within any State, or designated part of a State, the people whereof shall then be in rebellion against the United States, shall be then, thenceforth and forever free, and the Executive Government of the United States, including the military and naval authorities thereof, will recognize and maintain the freedom of such persons, and will do no act or acts to repress such persons, or any of them, in any efforts they may make for their actual freedom.

"That the Executive will, on the first day of January aforesaid, by proclamation, designate the States and parts of States, if any, in which the people, therein respectively shall be then in rebellion against the United States, and the fact that any State, or the people thereof, shall on that day be in good faith represented in the Congress of the United States by members chosen thereto, at elections wherein a majority of the qualified voters of such States shall have participated, shall, in the absence of strong countervailing testimony, be deemed conclusive evidence that such state and the people thereof are not then in rebellion against the United States."

Now, therefore, I, Abraham Lincoln, President of the United States, by virtue of the power in me vested as Commander in Chief of the Army and Navy of the United States, and as a fit and necessary war measure for suppressing said rebellion, do, on this first day of January, in the year of our Lord one thousand eight hundred and sixty-three, and in accordance with my purpose so to do, publicly proclaimed, for the full period of one hundred days from the day above the first above-mentioned, order and designate as the States and parts of States wherein the people thereof respectively are this day in rebellion against the United States, the following, to wit: Arkansas, Texas, Louisiana, except the parishes of St. Bernard, Plaquemines, Jefferson, St. John, St. Charles, St. James, Ascension, Assumption, Terre Bone, Lafourche, St. Mary, St. Martin, and Orleans, including the City of New Orleans, Mississippi, Alabama, Florida, Georgia, South Carolina, North Carolina and Virginia, except the forty-eight counties designated as West Virginia, and also the three counties of Berkeley, Accomac, Northampton, Elizabeth City, York, Princess Ann, and Norfolk, including the cities of Norfolk and Portsmouth, and which excepted parts are, for the present, left precisely as if this proclamation were not issued.

And by virtue and for the purpose aforesaid, I do order and declare that all persons held as slaves within the said designated States and parts of States are, and henceforward shall be free; and that the Executive Government of the United States, including the Military and Naval authorities, thereof, will recognize and maintain the freedom of said persons.

And I hereby enjoin upon the people so declared to be free, to abstain from all violence, unless in necessary self-defense, and I recommend to them, that in all cases, when allowed, they labor faithfully for reasonable wages.

And I further declare and make known that such persons of suitable condition will be received into the armed service of the United States to garrison forts, positions, stations, and other places, and man vessels of all sorts in said service.

And upon this, sincerely believed to be an act of justice, warranted by the Constitution, upon military necessity, I invoke the considerate judgment of mankind and the gracious favor of Almighty God.

In witness whereof, I have hereunto set my hand and caused the seal of the United States to be affixed.

Done at the city of Washington, this first day of January; in the year of our Lord one thousand eight hundred and sixty-three, and of the Independence of the United States of America the eighty-seventh.

Abraham Lincoln

By the President—William H. Seward
        Secretary of State

APPENDIX VII

The Gettysburg Address
November 19, 1863

Fourscore and seven years ago our fathers brought forth on this continent, a new nation, conceived in Liberty and dedicated to the proposition that all men are created equal.

Now we are engaged in a great civil war, testing whether that nation or any nation so conceived and so dedicated can long endure. We are met on a great battlefield of that war. We have come to dedicate a portion of that field, as a final resting place for those who here gave their lives that the nation might live. It is altogether fitting and proper that we should do this.

But, in a larger sense, we cannot dedicate—we cannot consecrate—we cannot hallow—this ground. The brave men, living and dead, who struggled here, have consecrated it far above our poor power to add or detract. The world will little note nor long remember what we say here, but it can never forget what they did here. It is for us, the living, rather to be dedicated here to the unfinished work which they who fought here have thus far so nobly advanced. It is rather for us to be here dedicated to the great task remaining before us—that from these honored dead we take increased devotion to that cause for which they gave the last full measure of devotion; that we here highly resolve that these dead shall not have died in vain; that this nation, under God, shall have a new birth of freedom; and that the government of the people, by the people, for the people, shall not perish from the earth.

Abraham Lincoln's Second Inaugural Address
March 4, 1865

Fellow-Countrymen:—At this second appearing to take the oath of the presidential office there is less occasion for an extended address than there was at the first. Then a statement somewhat in detail of a course to be pursued seemed fitting and proper. Now, at the expiration of four years, during which public declarations have been constantly called forth on every point and phase of the great contest which still absorbs the attention and engrosses the energies of the nation, little that is new could be presented. The progress of our arms, upon which all else chiefly depends, is as well known to the public as to myself, and it is, I trust, reasonably satisfactory and encouraging to all. With high hope for the future, no prediction in regard to it is ventured.

On the occasion corresponding to this four years ago all thoughts were anxiously directed to an impending civil war. All dreaded it, all sought to avert it. While the inaugural address was being delivered from this place, devoted altogether to *saving* the Union without war, insurgent agents were in the city seeking to *destroy* it without war— seeking to dissolve the Union and divide effects by negotiation. Both parties deprecated war, but one of them would *make* war rather than let the nation survive, and the other would *accept* war rather than let it perish, and the war came.

One eighth of the population was colored slaves, not distributed generally over the Union, but localized in the southern part of it. These slaves constituted a peculiar and powerful interest. All knew that this interest was somehow the cause of the war. To strengthen, perpetuate, and extend this interest was the object for which the insurgents would rend the Union even by war, while the Government claimed no right to do more than to restrict the territorial enlargement of it. Neither party expected for the war the magnitude or the duration for which it has already attained. Neither anticipated that the *cause* of the conflict might cease with or even before the conflict itself should cease. Each looked for an easier triumph, and a result less fundamental and astounding. Both read the same Bible and pray to the same God, and each invokes His aid against the other. It may seem strange that any men should ask a just God's assistance in wringing their bread from the sweat of other men's faces, but let us judge not, that we be not judged. The prayers of both could not be answered. That of neither has been answered fully. The Almighty has His own purposes. "Woe unto the world because of offenses; for it must needs be that offenses come, but woe to that man by whom the offense cometh." If we shall suppose that American slavery is one of those offenses which, in the providence of God, must needs come, but which, having continued through His appointed time, He now wills to remove, and that he gives to both North and South this terrible war as the woe due to those by whom the offense came, shall we discern therein any departure from those divine attributes which the believers in a living God always ascribe to Him? Fondly do we hope, fervently do we pray, that this mighty scourge of war may speedily pass away. Yet, if God wills that it continue until all the wealth piled by the bondman's two hundred and fifty years of unrequited toil shall be sunk, and until every drop of blood drawn with the lash shall be paid by another drawn with the sword, as was said three thousand years ago, so still it must be said, "The judgments of the Lord are true and righteous altogether."

With malice toward none, with charity for all, with firmness in the right as God gives us to see the right, let us strive on to finish the work we are in, to bind up the nation's wounds, to care for him who shall have borne the battle and for his widow and his orphan, to do all which may achieve and cherish a just and a lasting peace among ourselves and with all nations.

The Thirteenth Amendment to the Constitution

Section 1

Neither slavery nor involuntary servitude, except as a punishment for crime whereof the party shall have been duly convicted, shall exist within the United States, or any place subject to their jurisdiction.

Section 2

Congress shall have power by appropriate legislation to enforce the provisions of this article.

The United States Senate had passed an amendment to the Constitution prohibiting slavery on April 8, 1864. But the House defeated the measure in June 1864. After the consideration of legal questions, the House passed the Thirteenth Amendment on January 31, 1865, and it was sent to the states for ratification. The amendment was proclaimed in effect on December 18, 1865.

The Fourteenth Amendment to the Constitution

1. All persons born or naturalized in the United States, and subject to the jurisdiction thereof, are citizens of the United States and of the State wherein they reside. No State shall make or enforce any law which shall abridge the privileges or immunities of the citizens of the United States; nor shall any State deprive any person of life, liberty, or property, without due process of law; nor deny to any person within its jurisdiction the equal protection of the laws.

2. Representatives shall be apportioned among the several States according to their respective numbers, counting the whole number of persons in each State, excluding Indians not taxed. But when the right to vote at any election for the choice of Electors for President and Vice-President of the United States, Representatives in Congress, the executive and judicial officers of a State, or the members of the Legislature thereof, is denied to any of the male inhabitants of such State, being twenty-one years of age, and, citizens of the United States, or in any way abridged, except for participation in rebellion, or other crime, the basis of representation therein shall be reduced in the proportion which the number of such male citizens shall bear to the whole number of male citizens twenty-one years of age in such State.

3. No person shall be a Senator or Representative in Congress, or Elector of President and Vice-President, or hold any office, civil or military, under the United States, or under any State, who, having

previously taken an oath, as a member of Congress, or as an officer of the United States, or as a member of any State Legislature, or as an executive or judicial officer of any State, to support the Constitution of the United States, shall have engaged in insurrection or rebellion against the same, or given aid or comfort to the enemies thereof. But Congress may by a vote of two-thirds of each House, remove such disability.

4. The validity of the public debt of the United States, authorized by law, including debts incurred for payment of pensions and bounties for services in suppressing insurrection or rebellion, shall not be questioned. But neither the United States nor any State shall assume or pay any debt or obligation incurred in aid of insurrection or rebellion against the United States, or any claim for the loss or emancipation of any slave; but all such debts, obligations and claims, shall be held illegal and void.

5. The Congress shall have power to enforce, by appropriate legislation, the provisions of this article.

The Fourteenth Amendment was proposed to the legislatures of the states by the 39th Congress on June 13, 1866. It was declared to have been ratified on July 28, 1868.

Appendix XI

The Fifteenth Amendment to the Constitution

1. The right of citizens of the United States to vote shall not be denied or abridged by the United States or by any State on account of race, color, or previous condition of servitude.

2. The Congress shall have power to enforce this article by appropriate legislation.

This amendment was proposed to the legislatures of the states by the 40th Congress on February 26, 1869, and was declared to have been ratified on March 30, 1870.

The Civil Rights Act of 1875

An act to protect all citizens in their civil and legal affairs.

Whereas it is essential to just government we recognize the equality of all men before the law, and hold that it is the duty of government in its dealings with the people to mete out equal and exact justice to all, of whatever nativity, race, color, or persuasion, religious or political; and it being the appropriate object of legislation to enact great fundamental principles into law: Therefore,

Be it enacted, That all persons within the jurisdiction of the United States shall be entitled to the full and equal enjoyment of the accommodations, advantages, facilities, and privileges of inns, public conveyances on land or water, theaters, and other places of public amusement; subject only to the conditions and limitations established by law, and applicable alike to citizens of every race and color, regardless of any previous condition of servitude.

Known as the Second Civil Rights Act of 1875, the law was passed by Congress in an attempt to protect the civil equality of Negroes by giving them equal rights to public accommodations and to act as jurors. The first Civil Rights Act, passed over Andrew Johnson's veto in 1866, had been supplanted by the ratification of the Fourteenth Amendment to the Constitution (see Appendix X). This act was made unenforceable by two later Supreme Court decisions: the Civil Rights Cases in 1883 and the more significant *Plessy* v. *Ferguson* case in 1896, which negated this law's protections as well as diminishing those in the Fourteenth Amendment.

Bibliography

The literature of the Civil War is vast, so this bibliography is divided into subject areas. The books listed here include standard references as well as those more recently published books that take into account the research that has produced a significant revision of Civil War history. An asterisk (*) denotes a paperback edition.

General Reference

Bernstein, Iver. *The New York City Draft Riots.* New York: Oxford University Press, 1990. A scholarly account of the race riots that led to the deaths of at least 105 people.

Bierce, Ambrose. *Ambrose Bierce's Civil War*, ed. and with introduction by William McCann. *Washington, D.C.: Regnery, 1956. This famed American writer served in the war, and this book collects both his fiction and nonfiction accounts of the experience.

Boatner, Mark Mayo, III. *The Civil War Dictionary*, rev. ed. New York: David McKay, 1959; *Vintage. An exhaustive reference book, including sketches of some two thousand people, descriptions of battles, campaigns, weapons, and military terms.

Boller, Paul F., Jr. *Presidential Campaigns.* New York: Oxford University Press, 1984; *Oxford University Press, 1985.

Boritt, Gabor S. *Why the Confederacy Lost.* New York: Oxford University Press, 1992. Collected essays.

Carey, John, ed. *Eyewitness to History.* Cambridge, Mass.: Harvard University Press, 1987; *New York: Avon Books, 1990. Firsthand accounts of great moments in history, including several Civil War events.

Carroll, Peter N., and David W. Noble. *The Free and the Unfree: A New History of the United States*, 2nd ed. *New York: Penguin Books, 1988. Interesting Civil War material; an alternative to traditional American histories.

Carruth, Gorton. *What Happened When: A Chronology of Life and Events in America.* New York: Harper & Row, 1989; *Signet, 1991.

Catton, Bruce. *The American Heritage Picture History of the Civil War.* New York: American Heritage/Bonanza Books, 1960; *Boston: Houghton Mifflin. With photographs, maps, illustrations, and period documents, a solid overview of the war with an emphasis on military history.

————. *This Hallowed Ground.* Garden City, N.Y.: Doubleday, 1955; *Washington Square Press, 1961. A single-volume narrative history of the war; a standard in the field, though now slightly dated.

Craven, Avery. *The Coming of the Civil War,* 2nd ed., rev. New York: Scribner's, 1942; *Chicago: University of Chicago Press, 1957.

*Cunliffe, Marcus. *The Presidency,* 3rd ed. Boston: Houghton Mifflin, 1987. A thematic approach to the presidency, containing interesting views of Lincoln's performance.

Denney, Robert E. *The Civil War Years: A Day-by-Day Chronicle of the Life of a Nation.* New York: Sterling, 1992; *Sterling, 1994.

de Tocqueville, Alexis. *Democracy in America,* ed. J. P. Mayer. New York: Harper & Row, 1966; *HarperPerennial, 1988. A classic collection of observations on American life in the 1840s by a young Frenchmen who traveled about the country. (Many other editions are available.)

Dinnerstein, Leonard. *Anti-Semitism in America.* New York: Oxford University Press, 1994; *Oxford, 1995. A comprehensive analysis of prejudice against Jews in America, with several sections detailing the attitude of Americans toward Jews during the Civil War.

Donald, David. *Why the North Won the Civil War.* Baton Rouge: Louisiana State University Press, 1960; *New York: Collier, 1962. Examines the economic, military, diplomatic, social, and political factors in the victory.

Dupuy, R. Ernest, and Trevor N. Dupuy. *The Compact History of the Civil War.* *New York: Warner Books, 1993.

*Editors of Combined Books. *The Civil War Book of Lists.* Conshohocken, Pa.: Combined Books, 1993.

Foner, Eric. *Free Soil, Free Labor, Free Men.* New York: Oxford University Press, 1970; *Oxford, 1995. A scholarly account of the rise of the Republican party and its ideological commitment to free labor.

Foote, Shelby. *The Civil War: A Narrative,* Vol. 1, *Fort Sumter to Perryville.* New York: Random House, 1958. Vol. 2, *Fredericksburg to Meridian.* New

York: Random House, 1963. Vol. 3, *Red River to Appomattox*. New York: Random House, 1974; *Vintage. The classic narrative of the war.

Gragg, Rod. *The Illustrated Confederate Reader*. New York: Harper & Row, 1989; *HarperPerennial, 1991. A collection of personal experiences, eyewitness accounts, and facts about southern soldiers and civilians.

Hacker, Andrew. *Two Nations: Black and White; Separate, Hostile and Unequal*. New York: Scribner's, 1992; *Ballantine, 1993. This best-seller traces the deep racial divide in America to its historical roots and examines how it continues to play a role in American history.

*Hansen, Harry. *The Civil War: A Narrative*. New York: Mentor Books, 1961. A single-volume history, with an emphasis on the military aspects of the war.

Harwell, Richard B. *The Confederate Reader: How the South Saw the War*. New York: Longmans, Green & Co., 1957; *Dover, 1989.

Hofstadter, Richard, ed. *Great Issues in American History*, Vol. 2, *From the Revolution to the Civil War*. *New York: Vintage Books, 1958. This volume presents significant speeches and court decisions of the period with an excellent historical overview by the editor.

Hofstadter, Richard, and Beatrice K. Hofstadter, eds. *Great Issues in American History*, Vol. 3, *From Reconstruction to the Present Day, 1864–1981*. *New York: Vintage Books, 1982.

Holzer, Harold, ed. *The Lincoln–Douglas Debates: The First Complete Unexpurgated Text*. New York: HarperCollins, 1993.

Holzer, Harold, and Mark E. Neely, Jr. *Mine Eyes Have Seen the Glory: The Civil War in Art*. New York: Orion Books, 1993. A lavishly illustrated collection of Civil War artwork.

Keegan, John. *A History of Warfare*. New York: Knopf, 1993. Written by an acclaimed military historian, this book offers a sweeping overview of the place of war in human culture.

*Kennedy, Frances H., ed. *The Civil War Battlefield Guide*. Boston: Houghton Mifflin, 1990. Published by the Conservation Fund, a group committed to the preservation of Civil War sites, this is a travel guidebook to the sites of Civil War battles.

Kennedy, Paul. *The Rise and Fall of the Great Powers*. New York: Random House, 1987; *Vintage, 1989.

*Lawliss, Chuck. *The Civil War Sourcebook: A Traveler's Guide*. New York: Harmony Books, 1991. Includes a reference guide to battlefield tours and listings of sources of information on war memorabilia, collectibles, and publications.

Lowry, Thomas P. *The Story the Soldiers Wouldn't Tell: Sex in the Civil War*. Mechanicsburg, Pa.: Stackpole Books, 1994. An unusual but revealing and fascinating look at the secret life of soldiers and American morality and the spread of pornography and prostitutes during the war years.

*McEvedy, Colin. *The Penguin Atlas of North American History*. New York: Penguin Books: 1988. American history to the time of the Civil War, illustrated through a progressive series of maps.

McPherson, James M. *Battle Cry of Freedom*. New York: Oxford University Press, 1988; *Ballantine Books. One of the best single-volume histories of the war.

McPherson, James M., ed. *The Atlas of the Civil War*. New York: Macmillan, 1994.

*Malone, John. *The Civil War Quiz Book*. New York: William Morrow/Quill, 1992. A multiple-choice format provides a simple factual review of the war.

Masur, Louis P., ed. *"The real war will never get in the books": Selections from Writers During the Civil War*. New York: Oxford University Press, 1993.

Morison, Samuel Eliot. *The Oxford History of the American People*, Vol. 2, *1789 Through Reconstruction*. New York: Oxford University Press, 1965; *Mentor, 1972. One of three volumes, this standard history covers the Civil War period.

Nelson, Christopher. *Mapping the Civil War: Featuring Rare Maps from the Library of Congress*. Washington, D.C.: Starwood Publishing, 1992.

*Nevins, Allan, and Henry Steele Commager with Jeffrey Morris. *A Pocket History of the United States*, 8th rev. ed. New York: Washington Square Press, 1986. A single-volume overview of American history that includes an extensive section on the Civil War.

Pratt, Fletcher. *Civil War in Pictures*. Garden City, N.Y.: Garden City Books, 1955. Original correspondents' reports and war drawings from the front lines as they appeared in contemporary publications such as *Harper's Weekly*.

*Schlesinger, Arthur M., Jr., ed. *The Almanac of American History*. New York: Perigee Books, 1983. A chronological listing of events in American history with interpretive essays by leading historians and brief biographical sketches of major figures; a sound, useful reference volume.

Shenkman, Richard. *Legends, Lies & Cherished Myths of American History*. New York: Morrow, 1988. An amusing collection of correctives to some commonly accepted ideas about American history, including Civil War misconceptions.

Sifakis, Stewart. *Who Was Who in the Civil War*. New York: Facts on File, 1988.

Stampp, Kenneth M., ed. *The Causes of the Civil War*. *Englewood Cliffs, N.J.: Prentice-Hall, 1974.

Thomas, Emory M. *The Confederate Nation: 1861–1865*. New York: Harper & Row, 1979.

Wade, Wyn Craig. *The Fiery Cross: The Ku Klux Klan in America*. New York: Simon & Schuster, 1987; *Touchstone, 1988.

Ward, Geoffrey C., with Ric Burns and Ken Burns. *The Civil War*. New York: Knopf, 1990; *Knopf. The companion book to the award-winning public television series.

Wheeler, Richard. *Voices of the Civil War*. New York: T. Y. Crowell, 1976; *Meridian, 1990. A narrative of the war through the eyewitness accounts of men and women on both sides of the conflict.

Woodward, C. Vann, ed. *Mary Chesnut's Civil War*. New Haven: Yale University Press, 1981; *Yale University Press. Winner of the Pulitzer Prize in history; it gives a woman diarist's view of the Confederacy and the war.

Yanak, Ted, and Pam Cornelison. *The Great American History Fact-Finder*. Boston: Houghton Mifflin, 1993.

Zinn, Howard. *A People's History of the United States*. New York: Harper & Row, 1980; *Perennial Library, 1980. In this decidedly different, highly revisionist approach to American history (often told from the point of view of the losers), there are three excellent chapters dealing with the Mexican War, slavery, and the Civil War. Highly recommended as an alternative perspective to the more traditional styles of American history.

Biographies

Baker, Jean H. *Mary Todd Lincoln: A Biography*. New York: Norton, 1987; *Norton, 1989. A modern, well-written, and sympathetic account of the tragic life of Lincoln's wife.

Boritt, Gabor S., ed. *Lincoln, the War President: The Gettysburg Lectures*. New York: Oxford University Press, 1992. Essays by leading historians (including James M. McPherson, Kenneth Stampp, and Arthur M. Schlesinger, Jr.) analyze Lincoln's actions during the war.

Bowers, John. *Stonewall Jackson: Portrait of a Soldier*. New York: William Morrow, 1989; *Avon Books.

Davis, William C. *Jefferson Davis: The Man and His Hour*. New York: HarperCollins, 1991; *HarperPerennial, 1992. A thorough, modern, and balanced account of the Confederate president's life.

Donald, David Herbert. *Lincoln*. New York: Simon & Schuster, 1995. A best-selling, comprehensive biography that is widely considered the definitive life of Abraham Lincoln.

*Douglass, Frederick. *Narrative of the Life of Frederick Douglass: An American Slave*. New York: Signet Books, 1968. Written in 1845, this memoir of a slave who became one of the most influential men of his time is an important classic.

Duncan, Russell, ed. *Blue-Eyed Child of Fortune: The Civil War Letters of Colonel Robert Gould Shaw*. Athens: University of Georgia Press, 1992; *New York: Avon Books, 1992. The wartime correspondence of the young man who led the first black regiment, the 54th Massachusetts, which was depicted in the film *Glory*.

Evans, Eli N. *Judah P. Benjamin: The Jewish Confederate*. New York: Free Press, 1988; *Free Press, 1989.

Farwell, Byron. *Stonewall: A Biography of General Thomas J. Jackson*. New York: Norton, 1992; *Norton, 1993. A thorough, balanced life of one of the most legendary and complicated heroes of the South.

Fehrenbacher, Don E., ed. *Lincoln: Speeches and Writings, 1859–1865*. New York: Library of America, 1989. Lincoln's own words during his presidency.

Freeman, Douglas Southall. *Lee*, abridged by Richard Harwell. New York: Scribner's, 1934 (abridgment copyright 1961); *Collier, 1993. A one-volume abridgment of the Pulitzer Prize–winning four-volume life of the general.

Gooding, James Henry. *On the Altar of Freedom: A Black Soldier's Civil War Letters from the Front*. Amherst: University of Massachusetts Press, 1991; *New York: Warner, 1992.

Hedrick, Joan D. *Harriet Beecher Stowe: A Life*. New York: Oxford University Press, 1994. A full-scale Pulitzer Prize–winning biography of the pioneer abolitionist, feminist, and author of the novel *Uncle Tom's Cabin*.

Hurst, Jack. *Nathan Bedford Forrest: A Biography*. New York: Knopf, 1993; *Vintage, 1994. The life of a somewhat overlooked and notorious Confederate general, widely considered a military genius, who went on to help found the Ku Klux Klan.

Kunhardt, Philip B., Jr., Philip B. Kunhardt III, and Peter W. Kunhardt. *Lincoln: An Illustrated Biography*. New York, Knopf, 1992.

McFeely, William S. *Frederick Douglass*. New York: Norton, 1991; *Touchstone, 1992.

———. *Grant: A Biography*. New York: Norton, 1981; *Norton, 1982. A thorough and modern account of the general and later president.

McPherson, James M. *Abraham Lincoln and the Second American Revolution*. New York: Oxford University Press, 1990. A collection of essays about Lincoln by the Pulitzer Prize–winning historian.

Marszalek, John F. *Sherman: A Soldier's Passion for Order*. New York: Free Press, 1993; *Vintage, 1994.

Morris, Roy, Jr. *Sheridan: The Life and Wars of General Philip Sheridan*. New York: Crown, 1992; *Vintage, 1993.

Neely, Mark E., Jr. *The Abraham Lincoln Encyclopedia*. New York: McGraw-Hill, 1982; *Da Capo Press. A comprehensive, illustrated reference work about Lincoln featuring essays about his views, biographical sketches of his contemporaries, and appraisals of other Lincoln books.

———. *The Fate of Liberty: Abraham Lincoln and Civil Liberties*. *New York: Oxford University Press, 1991. This winner of the 1992 Pulitzer Prize for History by a leading Lincoln scholar explores the constitutional and political issues behind Lincoln's controversial suspension of habeas corpus.

————. *The Last Best Hope on Earth: Abraham Lincoln and the Promise of America*. Cambridge, Mass.: Harvard University Press, 1993. A compact biography of Lincoln as politician by one of the country's preeminent Lincoln scholars.

Nevins, Allan. *Frémont: Pathmarker of the West*. *Lincoln: University of Nebraska Press/Bison Books, 1992. A detailed life of the extraordinary adventurer and soldier who helped map the West, founded California, ran for president, and lost his Civil War command.

Nolan, Alan T. *Lee Considered: General Robert E. Lee and Civil War History*. Chapel Hill: University of North Carolina Press, 1991.

Oates, Stephen B. *Abraham Lincoln: The Man Behind the Myths*. New York: Harper & Row, 1984; *Meridian Books, 1985.

————. *With Malice Toward None: The Life of Abraham Lincoln*. New York: Harper & Row, 1977; *Mentor Books, 1978. A readable and realistic single-volume life of Lincoln.

————. *A Woman of Valor: Clara Barton and the Civil War*. New York: Free Press, 1994; *Free Press, 1995. A thorough life of the courageous woman who nursed soldiers during the war and went on to found the American Red Cross.

Paludan, Phillip Shaw. *The Presidency of Abraham Lincoln*. Lawrence: University Press of Kansas, 1994.

Robertson, James I., Jr. *Soldiers Blue and Gray*. Columbia: University of South Carolina Press, 1988; *New York: Warner, 1991. Adapted from letters, a re-creation of the common soldier's experience in the Civil War.

Sears, Stephen W. *George B. McClellan: The Young Napoleon*. New York: Ticknor & Fields, 1988; *Ticknor & Fields. A thorough biography of the controversial commander who later ran for president against Lincoln.

Thomas, Emory M. *Bold Dragoon: The Life of J.E.B. Stuart*. New York: Random House, 1986; *Vintage, 1988. A life of the audacious Confederate cavalry leader, one of Lee's chief lieutenants.

————. *Robert E. Lee: A Biography*. New York: Norton, 1995. The most recent assessment of the legendary Confederate hero.

Warren, Robert Penn. *John Brown: The Making of a Martyr*. Payson & Clarke, 1929; *Nashville: J. F. Sanders, 1993.

Wills, Garry. *Lincoln at Gettysburg: The Words That Remade America.* New York: Simon & Schuster, 1992. This Pulitzer Prize–winning best-seller analyzes the impact of Lincoln's brief but timeless speech.

Woodward, C. Vann, ed. *Mary Chesnut's Civil War.* New Haven: Yale University Press, 1981; *Yale University Press. This winner of the Pulitzer Prize in history shows how a famous diarist's view of the Confederacy and the war was considerably revised well after the war to conform to acceptable notions of the conflict.

Military History

Billings, John D. *Hard Tack & Coffee: The Unwritten Story of Army Life.* Boston: George M. Smith, 1887; *Lincoln: University of Nebraska Press/Bison Books, 1993. A classic account of the war from the enlisted soldier's point of view.

Boritt, Gabor S. *Lincoln's Generals.* New York: Oxford University Press, 1994. Essays on Lincoln's often stormy relationships with his commanders by Stephen Sears, Mark E. Neely, Jr., Michael Fellman, and other Lincoln scholars.

Catton, Bruce. *Glory Road.* Garden City, N.Y.: Doubleday, 1952; *Anchor Books, 1990. Volume 2 of the Army of the Potomac trilogy.

———. *Mr. Lincoln's Army.* Garden City, N.Y.: Doubleday, 1951; *Anchor Books, 1990. Volume 1 of the author's acclaimed Army of the Potomac trilogy; almost exclusively concerned with the military aspects of the war in Virginia.

———. *A Stillness at Appomattox.* Garden City, N.Y.: Doubleday, 1953; *Anchor Books, 1990. Volume 3 of the Army of the Potomac trilogy.

Coffin, Howard. *Full Duty: Vermonters in the Civil War.* Woodstock, Vt.: Countryman Press, 1993. An account of the unusual response of the small, poor, and remote northeastern state that sent more than 10 percent of its total population to the war.

Cornish, Dudley Taylor. *The Sable Arm: Black Troops in the Union Army, 1861–1865.* *Lawrence: University of Kansas Press, 1987. A classic account of the wartime service of black troops.

Davis, Burke. *The Long Surrender.* New York: Random House, 1985; *Vintage, 1989. An eyewitness reconstruction of the last days of the Confederacy, including the capture of Jefferson Davis.

———. *Sherman's March.* New York: Random House, 1980; *Vintage, 1988. An account of the notorious March to the Sea through Georgia.

Davis, William C. *The Commanders of the Civil War.* New York: Salamander Books, 1990.

———. *The Fighting Men of the Civil War.* New York: Salamander Books, 1989.

Dufour, Charles L. *The Night the War Was Lost.* *Lincoln: University of Nebraska Press, 1994. The author contends that the war was essentially over when New Orleans fell to the Union navy in 1862, leading to the economic defeat of the Confederacy.

Eisenhower, John S. D. *So Far from God: The U.S. War with Mexico, 1846–1848.* New York: Random House, 1989. A basic account of the Mexican War with an emphasis on its military aspects.

Furgurson, Ernest B. *Chancellorsville 1863: The Souls of the Brave.* New York: Knopf, 1992.

Hastings, William H., ed. *Letters from a Sharpshooter.* *Belleville, Wis.: Historic Publications, 1993. The vivid letters home from a young Union soldier who was employed in one of the most dangerous occupations.

Hattaway, Herman, and Archer Jones. *How the North Won: A Military History of the Civil War.* Urbana: University of Illinois Press, 1983; *Illini Books, 1991.

Hennessy, John J. *Return to Bull Run: The Campaign and Battle of Second Manassas.* New York: Simon & Schuster, 1993. A campaign history of the less famous but more bloody second battle at this famous sight.

Keegan, John. *The Mask of Command.* New York: Viking Penguin, 1987; *Penguin Books, 1988. A brilliant discussion of four military leaders, including an excellent chapter on Ulysses S. Grant.

*Macdonald, John. *Great Battles of the Civil War.* New York: Macmillan, 1988; Collier, 1992.

McPherson, James M. *Marching Toward Freedom: Blacks in the Civil War.* New York: Knopf, 1967; *Facts on File, 1994.

———. *What They Fought For*. Baton Rouge: Louisiana State University Press, 1994. Insightfully examines the motives of the enlisted men on both sides of the war.

Moe, Richard. *The Last Full Measure: The Life and Death of the First Minnesota Volunteers*. New York: Henry Holt, 1993; *Avon Books, 1994. A history of one of the most famed state units, a group of volunteers perhaps best known for its crucial charge during the battle at Gettysburg.

Ray, Delia. *Behind the Blue and Gray: The Soldier's Life in the Civil War*. New York: Lodestar Books, 1991; *Scholastic, 1994. Aimed at young readers, a well-documented and strikingly illustrated account of the average soldier's life.

Royster, Charles. *The Destructive War: William Tecumseh Sherman, Stonewall Jackson, and the Americans*. New York: Knopf, 1991; *Vintage, 1993.

Sears, Stephen W. *Landscape Turned Red: The Battle of Antietam*. New Haven: Ticknor & Fields, 1983; *New York: Warner Books, 1985. A best-selling account of the bloodiest battle in the war.

———. *To the Gates of Richmond: The Peninsula Campaign*. New York: Ticknor & Fields, 1992. A campaign history of McClellan's disastrous attack on Richmond in 1862.

Svenson, Peter. *Battlefield: Farming a Civil War Battleground*. Boston: Faber and Faber, 1992. A modern farmer writes about his home on a Civil War battlefield.

Tapert, Annette, ed. *The Brothers' War: Civil War Letters to Their Loved Ones from the Blue and Gray*. New York: Times Books, 1988; *Vintage, 1989. Revealing letters home.

Watkins, Sam R. *"Co. Aytch": A Side Show of the Big Show*. New York: Macmillan, 1962; *Collier. A classic soldier's-eye view of the war from a twenty-one-year-old Tennessean who fought in many of the major battles.

Waugh, John C. *The Class of 1846: From West Point to Appomattox*. New York: Warner Books, 1994. A chronicle of the numerous classmates from the U.S. Military Academy (Stonewall Jackson and George McClellan among them) who fought together in the Mexican War but faced each other as enemies in the Civil War.

Wert, Jeffrey D. *Mosby's Rangers.* New York: Simon & Schuster, 1990; *Touchstone, 1991. An account of the notorious Confederate "Gray Ghost's" much-feared cavalry battalion.

*Wiley, Bell Irvin. *The Life of Billy Yank: The Common Soldier of the Union.* Baton Rouge: Louisiana State University Press, 1978. Originally published in 1952, a classic account of life among the ranks of the average Union soldier.

————. *The Life of Johnny Reb: The Common Soldier of the Confederacy.* Baton Rouge: Louisiana State University Press, 1943. Originally published in 1943, a companion volume to the book above.

Williams, T. Harry. *Lincoln and His Generals.* New York: Knopf, 1952; *Vintage Books. A standard volume by one of the deans of American military history.

Slavery and African-American History

Bennett, Lerone, Jr. *Before the Mayflower: A History of Black America,* 5th ed. Chicago: Johnson Publishing, 1962; *New York: Penguin Books, 1988.

————. *The Shaping of Black America.* Chicago: Johnson Publishing, 1975; *New York: Penguin Books, 1993.

Berlin, Ira, Barbara J. Fields, Steven F. Miller, Joseph P. Reidy, and Leslie S. Rowland, eds. *Free at Last: A Documentary History of Slavery, Freedom and the Civil War.* New York: The Free Press, 1992.

Blockson, Charles L. *The Underground Railroad.* New York: Prentice Hall Press, 1987; *Berkley Books, 1994. Firsthand accounts of slaves' escapes to freedom.

Brandt, Nat. *The Town That Started the Civil War.* Syracuse, N.Y.: University of Syracuse Press, 1990; *Laurel, 1991. A narrative account of an 1858 incident in which a runaway slave was rescued from slavecatchers by the citizens of Oberlin, Ohio.

Brent, Linda. *Incidents in the Life of a Slave Girl,* intro. Walter Teller. New York: Harcourt Brace Jovanovich, 1973; *Harvest Books, 1973. The autobiography of a young former slave, first published in 1861.

Davis, David Brion. *The Problem of Slavery in Western Culture.* Ithaca, N.Y.: Cornell University Press, 1966; *New York: Oxford University Press, 1988.

A Pulitzer Prize–winning study of the traditions of slavery in Europe leading up to the Civil War.

*Dennis, Denise. *Black History for Beginners*. New York: Writers and Readers, 1984. A documentary in cartoon format.

*Douglass, Frederick. *My Bondage and My Freedom*. New York: Dover Editions, 1969. First published in 1855, the second of the great abolitionist's three autobiographies details his life as a slave.

Editors of Time-Life Books. *Perseverance: African American Voices of Triumph*™. Alexandria, Va.: Time-Life Books, 1993. Volume 1 of a series detailing the African-American experience, it deals with slavery and the Civil War years as well as the civil rights movement.

Egerton, Douglas R. *Gabriel's Rebellion: The Virginia Slave Conspiracies of 1800 and 1820*. *Chapel Hill: University of North Carolina Press, 1993. A scholarly look at the social and political history of the often overlooked slave revolts led by blacksmith Gabriel Prosser.

*Gates, Henry Louis, Jr., ed. *The Classic Slave Narratives*. New York: Mentor, 1987. Contains *The Life of Olaudah Equiano, The History of Mary Prince, Narrative of the Life of Frederick Douglass, Incidents in the Life of a Slave Girl.*

Harding, Vincent. *There Is a River: The Black Struggle for Freedom in America*. San Diego: Harcourt Brace Jovanovich, 1981; *Harvest, 1992.

Hurmence, Belinda, ed. *Before Freedom: 48 Oral Histories of Former North and South Carolina Slaves*. Winston-Salem: John F. Blair, 1990; *New York, Mentor, 1990.

Jordan, Winthrop D. *White over Black: American Attitudes Toward the Negro, 1550–1812*. Chapel Hill: University of North Carolina Press, 1968; *New York: Norton, 1977. A prize-winning classic that examines the nature of the creation of slavery and its role in America's development.

Kolchin, Peter. *American Slavery: 1619–1877*. New York: Hill and Wang. 1993. A modern scholarly look at the "peculiar institution."

Lamb, David. *The Africans*. New York: Random House, 1983; *Vintage, 1987. A contemporary journalist's view of Africa that discusses the history of slavery and its impact on the continent of Africa.

Litwack, Leon F. *Been in the Storm So Long: The Aftermath of Slavery*. New York: Knopf, 1979; *Vintage, 1980. Based on interviews with former slaves and slaveowners, a Pulitzer Prize–winner in history.

Mellon, James, ed. *Bullwhip Days: The Slaves Remember*. New York: Weidenfield & Nicolson, 1988; *Avon Books. Oral history of former slaves collected during the Depression by the Federal Writers Project.

Miller, John Chester. *The Wolf by the Ears: Thomas Jefferson and Slavery*. New York: Free Press, 1977; *Charlottesville: University Press of Virginia, 1991. A scholarly examination of Jefferson's contradictory attitudes about slavery.

Quarles, Benjamin. *Black Abolitionists*. New York: Oxford University Press, 1969; *Da Capo Press.

———. *The Negro in the Making of America*, 3rd ed. rev. New York: Macmillan, 1964; *Collier Books, 1987.

Ripley, C. Peter, ed. *Witness for Freedom: African American Voices on Race, Slavery, and Emancipation*. *Chapel Hill: University of North Carolina Press, 1993.

Rosengarten, Theodore. *All God's Dangers: The Life of Nate Shaw*. New York: Knopf, 1974. *Vintage, 1984. National Book Award–winning story of a tenant farmer who grew up in the society of former slaves and slaveowners.

———. *Tombee: Portrait of a Cotton Planter*. New York: William Morrow, 1986; *Quill. Life on a southern plantation before and during the war.

Stampp, Kenneth M. *The Peculiar Institution: Slavery in the Ante-Bellum South*. New York: Knopf, 1956; *Vintage.

Fiction

This list offers a handful of the best—or most notorious, in some cases—historical fiction about the Civil War. It is wise to remember that fiction is just that. Despite the best efforts of novelists to document their work, they are still creating characters, scenarios, and in many cases events. But many of these books bring the war to life as histories do not and are worth reading for that quality.

Crane, Stephen. *The Red Badge of Courage*. One of the first and still the greatest of all Civil War novels. Many paperback editions available.

Foote, Shelby. *Chickamauga and Other Civil War Stories.* *New York: Delta Publishing, 1993. A collection of short fiction by such writers as Stephen Crane, William Faulkner, Mark Twain, Thomas Wolfe, and Eudora Welty.

————. *Shiloh.* New York: Dial Press, 1952; *Vintage, 1991. A fictional version of the bloody two-day battle by the author of the acclaimed trilogy *The Civil War: A Narrative.*

Gurganus, Allan. *Oldest Living Confederate Widow Tells All.* New York: Knopf, 1984; *Ballantine Books, 1990.

Kantor, MacKinlay. *Andersonville.* New York: Harper & Row, 1955; *Signet. Pulitzer Prize–winning novel about the notorious Confederate prison.

*McSherry, Frank, Jr., Charles G. Waugh, and Martin Greenberg, eds. *Civil War Women.* Little Rock, Ark.: August House, 1988. A collection of short fiction about Civil War subjects written by women (including Louisa May Alcott, Kate Chopin, and Eudora Welty) both during and after the war.

Mitchell, Margaret. *Gone With the Wind.* New York: Macmillan, 1936; *Avon Books, 1973. "The mother of all historical romances" and one of the chief reasons for America's love affair with the Civil War.

Quinn, Peter. *Banished Children of Eve.* New York: Viking Penguin, 1994; *Penguin, 1995. A critically praised novel of Civil War New York focusing on the great Draft Riot.

Safire, William. *Freedom: A Novel of Lincoln and the Civil War.* Garden City, N.Y.: Doubleday, 1987; *Avon, 1988.

Shaara, Michael. *The Killer Angels.* New York: David McKay, 1974; *Ballantine, 1975. The basis of the film *Gettysburg* and perhaps the best novel about the Civil War.

Styron, William. *The Confessions of Nat Turner.* New York: Random House, 1967; *Vintage, 1993. A controversial prize-winning novel from the perspective of the leader of one of the great slave revolts.

Vidal, Gore. *Lincoln: A Novel.* New York: Random House, 1984; *Ballantine Books, 1985.

Young Adult Books

Beller, Susan Provost. *Medical Practices in the Civil War.* *Cincinnati: Shoe Tree Press, 1992. Written for students, it offers a brief overview of how medicine was practiced and how procedures changed during the war.

Chang, Ina. *A Separate Battle: Women and the Civil War.* New York: Dutton/ Lodestar Books, 1991. Part of a Young Readers' History of the Civil War.

Coit, Margaret L. *The Fight for Union.* Boston: Houghton Mifflin, 1961. A children's book that presents a very useful summary of the debate that preceded the Civil War.

Fleischman, Paul. *Bull Run.* New York: HarperCollins, 1993. The story of the war's first major battle told from the perspective of a variety of witnesses, both soldiers and civilians.

Freedman, Russell. *Lincoln: A Photobiography.* Boston: Ticknor & Fields, 1987; *New York: Clarion Books, 1988. An excellent young adult book, the winner of the 1988 Newbery Award, it is a useful introduction for adults as well.

Fritz, Jean. *Brady.* New York: Coward-McCann, 1960; *Puffin, 1987. A young adult novel about the Underground Railroad.

Lester, Julius. *To Be a Slave.* New York: Dial Books, 1968; *Scholastic/Point. The slave's-eye view of life in the words of survivors of slavery.

Lyons, Mary E. *Letters from a Slave Girl: The Story of Harriet Jacobs.* New York: Scribner's, 1992. A fictional version of an authentic 1861 autobiography.

Meltzer, Milton, ed. *Voices from the Civil War: A Documentary History of the Great American Conflict.* New York: Thomas Y. Crowell, 1989. A valuable reference book drawn from letters, memoirs, newspaper accounts, and speeches depicting life during the war.

*Murphy, Jim. *The Boys' War: Confederate and Union Soldiers Talk About the Civil War.* New York: Clarion Books, 1990. Award-winning account of the very young boys who saw action on both sides of the conflict.

———. *The Long Road to Gettysburg.* New York: Clarion Books, 1992. A young adult book that recounts the famous battle from the point of view of two teenage participants on either side.

Myers, Walter Dean. *Now Is Your Time: The African-American Struggle for Freedom*. New York: HarperCollins, 1991; *HarperTrophy, 1991. A young adult history of Africans in America, beginning with slavery.

Paulsen, Gary. *Nightjohn*. New York: Delacorte, 1993. The life of a twelve-year-old slave changes when another slave teaches her to read.

Rappaport, Doreen. *Escape from Slavery*. New York: HarperCollins, 1991. True accounts of runaway slave experiences.

Ray, Delia. *Behind the Blue and Gray: The Soldier's Life in the Civil War*. New York: Lodestar Books, 1991; *Scholastic, 1994. Aimed at young readers, a well-documented and strikingly illustrated account of the average soldier's life.

*Rinaldi, Ann. *Wolf by the Ears*. New York: Scholastic, 1991. A novel about the young slave girl who, in the author's view, might have been Jefferson's daughter.

Robertson, James I., Jr. *Civil War! America Becomes One Nation*. New York: Knopf, 1992. An up-to-date, readable history for young readers.

Sobol, Donald J., ed. *A Civil War Sampler*. New York: Franklin Watts, 1961. A collection of letters and documents from the period.

Index

About the Author

KENNETH C. DAVIS, the creator and author of the Don't Know Much About® series, was recently dubbed the "King of Knowing" by Amazon.com, and speaks often on national television and radio. His *USA Weekend* column is seen weekly by millions of readers. In addition to his adult titles, he writes the Don't Know Much About® children's series published by HarperCollins. He lives in New York City and Vermont with his wife and two children.

BOOKS BY KENNETH C. DAVIS

DON'T KNOW MUCH ABOUT MYTHOLOGY
Everything You Need to Know About the Greatest Stories
in Human History but Never Learned
ISBN 0-06-019460-X (hardcover)
ISBN 0-06-093257-0 (paperback)
Explores the great legends of human history and connects
them to historical events.

DON'T KNOW MUCH ABOUT HISTORY
Everything You Need to Know About American History but Never Learned
ISBN 0-06-008382-4 (paperback)
The million-copy *New York Times* bestseller, completely revised
and updated! With wit, candor, and fascinating facts, Davis
explodes long-held myths and misconceptions
of American history.

DON'T KNOW MUCH ABOUT THE BIBLE
Everything You Need to Know About the Good Book but Never Learned
ISBN 0-380-72839-7 (paperback)
Davis sets the panorama of the Scriptures against the historical
events that shaped them.

DON'T KNOW MUCH ABOUT THE CIVIL WAR
Everything You Need to Know About America's
Greatest Conflict but Never Learned
ISBN 0-380-71908-8 (paperback)
Davis sorts out the players, politics, and key
events of the Civil War.

DON'T KNOW MUCH ABOUT GEOGRAPHY
Everything You Need to Know About the World but Never Learned
ISBN 0-380-71379-9 (paperback)
A fascinating, breathtaking, and entertaining grand tour of the
planet Earth.

DON'T KNOW MUCH ABOUT THE UNIVERSE
Everything You Need to Know About Outer Space but Never Learned
ISBN 0-06-093256-2 (paperback)
Davis begins with 20th-century exploration and coverage of
celestial bodies, and concludes with an accessible exploration of
cosmic questions from the beginning of civilization to the present.